The Hunt for Earth Gravity

Top: Rose de Freycinet arrives in Timor. Painting by Jacques Arago. Bottom: Marooned in the Arctic. The caption reads "The Arctic Dandies during their residence on Melville Island, 1819–1820. Drawn by Captain Sabine and partly coloured by him in 1822, completed by Edward Noble his godson in 1906". Most of the 'dandies' seem rather under-dressed for temperatures that remained below −15°C from November until March. © 2015 Christies Images Limited

John Milsom

The Hunt for Earth Gravity

A History of Gravity Measurement
from Galileo to the 21st Century

 Springer

John Milsom
Gladestry Associates
Presteigne, UK

ISBN 978-3-030-09113-2   ISBN 978-3-319-74959-4   (eBook)
https://doi.org/10.1007/978-3-319-74959-4

Printed on acid-free paper

This Springer imprint is published by the registered company Springer International Publishing AG part
of Springer Nature
The registered company address is: Gewerbestrasse 11, 6330 Cham, Switzerland

... we come now to the other questions, relating to pendulums, a subject which may appear to many exceedingly arid
Galileo Galilei: Two New Sciences

Preface

There are proper physicists and there are exploration geophysicists. To proper physicists, their brothers and sisters in exploration are people of uncouth lifestyles and suspect intelligence who abuse the most fundamental and mysterious force in the universe (gravity—the one that they themselves still do not understand) by treating it as a mere tool for looking at rocks. Few such people have ever worried about the possible non-equivalence of inertial and gravitational mass (although Loránd Eötvös did, in 1890, and revolutionised gravity surveying by designing a practical torsion balance), and even fewer bother about the role of gravity in Special or General Relativity, or the possible use of quantum gravity to reconcile classical physics and quantum theory. The question of whether or not the Higgs boson exists and, if it does, whether it really does give mass to everything else, does not keep them awake at night. Instead, they take their instruments out into the 'field', which may be a real field, or a desert, or a forest, or an ocean, or a city street and, having mapped the changes in gravity to the best of their ability, they try to understand what they are being told about the rocks beneath their feet. It seems somehow appropriate that the universal constant of gravitation, which holds everything together, is known as 'Big G', while the local gravity field with which explorationists content themselves is merely 'little g'.[1]

---

[1] Referred to from here onwards simply as 'g'.

As the list of things that do not worry explorationists might suggest, the break with other physicists came only in the twentieth century, and with Einstein. Before him, theirs were shared histories, involving some astounding insights and some improbable characters. The people who investigated 'g' in the fifteenth, sixteenth, seventeenth, eighteenth and nineteenth centuries had interests far wider than mere gravity measurement, but this book is concerned only with their efforts to do this (and, in some cases, with the effects those efforts had on them).

Mixed in with these stories are some of my own memories. It may be presumptuous to talk about these in a book that figures giants such as Galileo, or to compare the trivial discomforts of modern fieldwork with the truly horrific challenges faced by the Frenchmen who, in 1730, went to South America to discover the shape of the Earth, but my hope is that this sometimes very personal approach can give people who know little about the Earth's gravity field some insight into the reasons why so many people have devoted so much of their time to its study during the past 500 years.

If this book has any readers, they may be people who know something about physics but little geology, or people who know geology but not physics or people with only a layman's knowledge of either. The pattern used tries to cope with this. The numbered chapters are the history and are generally in rough chronological order, although there are overlaps. Any dated sections within them are anecdotal and subject to the defects of my own memory (I was never a diarist), with positions determined by topics and not by chronology. Chapter 2 is entirely anecdotal and is out of sequence, because its aim is to give readers an early feeling for where the book is heading. A final section of 'Codas' (Chap. 14) is included for those who not only feel comfortable with graphs and equations, but would like to read about them.

Presteigne, UK   John Milsom

Acknowledgements

One of the more stressful aspects of writing a non-fiction book is the need to obtain permission to reproduce copyright material, and particularly illustrations. In my case, the load has been lightened by the generous responses and encouragement that I have received from many of the people and organisations contacted. In some cases, the end result was that I was provided with far better copies of the pictures than I had been able to find elsewhere. So, my very special thanks to Helena Ingham of Christies for the high-resolution image of the Arctic Dandies that forms half the frontispiece, to Valerie Shrimplin of Gresham College for the similarly excellent image of the original college, to the British Geological Survey for their pictures of the torsion balance in use in the 1930s, to John Noonan of Oil Search for photographs of an early LaCoste gravity metre in use in Papua in the late 1930s, to Michel van Camp for the pictures of his superconducting gravity meter and his son in use as a test mass and to Christopher Jekeli for his pictures of the Bell accelerometer and of a camper van containing an entire airborne gravity metre being driven on to a C-135. Ute Schiedermeier of the Siemens Historical Institute in Berlin confirmed my identification of William Siemens in the picture of him and his brothers, as well as giving permission for its use.

It is now hard to imagine attempting a project of this sort without the facilities provided by the Internet. Original sources which had in the past to be sought in the dusty recesses of scattered libraries are now available online. The decision by the Royal Society to digitise the whole of their Philosophical Transactions archive and to make it freely available has been especially important. And, as far as reproduction of extracts from recent material is concerned, I have been especially fortunate in the very relaxed attitude adopted by the Society of Exploration Geophysicists.

I also had helpers on the ground. Paola Marshall and the Braschi-Levi family from Bologna took photographs for me in Italy, Richard Dingley photographed the Oude Kerke in Delft, and my daughter Anna took the photograph of the Islington canal tunnel. My daughter Kate produced the drawings of Galileo's gears and the Chimborazo experiment. Where no other source is given, photographs, drawings and images are those that I have produced myself. All images of the gravity field were prepared using the Geosoft suite of programmes, and in marine areas are based on grids placed online by David Sandwell at http://topex.ucsd.edu/cgi-bin/get_data.cgi. Line drawings were prepared in CorelDraw.

Many people gave me encouragement along the way. Jason Ali, Gary Barnes, Mark Davies, Matthew Engel, Richard Howarth, Ed Lake, Ian Nash, Barry Oliver, Mike Rego and John Smallwood all read excerpts at various stages in the writing process and provided helpful and incisive criticisms. Ted Metcalfe helped me with Galileo and Magne Njåstad with the geography of Trondheim. And very special thanks are due to my two daughters, Anna and Kate, and especially to Marijana Dworski, bookseller extraordinary, who stuck with the project through thick and thin, and who also provided some of the books that I have used.

Contents

Introduction

From sea level near the North Pole to sea level at the equator, 'g', the Earth's gravity, decreases by about half of one per cent. Travel to Ecuador, take a trip inland, and climb to the top of Chimborazo, which is as far from the Earth's centre as it is possible to get with feet still firmly planted on the ground, and the overall decrease amounts to about two-thirds of one per cent. These are small differences, but modern gravity metres can measure 'g' to one part in a billion. In the future, they may become easier and quicker to use, and cheaper, but there would be little point in making them more accurate. They are already sensitive to changes of less than half a centimetre in their height above sea level.

The Seconds Pendulum

Galileo discovered many things about gravity, but it was left to a Dutchman, Christiaan Huygens, to do the maths and write down the equations that govern the motions of 'simple' pendulums, in which point masses are supported by weightless threads, and of the 'compound' pendulums that exist in the real world. One of his aims in doing so was to find the length of a pendulum that would beat seconds exactly, and what he also showed was that its length would be directly proportional to 'g'. From his time onwards until the beginning of the twentieth century, values of 'g' were routinely quoted in terms of this length.

The idea is simple, but there is room for confusion. The time taken by a pendulum to swing from one extreme to another and back again is known as its period and, for good mathematical reasons, this is considered by physicists to be its fundamental property. Early clock makers, however, were concerned with what was easily observable, and a pendulum is most easily observed when it is vertical. This happens twice in every period, and what has come to be universally acknowledged as the seconds pendulum has a half period, not a full period, of one second. Its length is very close to one metre, which is pure coincidence since the metre was originally defined as one forty-millionth part of the polar circumference of the Earth.

Units in Renaissance (and Later) Science

Anyone interested in the history of science has to learn to navigate a maze of units. In the history of gravity, lengths were measured not only in *braccia*, *toises*, *lignes* and the English, Rhenish, Roman and Royal (French) feet, but also in Galileo's own private *punti*, which nobody else used. The factors that convert one to another are usually known only to parts per thousand, but parts per million can be very significant in modern gravity measurements. Moreover, the accepted factors may not always be the right ones, in any particular case. Either Riccioli's measurement of the height of the Asinelli Tower in Bologna was wrong, by a considerable margin (which would call into question all his other work) or the Roman foot that he used was slightly different from the foot used in Rome.[2]

To add further to the confusion, translators have not always left well alone. The Tuscan *braccio* that was familiar to Galileo has on occasion been translated as cubit, and cubit as fathom. Any attempt to use the accepted conversion factors on these translated units must end in disaster.

Even where translation was not involved, uncertainties persisted well into the nineteenth century. Henry Kater, the originator of the reversible pendulum, found it necessary to specify the length of his '*pendulum vibrating in seconds in London*' according to 'Sir G. Shuckburgh's standard' (in which it was 39.13860 inches), General Roy's scale (39.13717 inches) and Bird's Parliamentary Standard (39.13842 inches). The differences amounted to several parts in a hundred thousand, in a science that even then was hoping for parts per million. Eventually, and presumably in despair, Kater gave up trying to measure 'g' and made a career out of defining standards of mass and length for the British government.

---

[2] See discussion in Chap. 14, Coda 2.

Time

Where time is concerned, things are easier. The difficulties faced by early sci-entists in measuring it were daunting, but the basic standard was in little doubt. A second is one-sixtieth of a minute, and a minute is one-sixtieth of an hour and an hour is one twenty-fourth of a day. It is true that, because the Earth orbits the sun but not the stars, there is a difference between the sidereal day, which is measured by the stars, and the solar day measured by the sun but this was well understood in the seventeenth century. Riccioli, who made the first respectable estimates of 'g', using pendulums as well as falling weights, had only to specify which sort of day he was using for every-one who was interested to understand.

There is one exception. The French Revolution introduced to the world the decimal second, which was equal to one-hundredth of a decimal minute which was equal to one-hundredth of a decimal hour which was equal to one-tenth of a solar day, and which was therefore equal to 1.1574 ordinary seconds. Even in revolutionary France, it was never popular and was quickly abandoned but as late as 1821, and seven years after the restoration of the French monarchy, it was still being used by French scientists when reporting the results of pendulum observations in France, Spain and the British Isles (e.g. Biot and Arago 1821).

Units for Gravity

The problems with units of length (and mass) all but vanished when the standardised version of the metric system, the Systeme International (SI), was adopted in 1960, but the gravity world was poorly served by the SI committees. All geophysicists should now be using units based on metres and seconds, and 'g', as an acceleration, should be measured in metres per second per second, often written as metres/sec^2 and officially as m s^{-2}. However, no special name was given to this unit, leaving the people who worked with 'g' on a daily basis to flounder about expressing its changes in terms of a 'practical' unit equal to a millionth of a metre per second per sec-ond, officially written as μm s^{-2} and requiring recourse to the special char-acter set on their word processors every time a value had to be written down. Some chose to use this unit but call it, ambiguously, the 'gravity unit' or 'g.u.', but many others preferred to stick to the previous standard with a memorable name, the Gal, equal to 1 cm/sec^2. The practical unit for geological purposes is one-thousandth of a Gal, officially written mGal but voiced as milligal (which is the way it is written in this book).

On the Earth's surface, 'g' is reasonably close to 10 m s^{-2}, or a million milligals, making it easy to think of the gravity effects of geology in terms of parts per million or ppm. Changes of a few tenths of a milligal can be important when looking for caves and cavities (Fig. 1), changes of a milligal or a few milligals when looking for mineable orebodies and of a few tens of milligals when defining the limits of sedimentary basins. Because there are ten μm s^{-2} to a milligal, the significance of features on a gravity map for which the units have not been specified may be in doubt by a factor of ten, and this can be a real cause of misinterpretation.

In recent years, the technology has advanced to such an extent that it is not just 'g' but its gradient that is being measured. For this, there is a practical unit, free of the prefixes that characterise the Systeme Internationale. It is the Eötvös unit, and it represents a change of just one milligal over a distance of 10 kilometres, or of one μm s^{-2} over a kilometre.

Fig. 1 The Islington canal tunnel, North London. Its effect on 'g', amounting to a few hundredths of a milligal, is just measurable by modern gravity metres on the road above it (*Photograph* Anna Milsom)

A Note for Obsessives

The phrase 'acceleration due to gravity' is a common one, and it therefore seems right and proper that 'g' should be measured in units of acceleration. The justification for doing so goes back to Newton, whose first law states that the acceleration of a body in free space is proportional to the force acting on it divided by its mass, and whose Law of Gravitation then implies that all masses in free space will receive an acceleration proportional to the gravity field. It is, however, arguable that this focuses attention on effects rather than causes, that the proper units should be of force divided by mass, and that the Gal should be defined as one dyne per gram and the SI unit as one Newton per kilogram. The numerical values would be unchanged.

All units, if used often enough, acquire a life of their own. When a boy racer gets his first car and dreams of whipping it up to (in Britain or America) a hundred miles an hour, he is not thinking of a hundred miles of road and the hour it would take him to drive down it. He is thinking 'fast'. Similarly, for the people who use it all the time, the milligal is not something to be thought of in terms of centimetres or seconds, and still less of 'seconds squared'. Much more simply, a hundred milligals means big, while a hundredth of a milligal is barely measurable.

Reference

Biot J-B, Arago F (1821) Recueil d'observations géodésiques, astronomiques et physiques executées en espagne, en france, en angleterre et en écosse. Courcier, Paris

1

The Beginning

There can be little doubt about one thing.

It all began with Galileo (Fig. 1.1).

He was, after all, the first person to show that the distances travelled by objects propelled only by gravity are proportional to the squares of the travel times. He was also the first to say that a weight on a string (a simple pendulum) always takes the same time to complete a swing, regardless of how far it swings and how heavy the weight, and to establish a relationship between this time and the length of the string. He thus pioneered both of the methods that have since been used to measure 'g'. Up until the middle of the 20th Century the most accurate way of doing this was to time a pendulum. More recently, it has been the rates of fall of objects in vacuum chambers that have been measured.

The Biographers

Most of the hundreds, or thousands, of books written about Galileo concentrate on his trial and the events that led up to it. Straightforward descriptions of the known facts compete with elaborate conspiracy theories that have him confessing to a lesser offence to avoid being consigned to the fire for a greater one. Dealing with this torrent of information is like wading into a river in full flood. There is a great deal of rubbish coming down. There are large gaps in the contemporary accounts, and much unsupported speculation in what has been written since. Thankfully, I am only trying to follow

© Springer International Publishing AG, part of Springer Nature 2018
J. Milsom, *The Hunt for Earth Gravity*,
https://doi.org/10.1007/978-3-319-74959-4_1

GALILAEUS GALILAEI PATRICIUS FLOR.
AET. SUAE
ANNUM AGENS QUADRAGESIMUM

Fig. 1.1 Galileo at forty, when he was making his experiments with Swing, Roll and Fall. 19th Century engraving by Giuseppe Calendi, based on a painting by Santi di Tito

the history of ideas about the Earth's gravity field, and the task of tracing Galileo's part in that story has been manageable. Of the recent authors, I have only really engaged with three: Alexandre Koyré, Stillman Drake and Arthur Koestler. These were writers with very different views. Koyré admired Galileo as a master of the thought experiment but scorned his lab techniques, while Drake saw him as the first great experimental scientist.

Drake was an interesting character in his own right. His lifelong obsession with Galileo took him from financial consultancy in California to the professorial chair at the University of Toronto that he occupied until his death. He was a prolific writer, the author or co-author of more than a hundred books and papers about Galileo, but he was no scientist. His greatest contribution was to learn 16th Century Italian and then spend long hours puzzling his way through the mass of surviving documents, including some two hundred sheets of chaotically semi-legible folio notes, that Galileo left behind and which, in three hundred and fifty years, no-one else had had the stamina to unravel.[1] These were not proper lab books or formal records of results but jottings for immediate use, made on any piece of paper or parchment that happened to be handy. They were not dated, and were not kept in any sort of order. The entries on any one sheet might have been made on widely separated dates, and on at least one occasion a scrap of paper from one sheet was pasted on to another.[2] Drake provided a path through this wilderness but in many cases his interpretations were mere guesses and some of his translations and explanations are incomprehensible. He was also highly partisan, always showing Galileo's actions in the best possible light and treating his science as beyond reproach. His final haul of real experimental results was pitifully small, but enough to counter the very negative views of Alexander Koyré, which at that time were generally accepted.

Koestler provided another perspective. He was clearly unable to decide whether he disliked the Catholic Church more or less than he disliked Galileo, and he gave neither an easy ride. Of Galileo he said that much of his fame rested on discoveries that he never made and on actions that he never performed, and he listed some of them. They included the inventions of the telescope, the microscope, the thermometer and the pendulum clock, and the discoveries of sun spots, the law of inertia and the parallelograms of forces and motions. It is, however, hardly Galileo's fault if he has sometimes received credit that he never claimed, and Koestler did have to admit that the man who even he described as an 'outstanding genius' had earned his place amongst the shapers of human destiny by founding the science of dynamics. When he quoted Newton's famous statement to the effect that 'If I have been able to see farther, it was because I stood on the shoulders of giants', he identified these giants as Kepler, Galileo and Descartes (Koestler 1959; p. 358).

It is, perhaps, being over-pedantic to point out that it was to kinematics, not dynamics, that Galileo made his most important contributions, and that when Newton made his statement he was talking about optics.

---

[1]Images of the folios can now be accessed, together with notes and text transcriptions, through the website of the National Library in Florence, http://www.imss.fi.it/ms72/index.htm or the Max Planck Institute for the Study of the History of Science; http://www.mpiwg-berlin.mpg.de/Galileo−Prototype/index.htm.

[2]Drake (1990), referred to in the text as *Pioneer Scientist*. Drake's standing at the National Library in Florence must have been very high indeed, since he persuaded the director to have the pasted strip removed so that he could read what was written underneath.

The Legends

Koestler also identified as mere myths events that cannot be proven to have either happened or not happened. He was, for example, adamant that when, in 1633, Galileo was forced by the papal court to deny that the Earth moved around the Sun, he did not add, under his breath, "*Eppur si muove*"—'and yet it does move'. But how would anyone (including Koestler) know? Whether or not you think it believable largely depends on your opinion of Galileo.

Koestler also said that Galileo never threw down weights from the Leaning Tower of Pisa (Fig. 1.2 centre), and there he has to be granted at least technical accuracy. If Galileo did take weights of different sizes up the tower, he would surely not have thrown them down. That would have made it very difficult to prove that they fell at the same speed. The whole point of such towers is that they are great places from which to drop things.

The tale of the tower is, of course, one of the legends by which Galileo is chiefly remembered, and there are always people who want to spoil good stories by claiming that they are mere inventions. Their duller and more mundane versions often seem depressingly plausible, but the evidence for this story being a fiction is actually weaker than the evidence for it being a fact. Did he really climb the tower and drop from it (perhaps) a cannon ball and a musket ball? No, say the sceptics, because if he had he would have recorded it in his notebooks. It is, they say, a tale that was first told by Viviani, and not circulated until long after Galileo's death.[3]

Is this convincing? Geologists are taught at the very start of their training that absence of evidence is not evidence of absence. It is fair to at least ask where Galileo would have written about such an event. In his letters to his favourite daughter in a convent? Unlikely, since she was not even born until eight years after he had ceased to live in Pisa, and in any case those letters were all destroyed by her abbess after her death (Sobel 1999). We have only her letters to him. A similar fate may have befallen much of his other correspondence, as former colleagues scrambled to distance themselves from a convicted heretic.

In his scientific notebooks? There are no notebooks, just loose sheets of scribblings. Moreover, what we do know about Galileo suggests that he would not have considered this a proper experiment, to be written down. For one thing, if he did do it, he would not have been the first. Simon Stevin had dropped a musket ball and a cannon ball from the conveniently tilted tower of the Oude Kerk in Delft (Fig. 1.2 left) in 1586 (Dijksterhuis

---

[3]For a relatively recent brief review of the arguments, see Segré (1989).

Fig. 1.2 The Three Towers. From left to right: The Oude Kerk in Delft, from which Simon Stevin dropped weights several years before Galileo may have done the same thing in Pisa (*Photo* Richard Dingley). The Leaning Tower of Pisa (*Photo* Warwick Mihaly). The Asinelli Tower in Bologna, used by Riccioli to make the first respectable estimates of 'g' (*Photo* The Braschi-Levi family)

1943), three years before Galileo was appointed to the chair of mathematics in Pisa, and there had been others. If Galileo knew of any of them, he would not have thought his own demonstration worth recording.

It is also not true that there is nothing in Galileo's writings to suggest that it happened. For most of his life he was locked in combat not with the church, but with Aristotle, who had died some two thousand years before. In his last book, *Two New Sciences* (Galilei 1638), he wrote.

> Aristotle says that "an iron ball of one hundred pounds falling from a height of a hundred cubits reaches the ground before a one pound ball has fallen a single cubit". I say they arrive at the same time. You find, on making the experiment, that the larger outstrips the smaller by two finger-breadths; … now you would not hide behind these two fingers the ninety-nine cubits of Aristotle, nor would you mention my small error and at the same time pass over his very large one.[4]

This does read as if it was not Galileo but someone else who made the demonstration, but *Two New Sciences* was written as a dialogue and the 'you' was Simplicio, an imagined Aristotleian disputant who had to be con-

---

[4]The quotations are from the translation from the Italian and Latin by Henry Crew and Alfonso de Salvio, entitled *Dialogues concerning Two New Sciences* and referred to in the text as *Two New Sciences*. The page numbers of the original Italian edition were inserted by the translators in their text, and these are given, separated by a right slash, after the page numbers of the translation.

founded. This was Galileo's favourite way of writing, and gives some insight into his thinking. A modern scientist is able to subject his theories to critical appraisal, first by his colleagues and then by his wider peer group. That route was not available to Galileo, whose critics would merely have repeated the words 'Aristotle said …'. He had to provide his own peer review. If there was no real person making the statement, then it is likely that he made the test himself. 'Two finger-breadths' sounds like observation, not theory.

Moreover, Viviani was not just any biographer. He was Galileo's last student, and his companion during the last four years of his life under house arrest. He was present when the old man died, and was the only one of his many biographers who had actually known him. As Galileo's sight failed, it was to Viviani that he dictated his final work. During the long years of confinement, their conversations must have wandered over many events that had not seemed worth writing about when they actually happened.

An even more convincing argument for the truth of the story comes from what we know of Galileo's character. If he did make such a demonstration, it would probably have been between 1589 and 1591, when he was teaching mathematics at Pisa University. His own writings, and the descriptions left by his contemporaries, all reveal a man who loved a good argument (as long as he won) and arguments about Aristotle must have been almost daily events during this time. How could he not, on at least one occasion, have decided to prove his opponents wrong with a simple demonstration? Viviani's description (Viviani 1654) suggests that he might have done it a number of times, because

> he showed that the speeds of bodies of different weights, moving in the same medium, were not in proportion to their weight, as described by Aristotle, but that they move at the same speed, this he demonstrated with repeated experiments made from the height of the bell tower of Pisa with the help of other teachers, philosophers and all the students.[5]

Viviani did not, as is known from comparisons with other contemporary accounts, get everything right, but his identifiable errors were mainly, and predictably, about dates. Mistakes of that sort would be expected in the ramblings of an old man reminiscing about events long ago. 'All the students' could not, of course, be strictly true, but who would expect it to be? It was certainly not intended to mean 'all the students in Italy', let alone in Europe,

---

[5]Excerpt translated by Ted Metcalfe.

so why should it be taken to mean, as some have argued, 'all the students in Pisa'? It is much more likely that it referred to all the students in a particular class, or taking a particular course. It is surely quite improbable that Viviani would have made all this up, without Galileo himself having said anything about it. It may not have happened in exactly the way described, but not all the things that people in their seventies remember are exactly true. That doesn't mean they are mere inventions.

Yet another story concerning Galileo that is now often dismissed as myth is that, as a bored teenager forced to sit through interminable services in Pisa cathedral, he used his own pulse to time the swing of a lamp hanging from the ceiling. Once again Viviani is the only source we have for this story but it has, in its unembellished form, a ring of truth. Dava Sobel (Sobel and Andrews 1998) talked of this as 'an early mystical experience', but Galileo was the least mystical of men, and the most straightforward version is likely to be the most accurate. When trapped with nothing to do, and nothing interesting happening, the mind wanders. It is entirely believable that a youthful Galileo would pass otherwise unproductive time in this way, and in *Two New Sciences* (p. 47/141) he showed that he thought such observations commonplace. And, after all, unless something of the sort had happened, why would he have begun experimenting with pendulums? It is much more difficult to accept Koyré's claim that Galileo made his great discovery by comparing the times of swing of pendulums of the same length, but first and foremost '*by hard mathematical thinking*' (Koyré 1953).

Koyré's conclusion is all the more remarkable because Galileo lacked the mathematical tools to treat the motion of pendulums, and the discussions of their motion in *Two New Sciences* are based around experiments and observations. The textbooks that Koyré scorned for repeating Viviani's story of the pulse and the chandelier at least had some basis in a near-contemporary text, however unreliable. Koyré had none, and his picture of Galileo sitting down at his desk and deciding what it was that he was going to think about mathematically that day is almost laughable. It may possibly be how he himself worked, but few, if any, scientists work like that. Science advances because someone becomes curious about something. There has to be a trigger, and it is just as likely to be a lamp swinging from a ceiling as anything else.

There is one other possibility, which would reflect less well on Galileo. Leonardo da Vinci had sketched a design for a clock using a pendulum many years earlier, and an Arthur Koestler might suggest that Galileo had known about this and that, in telling Viviani the story of the lamp, he was trying to establish his claim to originality, if not priority. But Leonardo's sketch does not necessarily mean that he had noticed the constancy of the times of swing. In all clocks, the energy needed to keep them going is supplied through devices known as escapements, and a typical escapement for a pendulum clock will only work if the swing is always almost the same. Leonardo might have based his idea for a clock (which was never built) on nothing more than that the same swing always takes the same time.

There is one more argument against the truth of the story, which to Koyré seemed conclusive. What is now pointed out as 'Galileo's lamp' was not there when he was a teenager. The cathedral guides have an answer to that, and one of which Galileo himself would have been proud. *Do you think that, before that, they worshipped in the dark?*

Galileo and Aristotle

In the satirical pamphlet Dialogue Concerning the New Star, Matteo, one of two argumentative peasants, is recorded as asking *What has philosophy to do with measuring anything?* The pamphlet was published in 1605 (the 'new star' being the object now sometimes known as 'Kepler's Supernova') and is generally accepted as the work of Galileo. It is easy to imagine him saying this, grumpily, in response to some particularly inane remark, and then stomping off, leaving no time for a reply. It is especially easy to sympathise because geologists also have been obstructed, on at least three important occasions, by 'philosophers' (i.e. theoreticians) who told the field observers, with absolutely certainty, that what they observed could not be true.[6] The 'philosophy' that Galileo, through Matteo, was talking about was the idea, grounded in the somewhat suspect writings of Claudius Ptolemy in the Second Century AD,[7] that the Earth was fixed in space and that the sun orbited around it.

Galileo had not only the followers of Ptolemy to cope with but, still more immovably, the followers of Aristotle. Why they had such a stranglehold on philosophy at the start of the 17th Century is something of a mystery. It is sometimes supposed that it was because they had the backing of the church, but there was no theological reason why this should have been so. Aristotle

---

[6]The first of the three was the conflict with the theoreticians of the Church who assigned the Earth an age of only 6000 years. Having (mainly) won that battle, geologists then had to contend with Lord Kelvin, who claimed that the Earth could not be more than 50 million years old—still nowhere near enough. Thirdly, they were faced with physicists who told them that the continents could not possibly have moved relative to each other, despite all the field evidence that indicated that they had.

[7]A modern view can be found in Newton (1977).

may have been an early monotheist but, having lived several centuries before the birth of Christ, he was by definition a pagan and therefore not, in the sight of the Church, a person deserving of any special respect. And while the ideas of an Earth that is fixed and a sun that rotates around it are firmly grounded in good solid common sense and observation, there was much in Aristotle that offended against both. Galileo spent much of his time pointing this out, and in doing so upset most of his fellow academics.

A good example of his approach can be found in *Two New Sciences*, which he had published following a trial that would have cured any sensible person of being controversial. He, however, evidently still enjoyed confronting paper opponents whose arguments he could destroy and who could not call on the services of the inquisition to back them up. Only a few pages into the book we find him renewing his old war with Aristotle over the motions of falling bodies. Rather than relying on experiments that he was by that time too ill to make, he based his attack on contradictions in his opponents' thinking. At its heart was a very basic question—what does it take for a collection of bits to be regarded as a single body? He himself did not have to answer that question, because, whether one body or multiple bodies, according to him it made no difference to their rate of fall. But the followers of Aristotle did have to give an answer, because they thought that a cannon ball and a musket ball would fall at very different speeds, and therefore had to be able to say at what speed they would fall if they were linked by a light but rigid rod.

Aristotle valued theory over observation. It seemed obvious to him that heavy objects <u>should</u> fall faster than light objects, and that their speeds of fall <u>should</u> be proportional to their weights, and so he wrote that it was so, despite what must have been almost daily experiences to the contrary. It is now almost impossible for us to even enter the mind of such a person, but it was for his unthinking followers that Galileo reserved his contempt. For the man himself he showed respect. He said that

> … we come now to the other questions, relating to pendulums, a subject which may appear to many exceedingly arid, especially to philosophers who are continually occupied with the more profound questions of nature. Nevertheless, the problem is one which I do not scorn. I am encouraged by the example of Aristotle whom I admire especially because he did not fail to discuss every subject which he thought in any degree worthy of consideration. (Two New Sciences 94–95/138)

Aristotle looked at the universe and speculated about its ultimate origin, and that was not a path that Galileo chose to follow. Rather, he contented himself with discovering the laws by which it operated. Why those laws existed was of less interest. That he was, throughout his life, an ardent Catholic must have helped shape this attitude, since to such a person the ultimate cause would always have been God. Scientists, to him, were in the business of discovering <u>how</u> God had arranged things, not why.

A Route Map

In tracing the history of Galileo's investigations of gravity, I have relied mainly on what he himself said in *Two New Sciences* and what Drake, in his various publications, said about the folio notes. The task would have been much easier had it been possible to follow him in supposing that Galileo discovered the square-law relationship between the distance travelled by an object in free fall and the time of fall by first studying pendulums, then relating pendulums to fall and only then relating fall to descents down inclined planes.

If this is true, it is rather odd that vertical fall was treated in *Two New Sciences* only as a special case of the Law of Roll that governs descents down inclined planes. Nor is the sequence the one that Drake himself followed in the first chapter of *Pioneer Scientist*. In this complex and in places almost incomprehensible account, Galileo is described as reaching his final enlightenment in a series of stages from which logical method and progression are entirely absent. It might be argued that to expect these things of a Renaissance scholar is unrealistic, but there is very little in Galileo's own writings or in what his contemporaries said about him that fails to strike a chord with the modern mind. To appreciate this, it is necessary only to compare the ease of translating his works (mainly in Italian, an innovation in its own right) into English with the near-impossibility of translating the Latin of his contemporary, Johannes Kepler.

It is not difficult to take the information assembled by Drake and construct a far more believable progression. It would be that:

1. Galileo notices (perhaps in Pisa cathedral—why not?) that things on strings swing more slowly when the strings are longer, and is sufficiently intrigued to investigate further.
2. He very quickly finds that the angle of swing does not affect the time of swing, as long as the angle is not too large. He wrongly, but under-

standably, attributes the longer times for larger angles to the effects of air resistance.

3. He also finds that the weight on the string does not affect the time of the swing, provided that is not so light that air resistance becomes important.
4. With a little more effort he finds that the time of swing is proportional to the square of the length of the string.
5. In trying to understand these results, he turns to the only thing he can think of that resembles the motion of a weight on a pendulum, and begins timing balls rolling down slopes.
6. By measuring over times that are simple multiples of one basic unit, he discovers the times-squared Law of Roll.
7. He makes further experiments with slopes using a water-clock to measure times, and discovers other relationships that suggest strong links to the motions of pendulums.
8. Realising that he now has two laws relating lengths or distances to the squares of times, and that these imply a constant ratio between the time taken by a body to fall vertically through a set distance and the time of swing of a pendulum of that length, he attempts to measure this ratio.
9. He sees a telescope for the first time and loses interest in everything else, only returning to Swing, Roll and Fall when he sits down, twenty years later, to write *Two New Sciences*.

Because the notes he left behind are incomplete, we will never know whether this was the route actually taken, but it does make sense. It does not, unlike the scheme proposed by Drake, imply that Galileo effectively discovered the times-squared Law of Roll twice, but on the first occasion failed to notice it.

Step 8 is important for what it can tell us about Galileo's skill as an experimenter. He did not know what the ratio should be, but we do, so we can use his answer to check his experimental accuracy. It turns out to be quite impressive but, to appreciate this, the experiments and their results have to be examined in detail.

The Pendulum

The time taken by a pendulum to swing from one extreme to the other and back again is known as its *period*. During a single period, the weight passes through every point (except the two extremes) twice, moving in opposite directions. The most accurate measurements of time and position are made when the string is vertical and the weight is moving at its greatest speed, and this happens twice in every period. It is for this reason that what came to be known as the 'seconds' pendulum was defined as having a half-period, rather than a full period, of one second.

There is no direct evidence of when Galileo discovered that the square of the time of swing was proportional to the length of the pendulum, but Drake's assumption that it would have been before the experiments on Roll is almost certainly correct. Pendulums are far easier to work with than either Roll or Fall, because they themselves can do the timing. In *Two New Sciences* Galileo not only stated the rule that

> … as to the times of vibration of bodies suspended by threads of different lengths, they bear to each other the same proportion as the square roots of the lengths of the threads ….

but described an experiment which is so simple that it must surely have been one of the first that he made

> For if I attach to the lower end of this string a rather heavy weight and give it a to-and-fro motion, and if I ask a friend to count the number of its vibrations, while I, during the same time interval, count the number of vibrations of a pendulum which is exactly one cubit in length, then … one can determine the length of the string …. (Galilei 1638, p. 96/139–140)

This is rather oddly set out, as a way of determining the length of a string (for which, one feels, there would have been many easier methods), but that is a consequence of Galileo's preferred way of writing his science, as arguments between participants of varying degrees of intelligence. There must surely have been many occasions on which he used the technique described, of counting the vibrations of two pendulums swinging simultaneously.

We have very little information on how Galileo carried out his pendulum experiments, but on Folio f151v[8] Drake found, scrawled across some written notes, a very rough sketch of what appeared to be two interlocking gears (Fig. 1.3a), and promptly demonstrated to the full his talent for making things much more complicated than they need be. While it is true, as he said, that running the string over a nail in a movable upright and anchoring it to a bench would allow the nail to be raised and lowered by gears and a

---

[8]The folio notes are referred to by f numbers, followed by v (for verso) or r (for reverse).

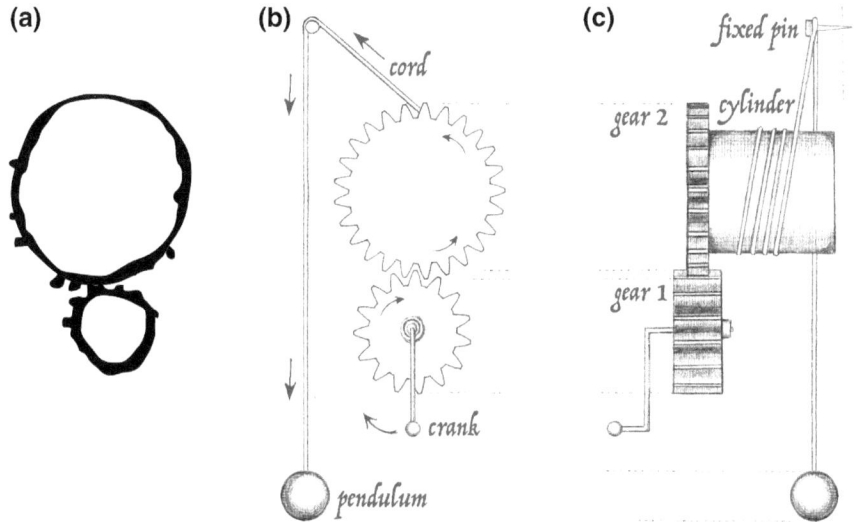

Fig. 1.3 A role for gears in pendulum experiments. **a** The essentials of Galileo's original sketch on Folio 151v. **b** and **c** Front and side views of suggested interpretation. The gearing allows finer adjustments to be made to the length of the pendulum than would be possible with a direct drive to the windlass, and it is possible that Galileo used such a system (*Drawing* Kate Milsom)

crank (Drake 1990, p. 15), the idea comes from a mind far more tortuous than Galileo's. It would be difficult to do, and quite unnecessary. If the basic purpose of the device was as proposed (and there is no strong evidence either way), then it would be much simpler for the larger wheel to be a windlass on to which the string could be wound or unwound. The smaller wheel, turned by a crank, would provide gearing to make small adjustments easier and more precise. The nail could stay exactly where it was.

The Making of a Scientist

One version of the pendulum story has Galileo timing the swinging lamp in Pisa cathedral with his own pulse and then rushing home, locking himself in his room and doing nothing but pendulum experiments for the next week. In others he is only a teenager, although most paintings of the event show a very mature person staring fixedly at a lamp. These accounts seem incompatible, let alone believable, but it is a fact that Galileo was involved with practical science from a very early age. It was all thanks to his father.

Vincenzo Galilei married Giulia degli Ammannati in the summer of 1562 and a little over seven months later Galileo, their first child, was born in Pisa. There is no evidence that the birth was premature, which suggests that the conception might have been. Socially, the marriage was a step up for Vincenzo, since Giulia's family was somewhat further in from the outer fringes of the nobility than the Galilei, but he may have lived to regret it. Then as now, marriage to a professional musician was not a passport to a secure or comfortable life, and although Vincenzo was a sufficiently skilled lute player and singer to attract powerful supporters, the interest of Renaissance patrons could disappear as quickly as it appeared. Guilia may have been permanently resentful of having married less well, and less prosperously, than she might. If there were any favourable comments made about her during her lifetime, they have not survived.

Vincenzo was much more than just a musician, he was a theorist and an experimentalist who made important discoveries concerning the physics of vibrating strings and vibrating columns of air. He was also a prickly and argumentative character, notorious for his heated attacks on his former teacher, Gioseffo Zarlino. His eldest son, who seems to have inherited almost all the most notable aspects of his character, both the good and the bad, almost certainly helped in some of the experiments. He was, at the very least, well aware of them, and in *Two New Sciences* his discussion of pendulums leads straight into a discussion of music and vibration that continues through to the end of the first 'day'. And, throughout his life, he followed closely the principles succinctly expressed by his father when he said, in his *Dialogue on Ancient and Modern Music*, that *they who in proof of anything rely simply on the weight of authority, without adducing any argument in support of it, act very absurdly.*

It was no part of Vincenzo's plan that Galileo should follow in his footsteps. With two daughters to dispose of, it was imperative that when the time came the family finances would be sound, and that meant placing the eldest son in a profitable profession. The one chosen was medicine, and it was with that in view that in 1581, at the age of seventeen, Galileo was enrolled in the University of Pisa. Once there, and in defiance of his father's wishes, he showed little interest in medicine and far more in mathematics. Four years later he returned to the family home, by this time in Florence, without a degree but with enough mathematics to earn a precarious living as a private tutor. It was not until 1588 that he achieved a sort of financial stability, with his appointment to a poorly-paid lectureship back in Pisa. Lecturers in philosophy at the time earned four to six times as much as lecturers in mathematics, which may have played its part in Galileo's life-long hatred of philosophers.

To have even reached this stage, Galileo had had to establish some sort of a reputation, and this he had done by giving public lectures and by designing things and making them. He was a skilled craftsman as well as in innovative scientist. Intriguingly, two of his lectures, and two of the best received, were concerned with the shape, location and volume of hell, as described in Dante's Inferno. It seems a strange thing for a serious scientist to do, but the boundaries of science had not yet been established.[9]

It was just as well that Galileo was able to supplement his stipend at Pisa, because in 1591 his father died, leaving him as the head of the family with the immediate responsibility for paying out the generous dowry owing to the husband of his recently-married sister Virginia. Fortunately also, in the following year he was able to leave Pisa (where he had made himself unpopular not only by arguing with almost everybody but also by ignoring and lampooning the university dress code) to take up a better-paid appointment at the University of Padua, in the Republic of Venice.

Galileo's eighteen years at Padua were, by his own account, the happiest of his life, and also the most productive. He invented a horse-driven pump for raising water and a device that he called a *geometric and military compass*, consisting of a jointed ruler that could be used to provide approximate solutions to a number of mathematical problems, and made money from the sales of both, and from their instruction manuals. He also began a long liaison with a Venetian commoner, Marina Gambia, which brought him a son and two daughters. He wrote extensively and argued intensively. And, during that time, he began to study objects falling under gravity.

The Inclined Plane

A pendulum swings from side to side, and a falling body drops straight down. Neither Drake nor any of Galileo's other biographers seems to have appreciated the magnitude of the conceptual leap needed to link the one directly to the other, and none of the surviving documents suggest that Galileo made such a leap. It is much more likely that he studied balls rolling down slopes because he recognised in their motion a similarity to the side-

---

[9]If they ever have been. One of my former university colleagues was much in demand for his lecture on the geology and plate tectonics of Tolkien's Middle Earth.

ways motion of pendulums, which he had already studied. Experiments with pendulums are easy but, in the days before Newton and Leibniz invented calculus, the theory behind their motion was mathematically challenging. Galileo may have begun experimenting with balls and slopes merely because the mathematics seemed likely to be simpler.

Even if this is true, there is no timeline. The end point, when he learnt of the telescope and switched from kinematics to astronomy, can be dated to 1609, but we do not have a start point, and he was a very busy man. There may have been a considerable gap between the main series of pendulum experiments and any experiments involving the Law of Roll, but once these had begun he made so many interesting discoveries that he devoted more than eighty pages of *Two New Sciences* to them (as against the mere twenty pages allocated to pendulums). Drake, despite believing that it was only after the discovering the Law of Fall that Galileo experimented with '*descents along planes*', began *Pioneer Scientist* by discussing three columns of numbers that he found scrawled in a corner of folio f107v and which he identified as the results of an early experiment with Roll. He offered no proofs, but his explanation is a very plausible.

The relevant part of the folio is shown in Fig. 1.4, redrawn to remove some extraneous material. The near-illegibility of Galileo's handwriting is preserved. Many of the numbers can only be deciphered because they are predictable or because they are repeated; the second of the columns merely lists the whole numbers from one to eight and the first, which is the most difficult to read and was almost certainly added later, lists their squares. It was written in the same hand but with different ink. There is no way of knowing how long it was before this column was added but, given the way that Galileo used his scraps of paper, it would probably have only been a few days before f107v disappeared under a pile of others.

The numbers in the third column are the crucial ones. They are, successively, 32, 120, 298, 526, 824, 1192, 1620 and 2104 and were interpreted by Drake as the total distances travelled by a ball rolling down a slope in the times listed in the second column. There is no actual proof of this idea, but they must surely represent something of the sort because, as discussed in more detail in Chap. 14, Coda 2, if each number is divided by 32 the result is a series of numbers very close to the series of squares from 1 to 64. The units can be assumed to be *punti*, the 0.94 mm gradations of Galileo's personal ruler, which was 60 *punti* long (Drake 1990, p. 9), because there are calculations to the right of the three columns that show that the distances

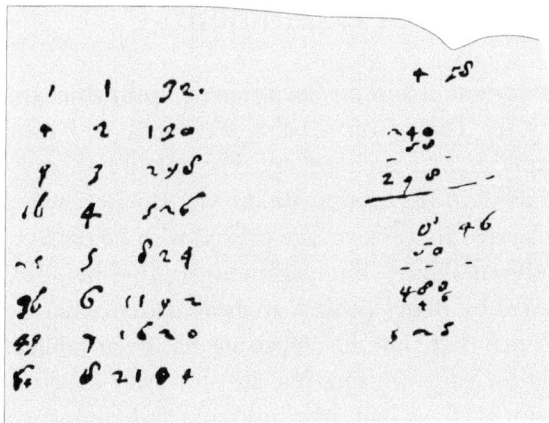

Fig. 1.4 The essentials of the top left hand corner of Folio f107v. The results are on the left, the calculations of the first two distances measured with the 60 *punti* ruler are on the right

were measured in 60-unit lengths. In some cases, these calculations provide more readable versions of the total distances than does the three-column listing. The maximum distance, of 2104 *punti*, would have been slightly less than two metres.

Taken together, the three columns look very much like records of a real experiment, but they are all the information we have. Galileo was clearly measuring time in intervals that were multiples of some basic unit, but we do not know what that unit was. He might have continued his (possibly apocryphal) use of his own pulse in Pisa Cathedral, but the human pulse is not an ideal clock. It is unlikely to be constant, even for a single individual, and is certainly not transferable from one person to another. For experiments in which distances were measured at fixed times there are many other possibilities, including pendulums. While there is no record of Galileo having ever used these for timing anything other than other pendulums, one of his colleagues in Padua is known to have developed a pendulum device that he called a '*pulsilogium*' for medical use (Sanctorius 1631), and where Galileo talks in his published work about a 'pulse', he might have been referring to this. His well-documented musical expertise could equally well have led him to use the vibrating strings that he and his father experimented with, or the beat-frequencies produced by two strings.

Experiments About Experiments

In 1947 Thor Heyerdahl and his five insanely optimistic companions drifted half-way across the Pacific on a balsa-wood raft and ushered in a whole new era of experimental archaeology. Heyerdahl's theory, which had the Polynesians arriving in the eastern Pacific via America, was probably wrong, and he never claimed that his voyage proved it to be correct. What it did do was answer, very effectively, one argument against his ideas that had been seen as conclusive by many people. It showed that what he was suggesting was possible. Since that time his approach has been replicated hundreds of times, not only for early voyages but also for early experiments. Inevitably, Galileo's experiments have had their imitators and, almost equally inevitably, in the forefront of these was Stillman Drake. He described what he did in a paper uncompromisingly entitled '*The Role of Music in Galileo's experiments*' (Drake 1975).

The scope implied by the title is very broad, but the paper actually dealt only with the experiment that supposedly produced the results recorded on f107v but which Galileo never described. To reconstruct it Drake cut a groove about six feet long in a block of hardwood, tilted it at a suitable angle and rolled down it a steel ball. In *Pioneer Scientist* he wrote that the slope angle was 1.7° and the basic time interval was 0.55 seconds, but it is disconcerting to find him admitting in '*The Role of Music*' that these were just guesses. This is a reminder of the need for caution when reading his work, but it is also true that he is almost indispensable. He identified the key entries in Galileo's labyrinthine folio notes, and no-one, surely, would want to repeat the long years that he spent in doing so.

Galileo's main difficulties when investigating the Laws of Fall and Roll arose from the need to measure very small intervals of time. With pendulums the problem was easily solved by counting the oscillations of different pendulums swinging simultaneously, and it was only when he began to study objects that were falling or rolling that he needed to do anything more complicated. Drake suggested that Galileo defined his basic unit by singing a song with a very strong beat, but while his arguments are plausible and he himself managed to obtain respectable results using *Onward Christian Soldiers*, there is nowhere any independent confirmation. Drake also suggested that strings or wires would have been placed across the slope in order to produce audible 'bumps' as the ball rolled over them, and even went so far as to argue for these having been lute frets. But lute frets would have impeded the ball if tight and given false readings if loose, and in his replica experiment Drake used rubber bands. It is just as likely that in Galileo's experiment the ball was halted by a sounding board that was moved until the time from release to audible impact matched the selected interval.

After their publication Drake's experiments were roundly criticised in a rather acrimonious correspondence in the pages of *Annals of Science* that was only brought to an end when the editor intervened by pointing out, rather crossly, that although the 'singing' hypothesis should have been the issue, none of the critics had actually mentioned it. Instead, the discussion had centred around a photograph, taken in Drake's absence by a photographer who had died shortly afterwards, of an experiment in which the timing was done electrically.[10]

There the matter rested. For the later experiments we have, instead of guesses, Galileo's own description.

> A piece of wood moulding or scantling about 12 braccia long, half a braccio wide, and three finger-breadths thick, was taken; on its edge was cut a channel a little more than one finger in breadth; having made this groove very straight, smooth and polished, and having lined it with parchment, also as smooth and polished as possible, we rolled along it a hard, smooth, and very round bronze ball. Having placed this board in a sloping position, by lifting one end some one or two braccia above the other, we rolled the ball … along the channel …. We repeated this experiment more than once … with an accuracy such that the deviation between two observations never exceeded one pulse beat. Having performed this operation … we now rolled the ball only one-quarter of the length of the channel; and having measured the time of descent, we found it precisely one-half of the former. Next we tried other distances … the times of descent, for various inclinations of the plane, bore to one another precisely that ratio which … the Author had predicted and demonstrated for them. (Galilei 1638, p. 96/212–213)

Here 'the Author' is Galileo himself, making a guest appearance in a work otherwise devoted to the imagined arguments between Simplicio, Sagredo and Salviati. He makes other appearances as 'the Academician'. Quite clearly, as far as the times-squared law was concerned, these were confirmations rather than investigations, since the distances were pre-selected in anticipation of the results. Because it required times to be measured to fractions of a single pulse, the experiment was timed using a water clock, which was also described.

---

[10]Editor's note accompanying MacLachlan (1982).

For the measurement of time, we employed a large vessel of water placed in an elevated position; to the bottom of this vessel was soldered a pipe of small diameter giving a thin jet of water, which we collected in a small glass during the time of each descent … the water thus collected was weighed, after each descent, on a very accurate balance; the differences and ratios of these weights gave us the differences and ratios of the times …. (Galilei 1638, p. 96/212–213)

For some reason these descriptions roused Koyré to a critical fury. The idea of an experiment based on rolling a bronze ball down a wooden groove appeared to him ridiculous. He thought the water clock described by Galileo inferior to the 'Roman' water-clock of Ctesebius (who was actually a Greek living in Alexandria long before Roman influence became significant there), and he ended his diatribe by concluding that Galileo's experiments were completely worthless (Koyré 1953).

Why he was so dismissive is a mystery. It is very clear from the discussions in *Two New Sciences* that Galileo obtained important information from his experiments, and there is no reason to suppose that his water-clock would have been any less accurate than its Alexandrian predecessor. There is no indication of the actual rates of flow, but Thomas Settle, the first person in modern times to make a serious attempt to reproduce the experiment (Settle 1961), used a tube that delivered water at a rate of about 20 cc per second and controlled the flow by placing or removing his finger from the inlet to the tube within the reservoir. This is not the obvious way of doing things, and he did not explain why he chose it, but he did say that at every stage where there was a choice he deliberately opted for the method that would produce the less accurate result, in order to give errors *every reasonable chance to accumulate*. He measured water volumes rather than weights and was less than scrupulous in measuring distances, but even so, after a number of 'training runs' to get into the rhythm of the experiment, he obtained results that were accurate to a tenth of a second.

Galileo, who was trying to be accurate, would surely have done as well, or better, especially if he used a better timing method. A major problem with a water clock is that errors are introduced when flow starts, because the flow pattern is being established and the conditions are not 'steady state'. As long as attention is on the reservoir and its outlet, there is no solution. Transferring attention to the collecting vessel changes all this. Steady-state flow can be established first and it then takes much less than a tenth of a second to put a collector in place to begin timing, or to remove it to end it. It is even quicker, and avoids the need to move the collector when partly full, if it has a lid that can be removed to start timing and replaced to end it This is a bit messy (when the water is not going into the collector it goes every-where else) but it does work. As with all methods, there are errors stemming from the observer's reaction times, but the delay at the start is roughly can-celled out by the delay at the end.

Despite all his efforts, Drake could find no evidence that Galileo ever attempted to relate any of his time measurements to the astronomical sec-ond but, inexplicably, he also felt able to state that the flow rate was about three fluid ounces per second '*very nearly indeed*' (Drake 1990, p. 12). This is equivalent to 1440 of the 'grains' in which Galileo weighed things, and to 60 cc per second, which is three times the rate in Settle's experiment and hardly seems consistent with flow 'via a narrow tube'. The tap in my kitchen has to be almost fully open to achieve this rate (but water pressure can be low on hillsides in Wales). A wide tube would have been needed, with the risk of water splashing out of a small collecting vessel, but with such rapid flow the weighing of amounts corresponding to a tenth of a second or less would have been easily within the capabilities of the balances available.

The Complexities of Roll

In the experiments described in *Two New Sciences* an important new element was introduced that was not present in the experiment of f107v. The angle of the slope down which the ball was rolled was varied, providing a whole new set of insights. The most important of these was the counterintuitive discovery that the time taken by a ball to roll down a slope that formed a chord of a circle with one end at the circle's lowest point did not depend on the position of the upper end, as long as it was on the circle (Fig. 1.5). That Galileo found this out must surely have been because he was think-ing about pendulums and Swing, and not about Fall, when he investigated Roll. We cannot now know whether he discovered the rule experimentally (Drake), and then worked out why, or deduced the result geometrically (Koyré), by one of the routes described on pages 190–191/222–223 of *Two New Sciences*, and then confirmed it by experiment. Either way, it was an impressive achievement, and it provided him with a rather direct link between Roll and Swing. He must have been delighted.[11]

A lesser man might have stopped there, but Galileo went further. He replaced the single chord with two chords forming a composite slope with the same start and end points (Fig. 1.5b), and obtained the even less intui-

---

[11]The mathematics underlying this discovery are discussed in Chap. 14, Coda 2.

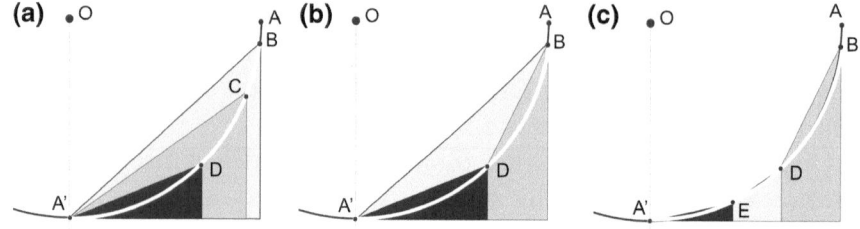

Fig. 1.5 The Laws of Roll. A' is the lowest point on the vertical quarter-circle AA'. **a** If B, C and D are points on AA', then the times taken by a ball to roll down the paths BA', CA' and DA' will be the same. **b** The time taken by a ball to roll down BD and then down DA' will be less than the time taken to roll down the straight path BA. **c** It follows that the time taken to roll down a segmented path between B and A' will decrease as the number of segments increases, until the path is effectively the arc BA' of the circle

tive result that the longer path was travelled in the shorter time. It was only after he had reached this result that he went astray. Lacking the mathematical tools for rigorous calculations, he guessed that the shortest time of all would be associated with a circular arc. He was wrong (the quickest path is down a cycloid curve, as described in Chap. 14, Coda 3), but not very wrong. For small swings, his guess was almost correct.

Nowhere in *Two New Sciences* did Galileo distinguish clearly between the laws governing balls that roll and balls that fall, and in many places he treated Fall as simply the extreme case of Roll. Once again his guess was wrong, because sliding friction is very different from rolling resistance and both are very different from the air resistance experienced by a body in free fall. Incorrect assumptions can, however, be fruitful, and although the idea may have led him to make some dramatically bad guesses about times of fall, it also prompted him towards some of his most important results.

Fall and Swing

Given his belief that free-fall was just a special case of roll, and therefore followed a times-squared law, it would not have taken Galileo long to conclude that there should be a fixed ratio between the length of a pendulum and the time taken by a falling body to travel the same distance. By bringing together results recorded on a number of different folios, Drake attempted to establish how this was measured. The first results he quoted (Drake 1990, pp. 12–14) were two timings of objects in free-fall over distances of

2000 and 4000 *punti* (about two and four metres). The amounts of water collected weighed in at 903 grains in the first case and 1337 grains in the second. Using the (rather suspect) conversion factors stated by Drake gives 0.63 seconds for the 2000 *punti* drop and 0.93 seconds for the 4000 *punti* drop. These very short times can be compared with the theoretical times of fall in a vacuum, which for 2000 *punti* would be about 0.62 seconds, and for 4000 *punti* about 0.88 seconds.

By themselves, the two results do not make a convincing case for a square law, but despite this Drake suggested that Galileo moved on immediately to find the lengths of the pendulums with quarter-periods (times to the vertical from the release of a displaced bob) corresponding to the 903 *grain* time-interval and half of the 1337 *grain* time-interval. His final values were 1590 and 870 *punti* respectively. If this process was purely one of trial and error, and if the water clock was being used, then the water would have had to be weighed after each 'run' and the pendulum length would then have had to be readjusted. It seems more likely that the falling mass and the pendulum bob were released simultaneously, and the pendulum length would have been altered until the thud of the mass hitting the floor coincided with the smack of the bob hitting the sounding board. This was the method used by Christiaan Huygens half a century later. It would have been tedious, but not as tedious as doing weighing after weighing after weighing. The folios suggest that some weighing was done, but only for confirmation.

Whatever the method, the ratios between the drop distances and the pendulum lengths were 2.52 and 2.30 in the two cases. The theoretical value, which is a constant regardless of pendulum length or gravity value, is equal to the constant 'π' divided by the square root of two, or 2.22. Galileo's results were thus very creditable, but did not have the almost ludicrously high accuracy that Drake claimed on his behalf.

Without the techniques that were developed seventy years later by Christiaan Huygens, Galileo could not explain the pendulum-length to drop-distance ratio, but without his work Huygens might never have arrived at the right equations. How much further Galileo might have taken his investigations into 'g' we can only guess. He did apply his acquired knowledge of the speeds attained by balls rolling down slopes to militarily important work on missile trajectories, but in 1609 a foreigner visiting Padua brought with him the first primitive telescope. At this point Galileo, the practical experimentalist and instrument maker, abandoned the sciences of motion for the science of astronomy. By the time he returned to 'g', when writing *Two New Sciences* some twenty-five years later, he must have forgotten many of the details of what he had done.

Tides and the Church

At first sight, the trial of Galileo by the Roman Curia in 1633 had nothing to do with gravity and everything to do with astronomy, but that is not the whole story. What got him into trouble was the publication of quite possibly his worst scientific idea; his Theory of Tides.

By 1616, Galileo was openly arguing in favour of the Copernican theory that put the sun, rather than the Earth, at the centre of the universe, but the attitude of the church to the idea had not yet hardened into outright opposition. For one thing, the Protestant versus Catholic tensions in central Europe were about to erupt into the Thirty Years War, and the Vatican had more important things to worry about than mere star-gazing. Drake, who, of all Galileo's biographers, is the one most interested in his science, has argued convincingly that it was not initially the church that wanted to crush the Copernican world view but the academic philosophical establishment, still rooted in the works of Aristotle. Galileo had spent much of his life upsetting this establishment by contradicting Aristotle, and it is more than likely that when he began to switch the places of the Earth and the Sun in his cosmology, the philosophers saw their chance. The first outright attack from a pulpit on Copernicus and, by implication, on Galileo was made not by cardinals or the Curia but by Thomas Caccini, a young Dominican friar on the make.

Since Caccini lacked both status and patronage, all might have been well, and the whole affair might have been quickly forgotten had not a riposte been published by a Carmelite theologian. This made things serious. The church was supposed to be united, it was already losing ground to heretics in northern Europe, and here were different monastic orders squabbling between themselves in a most unseemly manner. A leading Jesuit, Cardinal Roberto Bellarmine, was one of those who felt that something had to be done. He had been one of the judges who sent Giordano Bruno to the stake but he may have regretted this and become more flexible. At any rate, he delivered the Solomonic judgement that it was acceptable to use Copernican science as long as it was treated merely as a mathematical device for calculating the movements of planets, and not as a description of objective reality. Unwisely, Galileo sought a greater degree of clarity, and the affair ended with the issue of two contradictory documents, one a formal restatement of Bellarmine's position, the other rather stronger and supposedly representing the views of the Pope himself. This second document, however, had been prepared by a mere notary and was signed by nobody. In a final tidying up, the Carmelite defence of Copernicus was placed on the prohibited Index but, rather illogically, Copernicus' original book was merely 'suspended', pending 'correction'.

For sixteen years, nothing much more happened, but in 1632 Galileo decided to publish his '*Dialogue on the Tides*' a title that was changed to simply *Dialogo* at the request of the Church censors. The longer title '*Dialogue Concerning the Two Chief Systems of the World—Ptolemaic and Copernican*' only came into use much later. But, whatever its title, it was this publication that brought the inquisition down on Galileo's head.

In his tidal theory Galileo abandoned almost everything that had made him scientifically successful. It was based not on detailed measurements but on vague ideas that came to him, so it is said, when watching water slopping around in the bottom of a barge. Supposedly this all happened in 1595, and the barge was taking fresh water to Venice from its mainland possession of Padua. Inevitably, whenever the barge's speed or direction altered, the water inside moved about. When the barge grounded on a sandbar, the water first pushed up towards the bow and then ran back toward the stern, repeating this with decreasing force until it returned to a level state. In this, Galileo saw a mechanism for the tides that regularly sloshed up and down the Adriatic. Had his theory been correct, this might have been yet another fable about him that nobody believed, but since the theory was wrong, it seems that the story is acceptable.

In this tidal mechanism there was no role for the Moon. To anyone familiar with the Atlantic coasts of Europe, where the linkages between spring and neap tides and the Moon's phases have been taken for granted for thousands of years, this seems an incomprehensible error, but Galileo spent most of his life in inland cities (Fig. 1.6), and the seas that he did know were arms of the Mediterranean where the tidal ranges are almost trivial. Except to people who had chosen to live in a coastal swamp (which was, of course, exactly what his Venetian patrons had done), tides were of very little interest. Even in Venice, tides of more than a metre and a half are rare. Their consequences can be serious, but flooding is generally due to combinations of natural high tides and wind-driven storm surges.

Galileo certainly knew of theories that involved the Moon, because they were supported by Kepler, but this just made him angry. Kepler was, he admitted, a man of genius, but one who, to his mind, had fallen from grace because he had become '*interested in the action of the moon on water, and in other occult phenomena, and similar childishness*'. This rather uncharacteristic outburst (he was merciless in dealing with people for whom he felt only contempt, but he respected Kepler) is entirely consistent with Drake's view

Fig. 1.6 Galileo spent his whole life in northern Italy, moving between Pisa, where he was born, Florence and Padua. It was not an ideal background for someone wishing to study tides. His last visit to Rome was in 1633 for his trial. He then lived under house arrest in Arcetri, now part of metropolitan Florence but then a separate village, until he died, in 1642

of him not as a heretic but as a zealous Christian. At its heart is a total rejection of the astrological assumption that 'heavenly' bodies have some sort of influence on events here on Earth.

If it really was as early as 1595 that Galileo began to form his ideas about tides, he did little with them for another twenty-five years. When he did decide to publish, it didn't much matter that he was wrong. It did matter very much that his theory was incompatible with Bellarmine's polite fiction that putting the Sun at the centre of universe was a mathematical device unrelated to what was actually happening. That had kept everyone happy, but no mere mathematical short-cut could make the sea surge up and down the Adriatic twice a day. For this to happen in Galileo's theory, the Earth had to have at least two motions (orbital about the Sun and rotational about its own axis). There is irony in the fact that the true cause of the tides, in the gravitational attractions of the Moon and the Sun, could have worked just as well with the Earth motionless at the centre of the universe and the Sun and Moon in orbit around it. The numbers might have been wrong, but no-one in the early 17th Century had any idea about the numbers anyway.

To Urban VIII the *Dialogo* was a clear breach of the 1616 undertaking. There has been much speculation as to why this particular Pope, who had long been an admirer of Galileo and had granted him more audiences than most ambassadors, should have turned on him in the way that he did, but Galileo had not told Urban the full story of the events of 1616, which took place during the previous papacy, and it was this discovery that brought things to a head. It seems to have been a case of someone who, having mounted defence after defence of a person or organisation that is pushing the boundaries of the acceptable, eventually gives up and moves into opposition because something has been done that is simply indefensible

Why Galileo embarked on this course of action, so long after his initial tidal speculations and knowing that it might put an end to his easy relationship with the Vatican, is not clear. Drake's answer is that it was a last desperate attempt by a loyal Catholic to divert his church away from a rejection, for which he could see no theological justification, of the new sciences that were springing up all over Europe. On this interpretation, Galileo was quite deliberately provoking a confrontation that he thought, mistakenly, that he could win. This is at least easier to believe than that, as other authors have supposed, the trouble was actually caused by a hidden controversy over the communion wafers (e.g. White 2007). It is an idea based almost entirely on just two documents, the first of which is known as G3 and was discovered in the Vatican records in 1982 by Pietro Redondi. The second, known as EE291, is a report on G3 written by Father Melchior Inchofer in 1632 and is concerned not with the *Dialogo*, but with the earlier *Assayer*. In this work Galileo was, according to Inchofer, casting doubt on the doctrine of transubstantiation, the conversion of the wafer during the Mass into the literal body of Christ.

The argument is not convincing, because Galileo had a near-perfect defence. The acceptance of the possibility of miracles is basic to Christianity, to a greater extent, perhaps, than any other religion. For Catholics, if not for Protestants, the Communion witnesses the daily miracle of the transubstantiation. In *The Assayer*, Galileo was doing no more than describe the norm, the way things would be without divine intervention. Inchofer, evidently a rather stupid man, failed to notice that his own critique could be interpreted as denying the possibility of such intervention, and therefore as very heretical. Neither Galileo nor Urban were stupid men and could have easily dealt with him.

Another problem with this hypothesis is that it ignores the sheer venom with which academic disputes are so often conducted. Its proponents have been eager to condemn the Renaissance Papacy (and who, indeed, could defend it?), but in doing so they have underestimated the role played by the philosophical establishment (not particularly religious but strongly pro-Aristotle) in lobbying for Galileo's conviction. Even if he never did demonstrate their errors from the Leaning Tower, Galileo had made these people look foolish enough, on enough occasions, to have earned their permanent hatred. There is just as much evidence for the idea that the Church would, without their continued prompting, have preferred to leave Galileo alone as there is for the transubstantiation hypothesis. If it were correct, and the situation were as clear-cut as that implies, then it is surprising that three out of the eight judges voted for an acquittal.

Whatever the truth, there were, even after Galileo's conviction and sentencing, still churchmen who admired him, and were prepared to help him, and one of these, Archbishop Ascanio Piccolimini of Siena, had enough status to be allowed to oversee the first phase of his life under house arrest. The verdict came close to destroying Galileo mentally but it may have done a service to science, since it was while he was enjoying the Archbishop's rather courageous hospitality that he began writing *Two New Sciences*, his final testament. He continued the work after he had reached his final destination and begun the eleven years of confinement in his own home in Arcetri, where conditions were generally not too arduous but where he had little opportunity to do anything but think and write (and, after going blind, dictate). The 'New Sciences', which might be very loosely described as Kinematics and the Properties of Matter, were not really new, but represented most of those parts of his life's work that remained legal. Despite this, he had to look for a publisher in the Protestant Netherlands, because part of the Curia's sentence had been the proscription not only of everything he had ever written but of anything he might ever write.

In the end, the story of Galileo may simply be of a man who made so many enemies through flaws in his own character that the only way that his friends could find to protect him was to shut him up for good. He survived the publication of *Two New Sciences* by five years, dying, aged 78, in 1642. Later in the same year a sickly baby boy, not expected to survive and hurriedly christened with his father's name, was born prematurely to the wife of Isaac Newton, a recently deceased Lincolnshire farmer.

'Little g'

Was Galileo the first person to measure 'g'? And if not, who was?

Simple questions, but they do not have simple answers.

Galileo originally obtained the times-squared law from experiments with balls rolling down slopes and then extended it to objects in free fall. There are too many unknowns in such experiments for him to have deduced 'g' from them and, because he worked entirely with ratios, it is doubtful whether he ever made the attempt. Notoriously, in the *Dialogo* he had

Salviati say that a one hundred pound ball would fall 100 braccia in 5 seconds ('*una palla di ferro di cento libbre, la quale per replicate esperienze scende dall' altezza di cento braccia in cinque minuti secondi d'ora*'; Galilei 1638, p. 247), and it was this wildly inaccurate estimate that prompted the French mathematician Marin Mersenne to doubt that he had ever performed the experiment.[12] However the context suggests that this was a mere throw-away remark, not intended to be accurate, and it is even possible that the reference was to the *doppio braccio* of Fig. 1.7, rather than the single *braccio*. This would give a much better answer, but Galileo ultimately fails the test because he never attempted to express his experimental results in units that anyone else could use.

Was it then Mersenne, whose experiments with falling weights were '*repeated more than fifty times*', who was the first person to measure 'g'? He quoted a distance of fall in one second of 12 *pieds du Roy* (Royal feet, with an accepted length of 32.87 cm), so there is a calculable value. It implies a 'g' equal to 790 Gal,[13] a long way from the 980 Gal it should have been. For all his dedication and repetitions, Mersenne did not do very well.

A more convincing candidate is another priest, Giovanni Battista Riccioli, who completed his Jesuit novitiate in 1616. When, twelve years later and with his studies finished, he asked to be sent to China as a missionary, his request was turned down and he was assigned instead to teaching logic, physics and metaphysics in Parma and, in 1636, to a professorship in Theology in Bologna. His superiors may have decided to put him on this path because they were impressed by his scientific ability, but perhaps only because they doubted whether he would be able to cope with the physical demands of a missionary existence. Throughout his life he was described as weak or frail, and there is a suggestion that by the time his great work, the massive *Almagestum Novum*,[14] was published he was virtually crippled. He wrote in the *Almagestum* that

---

[12]'*Je doute que le sieur Galilee ayt fait les experiences des cheutes sur le plan puisqu'il n'en parle nullement, et que la proportion qui donne contredit souvent l'experience*': (Mersenne, quoted in Koyré 1953; p. 234). His further comment that '*one should not rely too much only on reasoning*', was one that Galileo would have been much more accustomed to delivering than receiving.

[13]One Gal is defined as an acceleration of one centimetre per second per second. See Introduction—Notes on Units.

[14]The title is a conscious referral back to Ptolemy's Almagestum, which it far exceeds in size. The section relating to the experiments with Fall is on pp. 384–387 of the second part of the first volume. Three volumes were planned but only the first was ever completed (and might, at more than 1500 pages, be considered quite enough). See also Graney (2012) and the more wide-ranging discussion of Riccioli's life and works in Graney (2015).

Fig. 1.7 The 'double braccio' and the metre on the wall of the town hall in Pistoia, near Florence. This is a relatively modern example, but in Medieval times such displays were common on buildings facing market squares, to serve as indisputable (even if very local) standards (*Photo* Paola Marshall)

When, later, I was given permission to read Galileo's Dialog, which the Holy Office had placed on the Prohibited Index, I found (on page 217 of the Italian or page 163 of the Latin version) that the rate of increase of distance with time that he measured followed the series of odd numbers 1, 3, 5, 7, 9, 11, etc. However, I suspected that there might be something wrong with his experiments, because in the same treatise (on page 219 of the Italian or page 164 of the Latin version) he says that an iron ball weighing 100 lb released from a height of 100 cubits reaches the ground in 5 seconds, yet my 8 ounce clay ball fell from a much greater altitude, (280 feet, which is 187 cubits) in precisely 26 strokes of my pendulum, i.e. in $4\frac{1}{3}$ seconds. I was certain that no perceptible error existed in my counting of time, and certain that the error of Galileo resulted from times not well calibrated against transits of the Fixed stars—error which was then transferred to the intervals traversed in the descent of that ball. Furthermore, I was scarcely believing that Galileo had been able to use an iron ball of such great weight, especially when he did not even name the tower from which he might have arranged for such a ball to be released.

And so, full of suspicion, I began in 1640 to make my own measurements ….

His doubts were evidently partly aroused by the difficulty he saw in getting so heavy a ball to a point from which it could fall a hundred cubits.[15]

---

[15]Riccioli, writing in Latin, used the cubit here, but in the *Dialogo* Galileo specified the distance as 100 braccia. In 1640 Galileo was still alive and in Arcetri, not impossibly far from Bologna, and could have been consulted, but direct contact with a heretic seems to have been considered a step too far.

In his own experiments with Fall he made the ground-level observations, leaving his younger and fitter colleagues to climb to the dropping point.

Published in 1651, the *Almagestum* includes a record of measurements made by Riccioli and some fellow Jesuits on the fall of 10-oz balls of clay from a range of levels of the tilted (and terrifying) Asinelli tower in Bologna (Fig. 1.2 right). The main advantage of this tower, apart from its great height, was that balls released from the upper platform would fall directly on to the pavement of a lower platform that was so wide that *six men can safely walk abreast* (although not, obviously, when the experiments were in progress). The falls were timed by pendulums that had half periods of one sixth of a second and were just over an inch long and also, simultaneously, by a musical beat. As an additional precaution, estimates were made by observers close to both the points of release and the points of impact, and then compared. The 'Roman foot' in which distances were measured has an accepted length of about 29.6 cm but a comparison of the known height of the tower in metres and the height as quoted by Riccioli puts his 'foot' closer to 30.1 cm (Graney 2012). If this latter value is correct, his 'g' was about 940 Gal, which is within 5% of the correct value and might considered the first really scientific estimate (see Chap. 14, Coda 2: the Asinelli Tower). It is all the more impressive because of Riccioli's initial doubts as to the truth of the times-squared law.

It was to be a very a long time before anyone did any better than Riccioli using Fall. This was partly because such measurements were inherently difficult until devices capable of timing accurately to thousandths rather than mere tenths of a second became available, but also because, once the theoretical ratio between the length of a pendulum and the distance a body would fall during a single one of its swings had been established, measuring the length of a seconds-pendulum became equivalent to measuring 'g'. Galileo made rough estimates of this length but, according to Drake, did not believe that there would be any point in obtaining an accurate value (Drake 1990, p. 24). Both Mersenne and Riccioli did think it would be useful and made the attempt. We do not know for certain how Mersenne went about it, but he may have used a water-clock. We do know what Riccioli did, because he wrote about it in the *Almagestum* with his usual attention to detail and Koyré (1953) has summarised the whole bizarre story for us. Fellow clergy were again enlisted to help, in this case by counting oscillations for hours on end, and on one occasion '*nine Jesuit fathers*' were persuaded to count, for a full 24-h solar day, the 87,998 oscillations of a pendulum three and one third Roman feet in length. A true seconds pendulum would have completed only 86,640 oscillations. Riccioli tried twice more, with diminishing numbers of Jesuits, and eventually, having exhausted them all, was reduced to making an untested estimate of 3 (Roman) feet, 3.27 in. This implies a value for 'g' of about 956 Gal using the standard conversion factor, or an impressive 972 Gal using the 'Asinelli Tower' factor.

The value of the Roman Foot varied through time as well as from place to place, and the conversion of Roman feet to centimetres based on the height of the Asinelli Tower is questionable. As far as is known, the first person to quote the length of the seconds pendulum in a unit with a reasonably stable value was Christiaan Huygens. He was certainly the first to derive a mathematical expression for the times of small swings of simple pendulums, the first to define the curve that the weight on the end of a truly isochronous pendulum would have to follow and the first to derive the equations governing the motion of real ('compound') pendulums that consist of rather more than ideal uniform and spherical weights supported by ideal weightless threads (see Chap. 14, Coda 3). In his instructions to the craftsmen who made his clocks (in a manuscript dated January 1660), he specified a length for the half-second pendulum of 9.5 in or 'pouce' which, if the widely accepted (although not consistently correct) conversion factor of 31.387 cm to one Rheinland foot is used, implies 'g' equal to 980.94 Gal (Howarth 2007). The now accepted value at the Paris Observatory is 980.928 Gal and since we do not know exactly where Huygens made his measurements, his estimate could conceivably have been closer to the true local value than the estimate made at the beginning of the 20th Century, after five years of concentrated effort, by Prussian scientists measuring the local value in Potsdam.

If so, it can have been no more than a happy accident. In quoting the length of his half-second pendulum to no better than the nearest tenth of an inch, Huygens was implicitly accepting the possibility of it being as little as 9.45 in long and as much as 9.55 in long. For what was to be just one component of a clock that, when eventually assembled, would be adjusted by actual observation to keep astronomical time, that was good enough. Translated into modern units, this would allow 'g' to be anywhere between about 975 Gal and 985 Gal, which neatly covers the entire range of values observable on the surface of the Earth. Huygens' note to his craftsmen did not require them to be in Paris. He would have given them exactly the same instructions had they been anywhere else.

References

Dijksterhuis EJ (1943) Simon Stevin. Nijhoff, The Hague

Drake S (1975) The role of music in Galileo's experiments. Sci Am 232:98–104

Drake S (1990) Galileo: pioneer scientist. University of Toronto Press, Toronto

Galilei G (1632) Dialogo intorno ai due massimi sistemi del mondo: Tolemaio e Copernicano. In: Alberi E (ed) (1842) Le opere de Galileo Galilei, Tomo I. Società Editrice Fiorentina, Firenze

Galilei G (1638) Discorsi e dimostrazioni matematiche intorno a due nuove scienze. Elzevir, Leiden. English edition: Galilei G (1914) Dialogues concerning two new sciences (trans: Crew H, de Salvio A). Macmillan, London

Graney CM (2012) Anatomy of a fall: Giovanni Battista Riccioli and the story of g. Phys Today 65:36–40

Graney CM (2015) Setting aside all authority: Giovanni Battista Riccioli and the science against Copernicus in the age of Galileo. University of Notre Dame Press, South Bend

Howarth R (2007) Gravity surveying in early geophysics: from time-keeping to the figure of the Earth. Earth Sci Hist 26:201–228

Koestler A (1959) The sleepwalkers. Hutchinson, London

Koyré A (1953) An experiment in measurement. Proc Am Philos Soc 97:222–237

MacLachlan J (1982) Note on R.H. Naylor's error in analysing experimental data. Ann Sci 39:381–384

Newton R (1977) The crime of claudius Ptolemy. John Hopkins University Press, Baltimore

Sanctorius S (1631) Methodi vitandorum errorum omnium qui in arte medica contingunt. Aubertum, Geneva

Segré M (1989) Galileo, Viviani and the tower of Pisa. Stud Hist Philos Sci 20:435–451

Settle TB (1961) An experiment in the history of science. Science 133:19–23

Sobel D (1999) Galileo's daughter. Walker & Company, New York

Sobel D, Andrews W (1998) The illustrated longitude. Fourth Estate, London

Viviani V (1654) Racconto Istorico dell Vita del Sig. Galileo Galilei. MS, Florence

White M (2007) Galileo antichrist. Weidenfeld & Nicholson, London

2

The Making of a Map

Three hundred years after Riccioli, 'g' was being routinely (and sometimes necessarily) measured, not to a few parts in a hundred but to one part in a hundred million. I was doing it myself.

Most lives are shaped by a small number of pivotal events. If something had not happened, or some decision had not been taken, those lives would have been unrecognisably different. For me, such an event took place in the mountains of Papua in 1965—and I wasn't even there. A geologist I had never met, working in the New Guinea section of the Australian Government's Bureau of Mineral Resources (the BMR), took a gravity meter out of the safety of the laboratory, and broke it.

Even worse, the accident went unrecognised and many hours of expensive helicopter time were used in making measurements that were completely meaningless. Never again, it was decreed, would a geologist be allowed to touch a BMR gravity meter, still less borrow one. If one had to be used, then a geophysicist would use it. And if, at that particular time, there was no geophysicist to spare in the gravity section, then an engineering geophysicist, already in place on the islands further north, would be taken away from the dam-site where he was working and be despatched to Eastern Papua.

That I was on the site of that never-to-be-built dam was because of an earlier event, over which I had at least had some control. Aged 16, I had begun to worry that maths, physics and chemistry were pushing me towards a life in laboratories or offices, instead of out in the open air where I wanted to be. I decided to switch to geography, because it included geology. The careers' master was suitably shocked. He pointed out that I was too heavy to be a cox and too light to pull on an oar, and that those were the only things for

© Springer International Publishing AG, part of Springer Nature 2018
J. Milsom, *The Hunt for Earth Gravity*,
https://doi.org/10.1007/978-3-319-74959-4_2

which the school's geographers were noted. He had, however, another solution. Only the day before, some pamphlets had arrived describing careers in geophysics, and they indicated that I could stay with maths, physics and chemistry and become a geophysicist. I now know that I could equally well have stayed with those subjects until university and become a geologist, but that is beside the point. One picture, in one pamphlet, made up my mind. It was of a geophysicist in a small motor boat, heading up a fast-flowing river under a rainforest canopy. That, I decided, was for me.

Ten years later, I was sitting in a small motor boat on one of the fast-flowing rivers that drain the mountain ridges of eastern Papua. It was all exactly as advertised.

1966 Bowutu Mountains

The foot that was pushing into the slope was only about a yard ahead of my nose. The toes gripped and compacted the mud. Stability established, the other foot went past, and did the same. Their owner walked the slope upright, carrying 30 kilo of rice. No European foot or shoulder was going to manage that; we went up the same slopes on our hands and knees. Crawled up two foot, slid down three foot, began again. It was hot, it was humid, each ridge was at least twenty metres high, and beyond each there was going to be another. The streams that ran between them were deep, and the leeches loved them, and us. Only a few miles away was the infamous Kokoda Trail, where the Australians and the Imperial Japanese Army had fought each other to an exhausted standstill just twenty-five years before. If there had been Japanese soldiers on the ridge above me, I would have stood up and begged them to shoot.

The first rule of field work is 'never go anywhere unless you are sure you can get back'. I have been foolish enough to have broken it on stupidly many occasions, and have suffered deservedly. I am just lucky that the consequences have at least been survivable. It is especially important to apply the rule if you have arrived in a helicopter, and are just about to wave it good-bye.

A quarter of an hour before that casual wave, the helicopter had collected us (a geologist, a geophysicist and a field-hand) from a rubber plantation at Ioma, in the Papuan foothills. From the airstrip, and beyond the layers of mist still hanging in the valleys, the peaks of the Owen Stanley Range had been visible, rising to more than four thousand metres, and only a few hundred yards away in that direction the rubber trees merged into the untouched forests. Somewhere in all that greenery was a narrow thread of track, cut by contractors at vast expense and ending in a helipad at the top of one of the Bowutu Mountains, the foothills to the main ranges. Somewhere also, and supposedly camped close to the pad, were thirteen carriers, big muscular Motu-speakers from the Southern Highlands. All we had to do was climb aboard the helicopter with hammer and gravity meter, meet the carriers on the peak, and stroll back down the track at our leisure. Everything we needed would be carried by someone else, in the long and cheerful line that would be following us. It was a gentleman's way to do field work.

It started well. The helipad was easy to find, and as we landed, we could see one of the carriers standing at the forest edge. With his help, the helicopter was unloaded, we waved, it left. It had made too much noise for conversation, and it was only after it had gone that we heard the news. It was not good. Half-way down the mountain the contractors had made a mistake, had gone off in one direction, realised their error, had back-tracked, had started again and had finished the job and left. Where the track split, our own carriers, with the one exception now standing before us, had decided to go no further. Being bush-hardened, forest-savvy Papuans of their generation, they had known exactly what to do. They cut another helipad out of the forest and waited for someone to drop out of the sky and tell them what to do next. With the helicopter on its way back to Ioma without us, that was not going to happen. What was going to happen was that we were going to have to load ourselves up with as much as we could of our impressive pile of cargo, and walk.

We began at about nine in the morning, and eight hours later reached the remains of one of the camps used by the contractors and decided to call it a day. It could have been worse. We may have been loaded, but at least it had been downhill. And, joy of joys as far as I was concerned, after three years in the deserts of central Australia, there had been no need to carry water. Not only is water plentiful in almost all parts of New Guinea, and decidedly overabundant in many, but the Papuans have a great respect for it. Pollution of water supplies was almost unknown in those days, and was certainly not going to be a problem on those uninhabited slopes. Even so, it had been a hard slog.

The next day began, very cold. The forest takes time to warm up, particularly at altitude, and if it is thick enough it never does, even at the equator. By six we were under way and by nine we had reached the delinquent carriers. Hugh Davies, my companion and later the first head of the Geology Department at the University of Papua New Guinea, was impressive in his address to them. Since it was all in Motu, I didn't understand a word, but the general drift was clear. Get up the top of the mountain (which it had taken us ten hours to come down), pick up all the cargo, and bring it back. We carried on walking downwards, rejoicing in having our packs carried by someone else. Five hours later, the four carriers sent up to the helipad caught up with us, just as we reached the gullies. It was my first, but not last, encounter with the sheer strength and stamina of the highlanders of Papua New Guinea.

1966 Jet Boat Survey

Two days on the mountain, and just five gravity stations. Walking the trails was not going to be the way to cover the country. The helicopter came back to Ioma and ferried us down the coast to where two of the BMR's fleet of Hamilton jet-boats were waiting for us. Puk-puk (Crocodile) and the ominously-named Fireball. Developed in New Zealand for the fast-flowing rivers that drain the Southern Alps, they were in their element in New Guinea. Water was sucked in through grids under the hull, accelerated by an impeller in a sealed pipe and ejected out the back at high velocity. The jet was steerable, and steered the boat. With no propeller to tangle in weeds or shatter on rocks, jet-boats can reach places that nothing else can. With sufficient speed, they can even bounce over gravel banks and carry on (Fig. 2.1).

In the boats we were using, the power for the water jets was supplied by massive six-cylinder petrol engines, the same engines that in those days powered the iconic Australian car, the Holden FJ. This was not entirely good news, as some petrol always leaked and collected in the bilges, where it eventually either caught fire or exploded. The boat named Fireball was slightly singed, having done this several times. I never saw it happen, but I heard what it was like. One of our geologists working on the Sepik River told me

> I thought I had two carriers with me, and I just had time to notice that they weren't there any more when there was a bang and I was in the water. The guys who had jumped ship pulled me out. The boat buried its nose in the bank. We dug it out and carried on. We didn't think it could do that twice in one day.

Each day on our much shorter rivers we headed upstream until they became too narrow for even our little boats, or until a waterfall blocked the way. We then came back downstream fast, usually leaking badly from having scraped its bottom on the river's bottom too many times. Coming down a river that we had only just been able to get up at top speed (25 knots through the water) could be interesting. The lowest speed that gave adequate steerage-way was about 10 knots, so some stretches were travelled at 35 knots

Fig. 2.1   Jet-boat 'Puk-Puk' after running into a muddy river bank at speed

and collisions with the banks were frequent. The evenings were usually spent working with fibre-glass, preparing the hull for the next day. By the time the job was done, we Europeans were exhausted and went to bed, but the carriers would sit up all night chatting to the villagers in Motu or Tok Pisin. I couldn't understand how that was possible with a shared vocabulary of less than a thousand words, but one of the old New Guinea hands who came with us explained it very well.

> If you only have a thousand words, it takes a long time to say anything.

How true. I still have a copy of my 1974 tax return, in English and Tok Pisin. It has a space marked *For office use only* or, alternatively, *Dispela ples em bilong long cuscus bilong inkam takis, em i wraitim tasol.* The cuscus is a middle-sized marsupial that does its best, although with rather less agility and energy, to fill the ecological niche occupied elsewhere by the tree sloth. It has enormous eyes and a perpetually startled expression. It is impossible to understand how its name came to be used as the standard Tok Pisin word for a government clerk.

The idea for the jet boat survey had come from Jack Thompson, a BMR Senior Geologist and New Guinea specialist. It was Jack who had planned and was now in charge of the Eastern Papua project of which it formed a part, and he had threatened to come out to see what we were up to. Since he had made no arrangements, we relaxed, assuming it was never going to happen. We remained relaxed right up to the moment that we came round a bend in a river to the unusual sight of a large European standing upright in a very small canoe (a trick I never mastered), attacking the rocky bank with a geological hammer. Jack had arrived. He had taken a mission Cessna flight to Popondetta and then walked through the forest for a couple of days. When he got to one of the rivers that he knew we would be using, he hired a canoe and settled happily down to doing geology until we arrived. He stayed with us a couple of days and then wandered off into the forest again. Of such stuff were the BMR's senior geologists made, in those days.

The next time I saw him he was back in Canberra, complaining bitterly about the way things were going in the new New Guinea. On his way back he had reached a river that could only be crossed by ferry (another dug-out canoe) and that was on the other side. The ferryman appeared when he shouted, but only to point to a large notice saying 'No wok Sundai'. Jack had, we were led to believe, persuaded him to work on that particular Sunday by using some of the more expressive elements of the English, Tok Pisin and Motu languages. The ferryman probably charged him double or triple rate, but since the standard pay for a day's work at the time was one shilling or one stick of tobacco, it probably didn't make too much of a dent in his budget.

The jet-boats were more fun that any survey vehicles I have ever used but, good as they were, when they reached a waterfall, they were stopped. To get further inland, we had to use helicopters, and they were fun too.

TPNG

Where were we?

The island of New Guinea has often been compared to a bird or a dragon (so, a dinosaur, either way), with its head facing west and its eastern tail eventually breaking down into strings of small islands that stretch east towards the Solomons. By the beginning of the 20th Century it had been divided into three parts. The western half was taken by the Dutch and included in the Netherlands East Indies, although ethnically (and geologically) very different from the other islands. In the south-east, the British, pushing up from Australia, had established the colony of Papua, but their expansion north had been checked by the presence of the Germans in the northeast and on the islands around the Bismarck and Solomon seas. At the end of the First World War, Germany was deprived of her empire and Britain took control of their New Guinea colony under a League of Nations mandate.

Although the whole of eastern New Guinea and its islands were then officially British, the two parts were governed separately until both were handed over to Australia after the Japanese were evicted in the closing stages of the Second World War. So it was that in the 1960s the combined colonies formed the Australian Territory of Papua New Guinea, generally known as TPNG. The earlier division into Papua and New Guinea had ceased to matter, except in one very important way. The trade language in Papua, used by the speakers of the myriad different languages to communicate with each other, was Motu. Pidgin English, the forerunner of modern Tok Pisin, was, anomalously, used only in the former German colony, because the long-serving governor of British Papua would not tolerate its use in his domain, regarding it as a bastardized baby-talk.

There he was wrong, or at least, not a linguist. Tok Pisin is a real language with a vocabulary taken largely from English with some local, German and even Samoan words, but with an entirely Melanesian grammar. One of the fascinating aspects of having visited the island at random times over more than forty years has been to hear it evolving from a simple 'pidgin' to a creole and then to a national language. The process has been extraordinarily rapid. My most recent attempts at speaking it were still understood, just about, but I no longer had any hope of understanding anything that was said to me. It had changed too much from the form in which I first learned it.

Linguistically, New Guinea is notorious in other ways. The figure of six hundred distinct languages is sometimes quoted, but this is a snapshot of a moving target, since the less spoken ones are dying fast. The Summer Institute of Linguistics used to send out its missionaries with instructions to translate the Bible into the language of the village where they eventually found themselves. This posed many problems (including that of presenting the parable of the sheep and the goats to people who originally had neither and are even now unable to recognise the difference), but was also often a race to complete the task before the last person who spoke the language died.

In 1960 few people could imagine a country called Papua New Guinea, with a seat in the United Nations and known as just PNG, but by 1970 most Australians had accepted the idea that independence was only a few years away. Thus my first major period of fieldwork in the country, from 1966 to 1968, also saw the twilight of the 'Rule of the Kiaps', the Australian

Patrol Officers, many still in their twenties, who administered the country in the time-honoured way of walking from village to village followed by long lines of carriers, collecting taxes, explaining the law and punishing those who broke it. One of the most common crimes was murder, and the approved punishment was a couple of years spent cutting the grass on the local airstrip. This was not, of course, done with a lawnmower. After landing at the remoter strips it was usual to be quickly surrounded by a dozen or more convicted assassins, all armed with yard-long bush-knives.

In those days the country didn't do roads (it still doesn't do them very much). Throughout most of it, you either walked or flew. There were plenty of Papuans who were quite blasé about aircraft and even helicopters but had never seen a motor car. There were still some tribes that had never seen a European and, like all field workers in the country, I was given a rudimentary training in how to handle a first contact (make no sudden movements and give gifts). Fortunately for everyone, I never had to use it

As one of its own parting gifts to the new country, the Australian government had decided to provide it with a complete set of geological maps, at a scale of about four miles to the inch. Whether this was a conscious decision or just happened, I don't know, but it was done under the guidance of the Geological Office in Port Moresby with abundant support from Canberra. Over the years, an extraordinarily effective mapping technique evolved, based on the use of small helicopters (Fig. 2.2). Geologists, usually in pairs, would be flown out of camp in the early morning and abandoned in a river bed somewhere. They would then walk along the river, mapping as they went, until they reached the pick-up point, either in the evening or a day later. This left the wide forest areas between the rivers unmapped, but rocks were hard to find there anyway, and one of a field geologist's most necessary talents is the ability to guess the rocks that can't be seen from observations in the sometimes pitifully small areas where they can be. The technique proved so successful that in less than ten years some 200,000 km^2 of country, largely covered in thick rain forest and almost completely devoid of roads, had been mapped geologically at a reconnaissance level.

It was easy to add gravity surveys. The helicopters were generally needed by the geologists only in the early morning and evening, leaving them available for geophysics during most of the day. The first time that this was tried was between 1966 and 1968, as part of Jack Thompson's Eastern Papua geological mapping project.[1] The terrain was challenging, but no more so than in most

---

[1] The geological results of the project are summarised in Smith and Davies (1971).

Fig. 2.2 The Papua New Guinea work-horse of the 1960s—the Bell 47 G3-B1

of the rest of the country. The peaks of the Owen Stanleys form the spine of the peninsula, and along their north-eastern flank the lesser ranges, including the Bowutu Mountains, are largely made up of heavy green and black rocks. Described as mafic and ultramafic because of their high content of iron and manganese, these rocks have long been recognised as more properly belonging either out in the oceans or in the uppermost mantle (the layer immediately below the crust). Their presence on mountain peaks on continents was a well-recognised geological problem,[2] and one way of investigating them was by measuring their effect on the gravity field.

The decade of the 1960 s was one in which not only PNG but the geological sciences changed for ever. At its start most of the geologists in Europe and North America believed that the Earth's continents were pretty much where they had always been. Ten years later, almost all of them accepted an Earth history of dramatic movement, punctuated by the formation of supercontinents that split apart and reassembled themselves in new shapes. Plate tectonics had arrived, but it had not yet made the presence of oceanic rocks on top of continental crust any easier to understand.

---

[2]Hugh Davies' geological account of these rocks can be found in his Stanford University PhD thesis and in Davies (1971).

The BMR

The sad thing about being very, very lucky in one's choice of a first job is that the good fortune is rarely appreciated at the time. It takes hard experience with other employers to hammer home the message that not all colleagues are congenial and not all bosses are sympathetic, imaginative and competent. I was extraordinarily lucky to have joined the BMR when I did, but I was not alone. In 1962, more than a dozen young geophysicists and geologists swore allegiance in London to Queen Elizabeth the Second, as queen also of Australia, and took the First Class boat trip from the UK to Melbourne. For some it was also their honeymoon.

Once there, they became Geophysical or Geological Officers (GOs). As public servantsof the Commonwealth of Australia, they found that, theoretically, they worked under the standard public service regulations. The 38-hour working week was achieved in strange ways, which included finishing each day at exactly six minutes past five. Also, there was no overtime. A field party working in the deserts of the Northern Territory was supposed to stop work at six minutes past five on Friday, spend two days staring blankly at the unchanging horizon, and begin their duties again at nine o'clock on Monday.

Fortunately the BMR was blessed with managers who knew when silly rules should be ignored. Norman Fisher, then Chief Geologist, had left Rabaul in 1942 fractionally ahead of the invading Japanese and sailed a small boat 600 miles to (temporary) safety in Papua. Lynn Noakes, who was to become his deputy, had remained in New Guinea for most of the war as a coast-watcher, radioing back details of the movements of Japanese ships in the narrow seas between the islands. These were people who were interested in seeing jobs being done, not in seeing pointless rules being obeyed. They expected their field parties to work seven days a week, and asked no questions when the GOs returned to Melbourne or Canberra, checked in their stores and then simply vanished for about the right number of days.

Other blind eyes were turned. The one thing that, in principle, you could not do throughout the public service was to lose any Commonwealth property. If something was damaged beyond repair, then that could be dealt with. A Board of Survey, consisting of at least two Commonwealth officers, could be convened and the offending item could then be solemnly sentenced to be buried, burned or otherwise destroyed. This process could be used unofficially to cover losses but was often just too much trouble, and the missing items were simply transferred to another field crew. After a couple of years of this sort of thing, most field parties would have needed an extra vehicle to hold all the non-existent stores that they had on their books, But, being Australia, help was always at hand. Sooner or later there would be a cyclone, a flood or a bush fire, and all the surplus items could then be safely written off.

Only occasionally did things come unglued, and sometimes with the best of intentions. Every new arrival was inoculated with the story of the great Port Keats disaster. Port Keats was a mission station in the far north-west of Australia, and for several months it had played host to a BMR airborne survey party. When they left, having no room on the aircraft for the massive kerosene refrigerator, the Party Chief donated it to the Mission. That sort of thing had been done many times before, a Board of Survey had been convened and the sentence had duly been pronounced, if not carried out. Unfortunately, this Party Chief was honest, and told the administrators what he had done.

A major incident was declared. No Commonwealth Officer is authorised to give away Commonwealth property. Reams of correspondence followed the tortuous paths from Melbourne to Port Keats and back. Clerks in offices in faraway Canberra gravely considered the consequences. Eventually an expedition was mounted to retrieve the fridge. It came back, at a cost of thousands of Australian pounds, and went into store, never to be seen again. Every junior GO knew the story, and took its lessons to heart.

When I first joined the BMR, I was put in the airborne section, which mapped changes in the Earth's magnetic, rather than gravity, field. I stayed there for two years and then spent another two years in the section that dealt with water resources and dam sites. If I had gone straight into gravity, I would have spent much of those four years, when not actually in the field, doing the sums the hard way, on a hand-cranked mechanical calculator. By 1966, the BMR had moved on, and the calculations were being done on a new thing called a computer, located in the Australian National University.

1966–1968 Eastern Papua Helicopter Surveys

New Guinea was still a strange and mysterious land in the 1960s, and most of the young GOs wanted to go there at least once, but for many one trip was enough, as they couldn't deal with what we called the ninety-ninety, ninety percent humidity and ninety degrees Fahrenheit. For me, after three and a half years of either having every last drop of moisture sucked out of my body in temperatures in the Centigrade 40°s (Northern Territory and western Queensland), or standing on a freezing mountain side in a never ceasing downpour (west coast of Tasmania), New Guinea, where I could guarantee being both warm and wet most of the time, seemed like paradise. Happily, when the BMR found someone who actually liked the climate, they kept on sending them back. By the end of 1966 the decision had been made to extend the Bowutu Mountains gravity survey to the rest of Eastern Papua, and to my mind there was only one suitable candidate.

There was a problem. The survey was going to be a long one, I had only been married three years, our first daughter was just one year old and I didn't want to miss out on that much family life. The BMR expected its junior GOs to spend long periods in the field, but made up for it by making it easy for them to take their families with them. However, the further they went, the longer they had to be away to be eligible for a family posting, and New Guinea was the end of a very long line. I wouldn't qualify, and the air fare alone would destroy our precarious finances. It was the chiefs, bless them, that found the solution.

In a gravity survey, the height above sea level of every point must be measured, as well as the gravity field. We used barometric altimeters for this, but barometric pressure is always changing, and during each survey loop there had to be someone sitting at base, reading an identical barometer every fifteen minutes. I was entitled to employ someone in PNG to do this (if I could find them).

It was suggested, quite unofficially, that my wife might like to do this job, unpaid. The pay she would have got would then cover her air fare. The presence of a one-year old child being carted from place to place in a government-chartered helicopter would be ignored. Somewhere there was a piece of paper that would make this dubious arrangement arguably official, but it was buried in an obscure file and all involved hoped it would never be needed. Everyone was happy, but it is hard to imagine any government organisation bending so many rules so far today.

The helicopter of choice in the 1960s was the Bell 47, then famous worldwide as the real star of the television series MASH. In TPNG the supercharged G3-B1 version was used, because of its performance at high altitude. This was very necessary in a place where there were mountains which, while only a few degrees from the equator, were high enough to have snowfalls. Even the G3-B1 became distinctly soggy in its flying characteristics above about 3000 m, but it was rare to have to go so high.

Helicopters provided the only possible way to cover the ground, but places where they could land were few and far between. In the forest-covered areas the options were usually limited to gravel banks in the rivers, and it was a work of artistry for the pilots to get down on some of these. A landing might involve dropping below the level of the forest canopy at one point along a river and then flying along it until a suitable spot for a landing was reached. On one memorable occasion, the helicopter had to be backed out the way it had come, because there was not enough room to turn round.

The broad grassy slopes that are also common in PNG looked easier at first sight but could also be problematic, since the coarse, spiky kunai grass grew high enough to reach the tail rotor. Even a single blade of grass could knock off one of the warning strike-tabs, rendering the aircraft legally and, if you were wise, actually, unflyable. To avoid this happening, the aircraft would sometimes hover just above the grass, allowing an adventurous observer to jump out, complete with gravity meter. Getting back again could be even more challenging. Sometimes the best the pilot could do was hover long enough for the observer to haul himself up on to one of the baggage trays, where he could sit while a search was made for a spot where the aircraft could land.

Sitting on the outside tray of a helicopter may sound risky, but was safe compared to another method developed later. For gravity surveys in shallow water, an underwater meter had been created that could be read remotely from a boat on the surface. It was heavy, but if all the waterproofing was stripped off, it could be carried in a G3-B1. The idea was that this 'dangle-meter' could be lowered into forest clearings too small for the helicopter to land, and the observer would take the reading remotely. I think that more than a hundred measurements were made by this method in the Sepik region before the DCA, the Civil Aviation department, discovered what was happening and put an end to it. A helicopter with engine failure, they pointed out, had a chance of landing safely as long as it had enough speed to keep the rotors turning, or enough height to reach that speed. If the engine failed when it was hovering at canopy level over a tiny clearing in the forest, there would be no speed and no height, and everyone on board would die. Someone did point out that in most New Guinea helicopter crashes everyone died anyway, but the DCA was unmoved. The dangle-meter was retired.

In Eastern Papua we had the same problems, but we were sometimes lucky. In one vast area of swamp we found only a single landing site, and that was on the wing of a B17, a Flying Fortress that had crash-landed there during the war (Fig. 2.3). There were bullet holes in various places, but the crew must have survived and at least tried to walk out, after smashing the bomb site. At any rate there was no sign of them, but everything else was pretty much intact. The helicopter pilot on that occasion was a small-arms

Fig. 2.3 An unconventional helipad. A B-17 'Flying Fortress' in the swamps east of Wanigela

enthusiast, and spent a considerable amount of time trying to get power to one of the gun turrets so that he could fire off a burst or two. Fortunately he never managed it, because he might well have sawn the helicopter in half. In the end he contented himself with unbolting the bomb winch, built for raising 500-pound bombs by hand. Back home in Lae he bolted it on to the front of his car, and never got stuck in the mud again.

When, in the year following the Bowutu Mountains survey, it was decided to expand the survey to the whole of the remainder of the Papuan Peninsula, it was also decided to go further, and include islands that ran out from the tip of the peninsula towards the Solomons. Extending the survey in this way raised a new problem. For flights over water, huge floats had to be attached to the skids on the G3-B1, in case of a forced landing, and that slowed it down considerably. The implications, in terms of speed and range, were sufficiently serious for Norman Fisher to call a meeting of all concerned to discuss the problem. *What would happen,* he asked, *if you had to land on the sea and didn't have the 'boats' on?*

Bruce Evans, helicopter pilot extraordinary and master of the great Australian one-liner, just looked at him.

She wouldn't even slow down.

Not literally true, of course, but the point was well made. We put on the floats and chartered a small trading vessel to run ahead of us, parking 44-gallon drums of aviation fuel on islands where they might be needed. It generally worked well, but on one occasion we arrived to find the drum breached and almost empty. The kiap on Woodlark Island later told us that some passing fishermen had found it and decided to fill their cigarette lighters. They weren't very good at it, and rather a lot was spilt. I did actually meet them a few months later, while they were cutting the grass on his airstrip.

The larcenous fishermen were from the Trobriand Islands, made famous as 'The Isles of Love' by the ethnically Polish but politically Austro-Hungarian anthropologist, Bronislaw Malinowski. He was on the islands doing some fieldwork when, at the start of World War I, he became an enemy alien but, rather sportingly, Australia interned him there for the duration so that he could continue his work (Malinowski 1929). Despite their reputation, and the fact that the unmarried girls wore only the local, and entirely traditional, version of the miniskirt that was then just setting swinging London aflame, the besetting sin of the islands was cricket. The game had developed independently of the MCC, and a village team would consist of all the able-bodied males, with the traditionalists still clinging to the curved bat and the two stumps. The fall of each wicket was marked by a ceremonial dance by the entire fielding side, and the games lasted until everybody, on both sides, had batted. Matches could easily last a couple of weeks, and at one stage had been declared illegal, thus freeing up the manpower to plant the gardens on which their lives depended.

Further out was Woodlark, separated from the much smaller Madau Island by a narrow, shallow, sinuous strait about 3 km long. We took the kiap on one of our flights across this and he told us, as we approached it, about the local superstitions, including the fact that nobody from either island, even though generally as happy in water as out of it, would ever swim across that channel. We then crossed it, getting an excellent view of the almost nose to tail assembly of sharks and rays parked in the warm water, and went on our way, shaking our heads over the strange beliefs of these unsophisticated village peoples.

Woodlark was also notable for its own reminder of World War II, in the form of a vast crumpled mound of aircraft (mainly Bell Airacobra fighters) in the jungle at the far end of the over-sized airstrip. Apparently they had all been lined up waiting to move on to somewhere where the fighting was hotter when the news arrived of the bombing of Hiroshima and Nagasaki. There being no foreseeable future use for all these expensive bits of metal, they were simply bulldozed out of the way and left there.

In the end the job was done. All the islands as far as the atolls beyond Woodlark in the northeast and Rossel in the southeast, were covered. Only Rennel, half-way to the Solomons, escaped. It was just too far.

1968: A Gravity Map for Eastern Papua

The first map of the gravity field of Eastern Papua was drawn by hand, on map sheets at a scale of four miles to the inch that covered an entire wall. The same information has been used to make the upper of the two maps in Fig. 2.4. Nothing done since 1968 has been included, and so there are no contours or shading where there are no islands, nor on the south-west side of the ranges, an area that was only mapped much later. A modern computer-contoured version that includes all the data collected since is shown below, but for almost six years, from the first trek down the Bowutu Mountains to the eventual defence of my thesis, the partial and hand-drawn map dominated my life. To no-one else can this be interesting, but the map itself can usefully serve to introduce some of the reasons why geophysicists might want to measure 'g'.

Almost every geological use of gravity measurements starts from the simple idea that dense rocks produce stronger gravity fields than light rocks, and the surveys in Eastern Papua spectacularly identified the places where the densest rocks were most abundant, which were not always the places where they could be mapped at the surface. The long gravity high labelled '1', which in the northwest more or less coincides with their outcrops continues in the east into swampy lowlands where only mud (and the odd B-17) is visible. The lower values of 'g' in the mountains south of the swamps show that the dense rocks in these places can only be a thin skin above much lighter ones on which they have been thrust from the north. This was the first and most obvious result of the survey, and it made a small contribution to understanding how these 'oceanic' rocks reached the places where they are seen today. It had, however, been anticipated, because a few years earlier a research student from the University of Tasmania had taken readings at a few places within the area of the gravity high.[3]

---

[3]St John (1965). The work was done in conjunction with the helicopter-supported teams that were making the first accurate maps of Papua New Guinea, and many of the measurements were made at the survey points on the peaks of its highest mountains, with correspondingly large gravity effects due the rugged terrain.

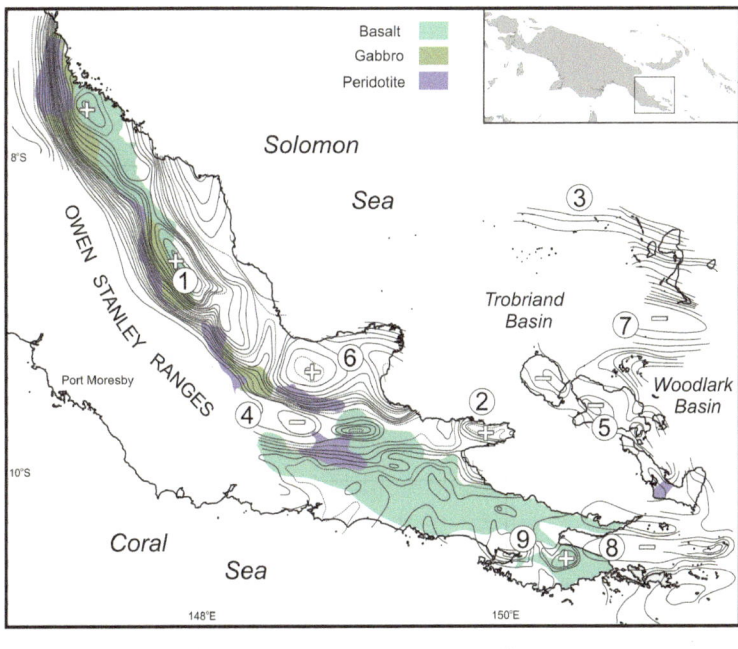

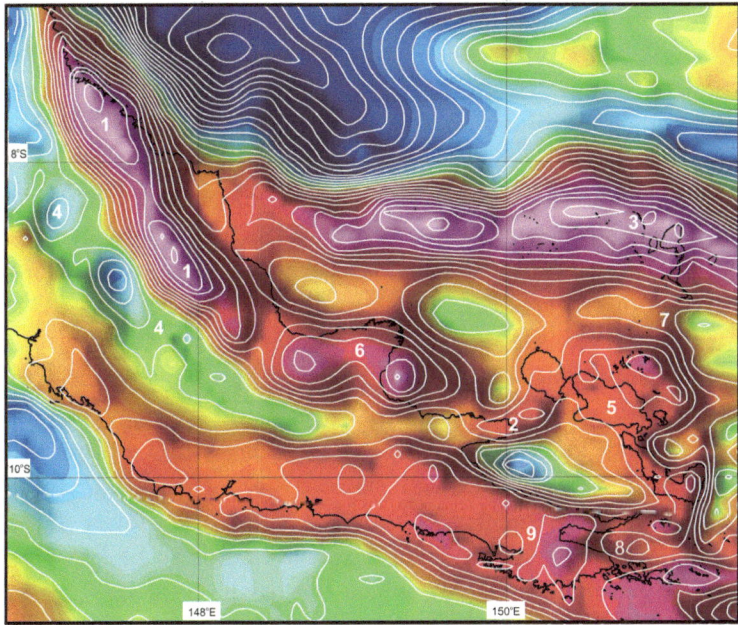

Fig. 2.4 Gravity maps of Eastern Papua, corrected for the gravity field of the ideal Earth at the heights and latitudes of the observation points, and for topographic masses above sea level. The upper map is based only on data collected before 1970, and also shows the distribution at the surface of the dense rocks of the Papuan Ultramafic Belt. The lower map makes use of all the data now available. Contour intervals are 10 milligal (roughly one hundred thousandth of 'g') for the upper map, 20 milligal for the lower

The university survey had not, however, covered the Cape Vogel Peninsula, which is marked by the number '2' and where the gravity high came as much more of a surprise. Most of the rocks there are coral limestones, now raised above sea level by very recent earth movements, but the few older lavas poking through were considered at the time to be very unusual rocks indeed. They contain a mineral known as clinoenstatite, which some specialists had claimed could not exist in the Earth's crust. Clearly it does, and similar lavas have now been identified in many other parts of the world, generally as parts of the belts of dense oceanic rocks which, when anomalously exposed onshore, are known as ophiolites. The ultramafic belt of Eastern Papua is a classic ophiolite, and it can now be seen as logical that the gravity highs it produces should continue on to Cape Vogel, but no-one at the time had thought that that would be the case.

The strongest gravity fields recorded in the survey were not over any part of the ophiolite, but far to the north, on the scattered islands of the Trobriand archipelago (marked by the number '3'). In this area the Earth's crust is thin, and the dense rocks of the upper mantle that come up to within a few kilometres of the sea floor beneath the Solomon Sea (which is one of the world's smallest oceans) are already beginning to make their presence felt by their effect on 'g'. The gradient marks the location of the transition from continent to ocean.

As well as gravity highs, the surveys defined some very definite gravity lows. The story behind the one marked '4' concerns continental crust. To a non-geologist, granite, which makes up much of continents beneath the sediment layers, may seem heavy, but it is much lighter than the rocks of the mantle, the next layer down. The continental core of the peninsula floats like an enormous iceberg on these denser rocks, and because it is relatively light the gravity field is low.

There is a special significance to the area of low gravity marked by a '5'. The presence in the rugged and mountainous D'Entrecasteaux Islands of what seems, both gravitationally and geologically, to be a displaced part of the continental core demands an explanation. A fuller understanding of what was happening there came only long after the gravity surveys were completed, with the recognition of what have come to be known as metamorphic core complexes in areas where the Earth's crust has been stretched and pulled apart, exposing the rocks of the deep crust. The islands are now considered to be amongst the best known examples of this process occurring in a region that is, geologically speaking, very young.

Gravity can be low for many reasons. Eastern Papua is host to the only active volcanoes on the New Guinea mainland and one of them, Mount

Lamington (just east of the '1' on the upper map), erupted in 1953, killing three thousand of the villagers who were living on its slopes without even recognising their danger. A BMR geologist, Tony Taylor, won the George Cross for staying on the mountain throughout the eruption, observing the progress of the glowing clouds of hot gas and ash as they cascaded down the slopes (Taylor 1958). Further east, in the region marked by a '6', there are two more volcanoes, named Mounts Victory and Trafalgar by 19th Century naval officers observing them from a safe distance. The molten volcanic magmas have pushed their way up through the ophiolite and, being lighter, have created breaks in the gravity high. The relationships between volcanoes and gravity fields are complex, and all over the world there are scientists who have devoted their lives to studying them. Two of those studies are described in Chap. 5.

Low gravity also characterises the area marked by a '7', between the Trobriand Islands to the north and the D'Entrecasteaux Islands to the south, and this low could have had commercial significance. Sedimentary rocks, the sources and hosts of hydrocarbons, are generally much lighter than either oceanic rocks or granites, and the measurements made on the scattered islands in this area revealed the presence of a deep sedimentary basin that later became a target for exploration for oil and gas. With the Deepwater Horizon disaster still fresh in mind, and the coral reefs that then seemed commonplace now globally endangered, I can be happy that neither was found, but in those days the environment seemed less fragile than it does today. The oil industry was the main sponsor of the development of modern gravity meters, described in Chap 9. A second 'basinal' gravity low is marked by an '8'. It coincides with the almost rectangular inlet known as Milne Bay, which is defined to north and south by steeply-dipping faults with very large displacements.

Even now, fifty years after the survey was completed, not all the patterns that were mapped have been explained. The origin of the paired high and low marked by the number '9' is still a mystery. The rocks at the surface are basaltic lavas, as they are almost everywhere else in that easternmost part of the peninsula, and show little variation. Something strange is happening at greater depths, but no later work has given even a clue as to what that is. Comparison of the ways in which these smaller features are presented on the two maps demonstrates also that hand contouring can sometimes give a clearer picture of what is going on than contouring by computer.

Bizarrely, the studies of the effect of mountains on gravity carried out by insanely dedicated scientists in the 18th Century were to be, in the 20th Century, the trigger for the development of ideas about the geological events that formed the Woodlark Basin, a blank space at the eastern edge the upper map. Their tale is told in Chaps. 5 and 6, and in Chap. 12 the story is continued to explain how it is that the water covered parts of the globe are now the parts of it where 'g' has been most completely mapped. The blank spaces on modern global maps are all on land, and testify to the difficulty, even now, of getting to some places.

Sadly, it is not always physical obstacles that stand in the way. All too many of the places in which I once worked peacefully and safely are now too disturbed and too dangerous to even think of visiting.

References

Davies HL (1971) Peridotite-gabbro-basalt complex in Eastern Papua: an overthrust plate of oceanic mantle and upper crust. BMR Bull 128, Canberra

Malinowski B (1929) The sexual life of savages in north-western Melanesia. An ethnographic account of courtship, marriage, and family life among the natives of the Trobriand Islands, British New Guinea. Eugenics Publishing, New York

Smith IEM, Davies HL (1971) Geology of Eastern Papua. GSA Bull 82:3299–3312

St John VP (1965) The gravity field in New Guinea. Ph.D. Thesis, University of Tasmania, Hobart

Taylor A (1958) The 1951 eruption of Mt Lamington, Papua. BMR Bulletin 38, Canberra

3

The Astronomers

The beginnings of modern gravity mapping can be traced back to Galileo, but there is another ancestral line. The first steps towards supposing that maps of 'g' would be useful were taken by astronomers, and their story begins with the overturning of the Ptolemaic universe and leads up to the idea of gravity as a universal, but variable, force.

Copernicus

It was Galileo's interest in astronomy that eventually got him into so much trouble. It was, to say the least, unwise of him to champion so enthusiastically the ideas of an obscure Polish churchman who had died almost a hundred years earlier in a tower at Frauenberg (now Frombork) in the bleak flatlands along the southern shores of the Baltic. Then in Poland, now once again in Poland, the area has been fought over by Teuton and Slav for a thousand years. When I visited it in 1960 the town and much of the cathedral were still in the state that they had been left by the retreating Wehrmacht and the advancing Red Army. Nothing but tumbled heaps of broken red bricks.

Much has since been rebuilt, and there is a museum dedicated to Nicolaus Copernicus, the churchman in question. He would have spoken more German than Polish, but the language in which he wrote and may have thought was Latin. In Arthur Koestler's 'The Sleepwalkers' he appears as the sleeper *par excellence*, barely aware of what he was doing.

© Springer International Publishing AG, part of Springer Nature 2018
J. Milsom, *The Hunt for Earth Gravity*,
https://doi.org/10.1007/978-3-319-74959-4_3

This is almost certainly unjust. Koestler was an exceptionally gifted writer (especially considering that English was his fourth or fifth language) but always highly partisan. He was an enthusiast for Kepler, he hated Galileo and he despised Copernicus. He may have allowed himself to be too much influenced by a contemporary comedy entitled *The Foolish Sage*, which caricatured Copernicus as an unlikeable, God-obsessed astrologer whose life's work remained for ever hidden. The real Copernicus, as well as fulfilling his ecclesiastical duties (which were financial and administrative rather than religious, and certainly not negligible), seems to have lived life to the full. The earliest portrait of him that still exists (Fig. 3.1) show a man with an interesting, if not handsome, face and a quizzical expression. He might, one feels, have been a good host or dinner table companion. He certainly had a very human side, and in later life anticipated the astronomers who followed him by getting into trouble with the church for a too close relationship with his housekeeper. Galileo, Tycho Brahe and Kepler were all criticised in their turns for selecting partners many rungs beneath them on the social ladders.

Fig. 3.1 Nicolaus Copernicus. The painting shows him as a young man but may not have been completed until after his death. By an unknown artist, it now hangs in the Town Hall in Torun, and has been the basis of many subsequent portraits

Koestler's prejudices sometimes took him to bizarre extremes. He devoted three completely irrelevant pages in *The Sleepwalkers* to imagining obscure and discreditable motives for Copernicus' description of Frombork as being on the Vistula rather than the shore of the Frisches Haff, but some arms of the Vistula do indeed feed into the Haff, and the largest of these, the Nogat, was the main channel until the 14th Century. That was why the Teutonic Knights chose its banks for their fortress capital of Marienburg. As if to deliberately prove Koestler wrong, Frombork's present citizens, who are Poles, refer to the Haff as the *Zalev Wiślany*—the Vistula Lagoon.

Copernicus would have known this geography, because he was a significant cartographer in his own right,[1] at a time when maps of Prussia were highly political. His first independent work is thought to have been a map of Warmia (Ermeland) and western Royal Prussia commissioned by Lucas Watzenrode, Prince-Bishop of Warmia, his uncle and patron. It was intended for use at a conference in Poznań instigated by the Teutonic Knights in an attempt to recover land lost in 1466, and it is said that when Copernicus accompanied his uncle there he took the map with him. This was wise, because during his absence his rooms were searched for it, fruitlessly, by a Teutonic spy. He later made other maps, of Warmia in 1519 to help his uncle in a border dispute with the town of Elbląg, and of the whole of Prussia in 1529 for Mauritius Ferber, who succeeded Watzenrode as Prince-Bishop. None have survived, but maps produced in collaboration with Rheticus and Henryk Zell (Fig. 3.2) were based upon them.

Another of Koestler's idiosyncrasies was to generally refer to Copernicus as 'Canon Koppernigk'. It is true that this was one version of his name, among many others in an era when spelling, particularly of names, was somewhat optional, but it was not the one under which he published. Why use it? Because it is slightly comic. There is a touch here of Churchill's practice, employed in a much better cause, of referring to Adolf Hitler, whenever possible, as 'Herr Schickelgruber'.

As a young man, Copernicus practised medicine, at least on his closer relatives. Later he had a military career of sorts, and in 1520 took a leading role in the defence of Allenstein (now Olsztyn), which was then a largely German town but part of Warmia and allied with Poland against the Teutonic Knights. It went on to become the second city in German East Prussia, and was, like Frombork, little more than a pile of rubble in

---

[1]Much of the primary material concerning Copernicus, including the portrait in Fig. 3.1 and the map in Fig. 3.2, can be accessed via the Copernicus Academic Portal, 'Nicolaus Copernicus Thorunensis' (Copernicus.torun.pl).

Fig. 3.2 Detail from a map attributed to Henryk Zell based on an original produced in about 1540 by Rheticus and Nicolaus Copernicus. Showing parts of East Prussia and Warmia (Ermeland), it was included in an atlas published by Abraham Ortelius in Antwerp in 1573 and a copy is now in the Copernicus Library in Torun. Frombork (Frauenburg) is close to the centre, on the shore of the 'Frisch haff'

1960. In the 16th Century it should not even have been necessary for it to be defended, because the Knights had been decisively beaten by the Poles and Lithuanians at Tannenberg in 1410, but they carried on making trouble along the Baltic coast for another hundred years. They then converted *en masse* to Lutheranism and settled down to becoming rich and, by all accounts, tight-fisted landed gentry.

Alongside this military episode, and in an interesting anticipation of Newton's much later role in running the Royal Mint, Copernicus for some years advised the governments of both Prussia and Poland on the reforms needed to stabilise their currencies. His writings on money were centuries ahead of their time and the first statements of some of the key ideas of modern economics. In 1526 in an essay entitled *Monetae cudende ratio* he set out a version of what is now known as Gresham's law, which states that bad money drives out good. He noted (Czartoryski 1985) that

> … maybe someone will argue that cheap money is more convenient for human needs, forsooth, by alleviating the poverty of people, lowering the price of food, and facilitating the supply of all the other necessities of human life, whereas sound money makes everything dearer, while burdening tenants and payers of an annual rental more heavily than usual. This point of view will be

applauded by those who were heretofore granted the right to coin money and would be deprived of the hope of gain.

Clearly the voice of a man with a rather better grasp of economics than many of the inhabitants of today's Wall Street and the City of London.

None of these things would, of course, have allowed Copernicus more than a footnote in history. It is for his astronomy that he is remembered, and it is true that as an astronomer his record of actual observations is slight and his instruments were probably not of the best. He seems to have relied heavily on the (generally remarkably accurate) observations by Arab scientists, which he probably found in Venetian libraries during the period, from 1501 to 1503, that he spent studying medicine in Padua. His interest was in the mathematics and he seems to have been one of those (perfectly respectable) scientists who make observations only as and when they need them to test their hypotheses. One can imagine him working for weeks to establish what he thought should be happening to one of the planets, and then climbing his observation tower to see if he was right. It would often have been a very long wait. Frombork is not a place anyone would pick for an observatory, given a choice. It lies low and is often covered by fog that rolls in from the Baltic.

Copernicus' persuasiveness, to those open to persuasion and able to handle his arguments, lay in his mathematical brilliance, displayed in the great work *De revolutionibus orbium coelestium* that was published just before he died in the spring of 1543. However, he had first set out his 'heliocentric' replacement (Fig. 3.3) for the Earth-centred system of Ptolemy in a pamphlet now know as the *Commentariolus* that he circulated to a few friends in 1514. Surviving only in later copies, it was there that he first presented the seven principles that formed his theory. These were that:

1. *There is no one single centre to the celestial spheres or to the orbits of heavenly bodies*
2. *The centre of the Earth is not the centre of the universe, but only of itself, and of gravity*
3. *The Sun is in the middle of the orbits of all the heavenly bodies, and the centre of the universe is therefore near the Sun*
4. *The distance between the Earth and the Sun is negligible when compared to their distance from the stars*
5. *The apparent movements of the stars are due to the movement of the Earth*
6. *What seems to us to be the motion of the Sun is actually due to the motion of the Earth around it, as with any other planet*

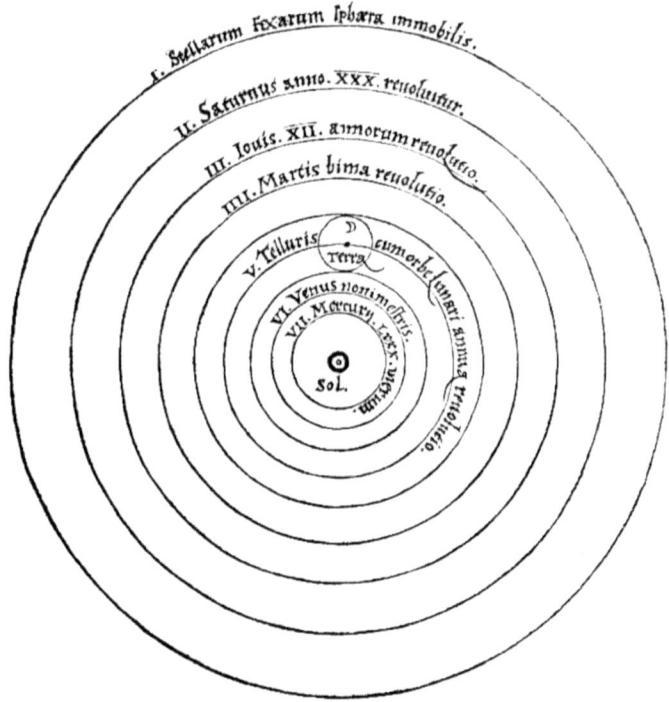

Fig. 3.3 The Copernican system, from De revolutionibus Orbium Coelestium. The Earth is the only planet shown as having a satellite

7. *The apparent retrograde motions of the planets are due to the motion of the Earth. The motion of the Earth is sufficient to explain the variations in the apparent motions of the other heavenly bodies.*[2]

At this stage these propositions were presented without any mathematical back-up. The theory was flawed because, like many later workers, Copernicus was obsessed with the idea of circular orbits, but he came achingly close to getting things right when he avoided placing the Sun at the exact centre of the universe. Sadly, the idea of circular purity was too deeply ingrained for him to take the final step and suggest an ellipse.

Given the accepted wisdom of the time, this was a staggering series of insights that went far beyond simply placing the Sun, instead of the Earth,

---

[2]See http://copernicus.torun.pl/en/archives/astronomical/1/?view=transkrypcja&lang=latina for the Latin original. I have translated 'firmamentum' as 'the stars' and 'mundus' as 'universe'.

at the centre of the universe, but Koestler unkindly dismissed *De revolu-tionibus* as 'the book nobody read'. It was this description and an encounter with a heavily annotated copy from the original printing that set Owen Gingerich, a Harvard professor, off on a lifelong investigation of his own. He had in his hands one copy that had very clearly been read exhaustively by its original owner, but what about the others? He found, in his often successful attempts to track them down, that of the roughly half of the thousand copies of the first two printings that have survived, an impressive number have annotations (Gingerich 2004). These are, as Koestler's defenders have pointed out, mainly in the sections dealing with mathematics rather than astronomy, but that may be no more than a consequence of their greater suitability for marginal notes.

No part of this story suggests that Koestler, in heading one of his chapters 'The Timid Canon', was being either fair or accurate. Indeed, he wanted to have it both ways. One the one hand, he berated Copernicus for his cowardice in not publishing, on the other he attacked him for the errors and inconsistencies in what he did publish. A more sympathetic biographer might have suggested that it was his awareness of the problems that delayed publication, and it was only eventually done at the urging of his pushy young disciple, Rheticus.[3] Indeed, it is possible to sense in '*De revolution-ibus*' a steadily decreasing confidence in the heliocentric idea. To have the planets, including the Earth, circling around the Sun should have simplified everything, but with circular orbits the theory could only be made to match observation by resorting to the Ptolemaic device of epicycles riding on cycles. Ultimately, there were even more epicycles in the Copernican system than in Ptolemy's (although they were much smaller). Small wonder there were so few observations. They must have been perpetually disappointing.

De revolutionibus was edited and printed in Nuremberg. If Copernicus ever saw a copy, it would have been on his death bed (and there is a legend that he did, and then died). It is unlikely that he ever knew that the editor, a Lutheran preacher called Andreas Osiander, had added an unsigned preface of his own, which said (among other things)

> … it is the duty of an astronomer to compose the history of the celestial motions through careful and expert study. Then he must conceive and devise the causes of these motions or hypotheses about them. Since he cannot in any way attain to the true causes, he will adopt whatever suppositions enable the motions to be computed correctly

---

[3]The role of Rheticus in persuading Copernicus to publish is a central theme of Sobel (2011).

… The present author has performed both these duties excellently. For these hypotheses need not be true or even probable. If they provide a calculus consistent with the observations, that is enough

… For this art, it is quite clear, is completely and absolutely ignorant of the causes of the apparent [movement of the heavens]. And if any causes are devised by the imagination, as indeed very many are, they are not put forward to convince anyone that they are true, but merely to provide a basis for computation.

However, since different hypotheses are sometimes offered for one and the same set of observations, the astronomer will take as his first choice that hypothesis which is the easiest to grasp. The philosopher will perhaps rather seek the semblance of the truth, but neither of them will understand or state anything as certain, unless it has been divinely revealed to him

… Let no one expect anything certain from astronomy, which cannot furnish it, lest he accept, as the truth, ideas conceived for another purpose, and depart this study a greater fool than when he entered.[4]

Here Osiander is anticipating the compromise that, in the next century, was to be seized upon by Cardinal Bellarmine, which was that the sun-centred universe should not to be taken literally but should be regarded merely as a convenient fiction that allowed calculations to be made more easily. Opinions differ as to whether this was a betrayal of Copernicus or a clever, although ultimately unnecessary, way of protecting him. Whatever the motive, it did provide a 'Get out of Jail' card (in some cases quite literally) for some of his later followers, and it gave Galileo twenty extra years of freedom. It is unlikely that anyone actually believed in it, but the appearance of believing was enough to allow the more liberal churchmen (but not Martin Luther) to turn a blind eye.

An Astronomical Revolution

While Koestler went much too far in dismissing *De revolutionibus* as almost completely unread, it is true that its initial impact was small. There were good reasons for this. The celestial scene during Copernicus' lifetime, and for ten years after his death, was rather boring. The planets moved about, in ways that had become more or less predictable, and the fixed stars remained—fixed. Halley's comet made an appearance in 1531, but may have

---

[4]Translation from Oster (2002).

been as unspectacular as it was in its most recent fly-past in 1986. Nobody much seems to have noticed. The incentives for mass interest in star-gazing simply did not exist.

All that changed in the next hundred years, when there were four major comets and two exceptionally bright supernovas (Fig. 3.4). These presented real challenges to the Ptolemaic universe in which the visible stars and planets were supposedly mounted on hollow and invisible crystal spheres that rotated peacefully in entirely predictable ways. The appearance of a supernova, which occupied a fixed place amongst the constellations for as long as it lasted but then faded away, was incompatible with the idea of a stellar sphere in which nothing ever happened.

Comets presented a different problem. They were certainly not stars, but where were they located? The popular view that they travelled within the Earth's atmosphere became untenable once astronomers had found ways of estimating their distances from Earth, and had shown that they must be accommodated in the space between the lunar and stellar spheres. Worse still, there was disturbing evidence that they were able to pass through some of the supposedly solid spheres on which the planets were mounted. In the

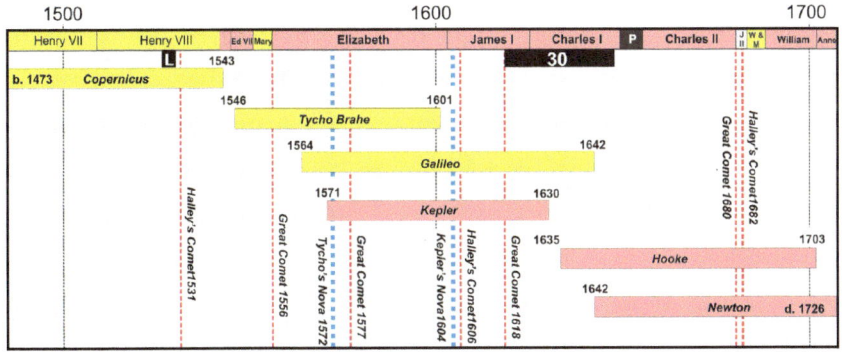

Fig. 3.4 The 150 years following the death of Copernicus were marked by an unusual number of spectacular astronomical events. Blue vertical dashed lines indicate supernovas, red dashed lines major comets. It was also a turbulent time for western Christendom, as demonstrated by the alternations in the religious leanings of the rulers of England. For both the monarchs and scientists, red rectangles indicate Protestants and yellow rectangles indicate Catholics, with Henry VIII changing allegiance late in life and Charles I, as ever, hard to pin down. The black rectangles marked 'P', 'L' and '30' identify, respectively, the period of the Puritan-dominated Commonwealth, the period between Martin Luther's publication of his 'Ninety-five Theses' and his excommunication, and the period of the Thirty Years War (1618–1648). The latter devastated much of Central Europe and was a major factor in the later life of Johannes Kepler

late 16th Century the night sky became something worth studying, and not just for the casting of horoscopes. The six-year old Kepler would not have been the only child whose interest was fired by being 'taken to a high place' by his mother to see the comet of 1577 (Jardine 1999). It was almost the only good thing she ever did for him.

The Instruments

It is impossible to get very far in understanding the astronomy of the time without considering the instruments and methods that were being used. The most basic, and oldest, of these was the sundial. The position of the shortest shadow defines the north-south direction (the meridian), and the direction of the sun at other times, and hence the time of day, is shown by the position of the shadow on a circular scale graduated in hours. This same scale could equally well be graduated in degrees and, for a conventional sundial with a horizontal circle, this would give the angle known as the *right ascension*.

The sundial uses shadows because the sun is normally too bright to be viewed by the unprotected eye, but it is easy to extend the principle to stars, which can be viewed directly. All that is needed is a sighting bar. It is also possible to use a vertical scale and measure the elevation of the sun or stars as an angle made with the horizontal, which is known as the *declination*. The instruments used to measure declinations are known (depending on the fraction of a circle occupied by the scale bar) as octants, sextants and quadrants.

In principle the angular distance between two stars could be obtained by measuring their individual declinations and right ascensions and then making some calculations in spherical geometry. There is, however, an easier way, using a very basic instrument known as a cross-staff or radius. This consists of a sighting arm a few feet long, along which can be slid a shorter arm mounted at right angles and projecting equal distances on either side. All the observer has to do is point the sighting arm towards the two selected stars and slide the cross-arm until there is a star at each end. Normally there would be a graduated scale along the sighting arm from which the angle between the two stars could be read directly.

The most important use of the cross-staff was for measuring *parallax*, or lack of it. A search in a modern encyclopaedia or on the internet for a definition of astronomical parallax leads to descriptions of the way in which the distances to the closest stars can be estimated by the changes in their angular positions as seen from opposite sides of the Earth's orbit around the

Sun. That is not what was meant in the 16th Century, when the idea of a moving Earth was accepted by very few, and the changes would, in any case, have been far too small to be measured with the instruments available. Parallax then was the change in angular position, over a period of only a few hours, of objects in space against the background of the fixed stars, and only the distances to the Sun, Moon, Mercury, Venus and some comets could be obtained by this method with the instruments available. The idea is most easily explained in terms of the geocentric Ptolemaic universe in which most astronomers then believed.

Figure 3.5 shows a vastly oversimplified geocentric universe, consisting of a fixed Earth, a stellar sphere rotating around it and a single object, C, that is between the two but rigidly attached to the stellar sphere by an invisible rod. At the moment represented by the star positions S_1 and S_2 the object, as seen from Earth, sits exactly between them, but two hours later, when the stellar sphere has rotated by 30° bringing the stars to S_1' and S_2', its position relative to them would have changed noticeably. The model of the universe was wrong but the estimates of distance were, within the limits of accuracy, correct.

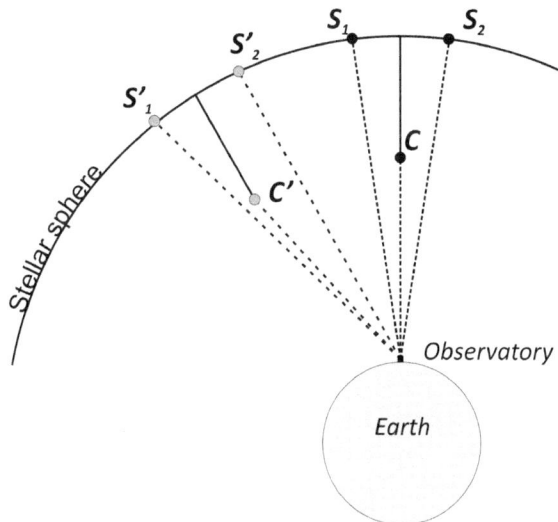

Fig. 3.5 Astronomical parallax in the 16th Century geocentric universe. The stellar sphere, on which the stars are located and to which the comet C is rigidly attached, has its centre at the centre of the Earth. Initially the comet appears to be exactly half way between S_1 and S_2. Two hours later, after the stellar sphere has rotated through 30°, the three bodies will be in the new positions S_1', S_2' and C' and the comet will no longer appear to be midway between the two stars

Things were not, of course, quite so simple, because comets, planets and the moon have their own motions, which could only be estimated by repeated observations made at the same times night after night. The better the instruments, the better the observations, and no-one ever built better instruments or made more accurate observations with the naked eye than a Danish aristocrat called Tycho Brahe.

Tycho Brahe

Tycho was born a few years after the death of Copernicus, in Skåne, on another shore of the Baltic. His birthplace is now in Sweden, not Denmark, but there has been no movement among the Swedes to claim him for their own. His father was a Brahe, his mother was a Bille and, as all lovers of *Hamlet* will be delighted to hear, his family tree included both Rosenkrantzes and Gyldernstiernes. He could not have been better connected. Almost a quarter of the 16th Century appointees to the Rigsraad, the tiny Council of State that effectively ruled Denmark through the king, came from those four families.

Tycho had a remarkable life,[5] and it began remarkably. When only two years old, he was kidnapped by his uncle Jørgen Brahe, who was childless and who presumably had reasons for thinking (correctly) that he would remain so. The true parents seemed to have objected only briefly, before devoting themselves, with considerable success, to producing replacements. The case was certainly unusual, but fostering of young nobles was a well-established Nordic tradition and had its advantages. For Tycho the move was crucial, because the Brahes were soldiers and statesman, and his four brothers all followed family tradition by being 'educated' as pages in foreign courts. Jørgen, although conforming to the Brahe military norm and ending his life as a vice-admiral, was married to Inger Oxe, and the Oxes considered studying in foreign universities more useful than fighting in foreign wars. Tycho was rigorously schooled and, fatally, was encouraged to think.

In 1565, at the age of nineteen, his life changed again. His foster father died, in circumstances that were romantic, if slightly unclear. The Danish fleet had just returned from a battle with the Swedes that they claimed as a victory (as did the Swedes) and the king, Frederick II, being drunk in celebration, fell into the water under a bridge that linked Copenhagen to one

---

[5]Described most comprehensively in Thoren (1990).

of the smaller islands. Accounts differ as to whether Jørgen (also probably drunk) fell in with him, or fell in trying to rescue him. Whatever the truth, and even though it was close to midsummer, the vice-admiral caught a chill and died of pneumonia a few days later. It is possible that Frederick's life-long indulgence of Tycho (a man who habitually pushed toleration to its limits) arose from this episode. Not only had he been robbed of the man he regarded as his father, but possibly also a very large fortune. At the time of his death, Jørgen had been about to make him his heir.

After Jørgen's death, Tycho returned to his natural parents, but it was too late for him to become a typical Brahe. He had no military experience and his interests lay elsewhere. He had bought his first astronomical instrument, a cross-staff, while studying in Leipzig, and had it modified to meet his own already exacting standards. Lack of military training has, however, never stopped adolescent males from fighting and a year later he lost most of his nose in a drunken duel with another Danish student in Wittenberg. Accounts differ in detail, but the quarrel certainly began with an argument involving the stars. In one version it was merely as to which of the two combatants was the better mathematician, but another has Tycho being derided for having cast a horoscope predicting the death of the already dead Turkish sultan (if he didn't know of the death, this seems like a very impressive near-miss). Some parts of this story were confirmed by two 20th Century exhumations of Tycho's body, which showed that most of the bridge of his nose was missing. Traces of metal near where it should have been suggested a replacement made of copper or brass, but did not disprove the legend of the man with the golden (or silver) nose. There may well have been one nose for important occasions and another one, lighter and cheaper, for everyday use. If there ever was a nose made of precious metals, it is not surprising that it was not buried with him.

Another result of the duel was a lifelong interest in medicine. It was probably only due to some unusually skilled doctoring that he did not die of infection.

Tycho's obsession with astronomy continued after his noseless return home, to the horror but eventual resignation of his family, and in about 1571 he followed sound astronomical tradition by beginning a liaison with a woman who was very much his social inferior. According to Jutish Law, any woman who lived with a man for three years, eating, drinking and sleeping with him and possessing the keys to his house (preferably worn openly, although it is not clear if this was absolutely necessary) was regarded as his wife, but if he was a noble and she a commoner, she could only be considered *slegfred*. Any children would be legitimate but could not inherit nobility. This sort of arrangement was common enough in Danish aristocratic circles, and Tycho's was unusual only in that it lasted until he died. Kirsten Jørgensdatter outlived her husband by several years, during which time she purchased a rural estate in Bohemia.[6]

It would be nice to think that it was a love match, and it may have been. After his arrival in Prague, where he died, Tycho went to the trouble of becoming a citizen of the Empire, thus placing the status of his wife and children beyond the reach of the medieval laws of Denmark. But love does not exclude the possibility that he also realised, consciously or unconsciously, that the daughter of a local pastor (as Kirsten was described in some accounts) would be less likely than a fellow aristocrat to object to his astronomical, alchemical and medical obsessions. The Viking sagas would have been warning enough of the ways in which ladies of the Nordic upper classes behaved when they felt that they were not getting sufficient attention.

In 1572 a stellar nova appeared in Cassiopeia. Tycho observed it and in the following year published a short monograph that ultimately led to its being known as 'Tycho's Star' (later he was to publish a monograph on the Great Comet of 1577, ensuring that it would sometimes be known as Tycho's Comet). His calculations proved conclusively, to the very few people who could follow them, that the nova lay beyond the orbit of the moon, in a region where nothing new was supposed to happen, and made him famous. By 1574 he was not only giving lectures on astronomy at the University of Copenhagen but his work had been noticed beyond Denmark, and he set out on a tour of the German observatories. On his return, Frederick, fearing that this now prominent scientist might leave Denmark, and possibly still feeling guilt over Jørgen Brahe's death, offered him the island of Hven, in the Øresund that separates Denmark from Skåne, together with the funding to build an observatory, if he would stay. The offer was accepted.

Tycho never did anything by halves, nor was he ever afraid of spending money (especially other people's) to get things as he wanted them. It has been estimated that in one of his years on Hven he worked his way through one per cent of the entire revenues of the Danish state. The Uraniborg, his great observatory cum castle, was built on a greenfield site to his own designs, complete with ornamental gardens in which he grew medicinal herbs (a personal interest since the loss of his nose), a private zoo (the

---

[6]Kirsten was buried next to Tycho in the Teyn Church in Prague. Letters between her and her sons and daughter written after Tycho's death (English translations in the 'Epilogue' to Thoren 1990) suggest a strong and united family.

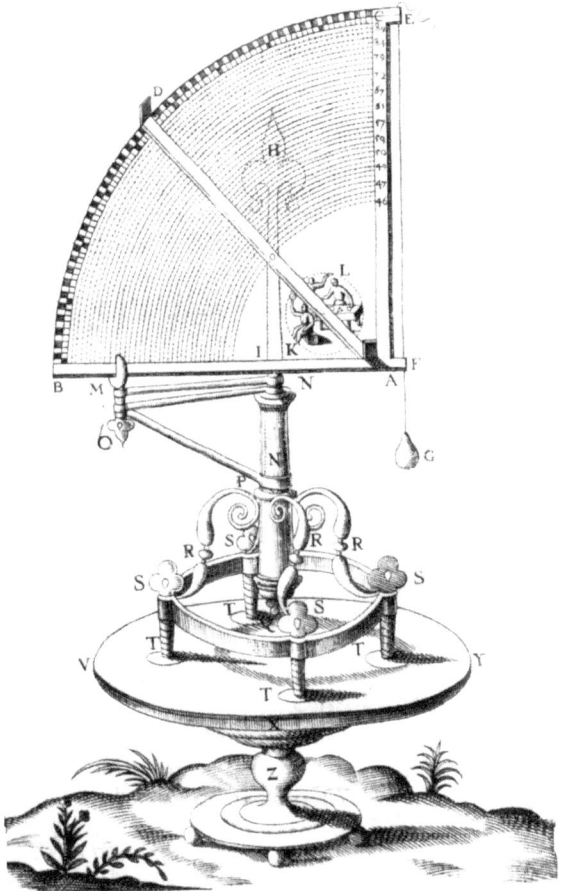

Fig. 3.6 The small quadrant (quadrans minor). From an illustration in Astronomia Instauratae Mechanica. The 90° arcs are transversal scales that allowed angles to be measured to better than one minute of arc. The letters refer to the detailed instructions for use provided in the accompanying text. In practice, this instrument, with a radius of only about 50 cm, was too small to produce results of the quality demanded by Tycho

elk got drunk one night, fell down some stairs, broke its leg and had to be destroyed) and even a prison (which was eventually to be a major cause of his downfall, due to his habit of illegally incarcerating his tenants). He designed new instruments (Fig. 3.6) and had them built, he made, or had his assistants make, observations on almost every night when it was possible to see any stars, he had his own printing press and he ran a school for young astronomers. He entertained royally, and royalty, and never ceased to lec-

Fig. 3.7 The Uraniborg, from a copper etching in Blaeu's Atlas Major of 1663. At the top of the building were eight bedrooms for assistants, and below them four observatories (with conical roofs). The ground floor was occupied by a library, a kitchen, a main dining room and three spare bedrooms. The alchemical laboratory was in the basement. The building took almost five years to complete, from early 1576 to late 1580

ture his visitors on the absolute necessity for accurate observations. When he found that winds shook the extravagant (and possibly somewhat jerry-built) structure of the Uraniborg (Fig. 3.7) so much that his observations were affected, he built an entire and lavish second observatory, the Stjerneborg, that was almost completely below ground.

Tycho's contributions to astronomy were enormous. He not only designed and built instruments of unprecedented accuracy (and size) and described them in his great work *Astronomiae Instauratae Mechanica* but, in a profound break from the then current practice, he calibrated them and repeatedly checked their accuracy. Like Galileo, he had no satisfactory way of measuring time and had to measure the positions of the planets in relation to the stars, which was much more difficult than just measuring their

Fig. 3.8 The great mural quadrant in Uraniborg, from the 1602 edition of Astronomia Instauratae Mechanica. The graduated scale, a form of transversal that allowed angles to be measured to better than a minute of arc without the need for the accurately engraved 90° arcs used in earlier and smaller instruments, was mounted on a wall oriented very precisely N-S. The accompanying text gives the radius of the arc as 5 cubits, or about 3 m, which suggests that it and Tycho (who is pointing to the small window through which the observations were made) are drawn to approximately the same scale, dwarfing his helpers and assistants. Even the observer making the measurements, although almost Tycho-size, is squashed in against the right-hand edge of the picture. The dog at Tycho's feet may be one of the pair of English mastiffs presented to him by the future James I of England (VI of Scotland)

positions at known times, and this led to the discovery of orbital anomalies that had never before been noticed. Without these comprehensive observations Kepler would never have been able to recognise that planets move in elliptical orbits. Tycho was also the first person to notice, and make corrections for, atmospheric refraction, because he was the first person to have instruments accurate enough for it to be noticeable. The observations made by previous astronomers had been accurate to perhaps a quarter of a degree, but Tycho's were routinely accurate to about two arc-minutes (one-thirtieth of a degree), and the best were accurate to about a quarter of that (Fig. 3.8).

Tycho's greatness as an observer and instrument designer is undisputed, but his achievements as a theoretician (which included a radically new understanding of the motion of the Moon) are less widely recognised. In part this is because he never discarded the idea of a geocentric universe. He gave various reasons for rejecting the heliocentric theory, but underpinning them all was his inability to abandon Aristotelian physics and its notion of absolute place. In his mind, heavy bodies fell to their natural place, the Earth, which was the centre of the universe.

Also, heliocentricism had a number of advantages. It produced a better match with lunar observations, based on smaller epicycles, and a ready explanation for the retrograde motions of the planets, and Tycho attempted to produce a system that would combine the best of both theories. He kept the Earth at the centre of the universe, so that he could retain Aristotelian physics, and made the spheres of the Sun and Moon and the fixed stars revolve about the Earth. The planets, however, he placed in orbit about the Sun, and the comet of 1577 was assigned a path between the orbits of Venus and Mars.

The Conjunction of the Bear

Tycho, dying, supposedly said, again and again, '*Let me not seem to have lived in vain*'.[7]

If he could have seen 300 years into the future, he might have been satisfied. It is not everyone who has given his name to a crater on the Moon, a crater on Mars and an astronomical observatory (in Copenhagen). Yet he would also have had to admit that without Johannes Kepler, the man who recorded his dying words, he might have been remembered only as a curiosity, almost the last of the 'naked-eye' astronomers who struggled to observe

[7]Quoted by Kepler in Astronomia Nova, Chap. 6 (see Koestler 1959).

the skies in the days before the telescope changed their science for ever. It was Kepler who completed his work, who published it and, being the man he was, gave the credit where it was due. There would have been little reason for either of the pair to be much remembered without the other, yet their collaboration lasted for less than two years. That it happened at all was something of a miracle, and one that owed much to Tycho's greatest enemy. The feud that dominated the last years of his life began in 1584 when Erik Lange, an old friend and fellow aristocrat (who eventually married his sister Sophie) brought to Hven a young astronomer on the make, Nicolaus Reimers Baer, otherwise known as Ursus the Bear.[8]

The Bear had made a remarkable ascent from the very bottom of the social ladder (as a pig-herding peasant) to employment first with Heinrich Rantzau, a noted scholar and governor of Holstein, and then with Lange. He must have been very bright indeed, since he taught himself both maths and Latin and by the time he came to Hven he had already written two books, one a Latin grammar and the other a treatise on surveying. Both were dedicated to Rantzau. Quite why Lange brought him along is not known, but he might have been hoping to pass him on to Tycho. As the next fifteen years were to show, the Bear was not a comfortable employee.

At first all went well. The Bear composed a poem in Tycho's honour, and may have done some paid work for him. That, however, did not last and Tycho, always paranoid where his work was concerned, began to suspect that there was more behind the questions he was being asked than mere casual interest.

There are two different versions of what happened next. Tycho's is the more coherent. The story that he, many years later, asked Lange's secretary to confirm in a notarised statement, was that he decided that the Bear needed watching and had one of his students share his sleeping quarters. While the Bear slept, the student felt in one of his pockets and found in it notes on Tycho's hybrid astronomical system, then not fully developed.

When, in the morning, the Bear discovered that his notes were missing, he accused Tycho of theft, an unwise move for a commoner when dealing with nobility. He was summarily dismissed by Lange (who might have been glad of the excuse) and left Hven under a cloud. He is next heard of as a tutor in the court of a minor Pomeranian noble, and that should have been the end of him, but this Bear was not easily penned. He managed to climb back, securing a position at the court of Wilhelm IV of Hesse-Kassel,

---

[8]The main facts given here concerning the conflict between Tycho and Ursus have been taken from Thoren (1990).

and in 1588 he published a volume entitled *Fundamentum Astronomicum*. In 16th Century science, even more than in today's, it was publish or perish, with starvation an occupational hazard. A publication was always, in part, a means to finding a patron, and the accepted way of increasing its chance of doing so was by dedicating it to the rich and the powerful. The *Fundamentum* had dedications to no fewer than thirty-eight such people. Tycho was not among them but the book contained a detailed description of a hybrid system of the universe very similar to his. It resembled even more closely the system as it had been four years earlier, when the Bear was on Hven and before Tycho realised that the orbit of Mars could not intersect that of the sun.

Tycho erupted. He was, perhaps, especially sensitive to claims on priority because his own claim was none too secure. A similar system had been proposed a thousand years earlier, by Martianus Capella, and had been adopted by Paul Wittich, a German astronomer who had also visited Hven. Wittich was dead, and so unable to object, but Tycho was very much alive. His counter-attack began with the publication of a series of his astronomical letters that testified to the early date at which he was developing his ideas. The Bear was mentioned only occasionally, and dismissively.

It was into this swamp of accusation and counter-accusation that Kepler splashed puppy-like, with his tail wagging. In November 1595, after a scant three-day read of the *Fundamentum*, he dashed off a letter to its author, addressing him in the most fulsome terms. Freely translated (and Kepler's style when in full flood is almost impossible to render literally), it went as follows:

There are in distant countries people who, being themselves unknown, write letters to people that they do not know; how strange are men. Your fame as the foremost amongst mathematicians is like that of the sun compared to the lesser stars. Even so, the more I praise you, the more all learned men will praise you, scorning in their judgement the opinions of the arrogant to agree with those of a modest young man. Since, therefore, it has been from your books that I have gained what knowledge I have of Mathematics, I have thought it right, and not something to be treated lightly, to consult with you on these difficult matters. If you approve of what I have written, I am blessed and my closest approach to happiness would be to be corrected by you. I value your judgement. I love your hypotheses, even though it is also impossible to esteem too highly the work of Copernicus, which I have saluted in these verses

What is the world but God's creation? Whence came God's numbers, the laws that rule the heavens? Why should there be six cycles, into which the orbits fall? Why are Jupiter and Mars so far apart? Believe that it is revealed in the five solids of Pythagoras

For between the orbits of Saturn and Jupiter we can inscribe a cube that just touches the sphere of Saturn and is just touched by the sphere of Jupiter. Between Jupiter and Mars we can inscribe a tetrahedron, between Mars and the Earth a dodecahedron, between the Earth and Venus an icosahedron and between Venus and Mercury an octahedron. Neither mathematicians nor metaphysicians can change the order of these bodies. The paths of the inner planets fit perfectly. The ratios of their distances are better known. There are deviations from the rule in the paths of the outer planets, just as a ray of light diffuses as it travels further from the source. Just so much can be derived from Copernicus. In fact, it follows from application of the sine and cosine laws that the differences cannot be greater than 12' for Saturn, 25' for Jupiter, 1°45' for Mars, 1° for Venus and 4' for Mercury.

I write no more and await your judgement, but must record my gratitude to this noblest of young men, D Sigismundi V Vagani, at whose prompting I have written and who is the means by which I have this most recent opportunity of communicating with you.

Farewell to you, the glory of Germany is in our knowledge of the stars.
Gratz, 15 November 1595.[9]

Most commentators have, understandably, concentrated on the embarrassingly obsequious first part of this letter, but what comes after the 'poem' is more remarkable. There Kepler is describing, and effectively giving away to a possible rival, his idea of a link between the orbits of the six known planets and the five regular Pythagorean (or Platonic) solids. The theory was wrong, but it was to dominate his scientific thinking for the rest of his life. The Bear, a serial plagiarist, might well have stolen it had it been compatible with his own geo-heliocentric model (and had he been able to understand what, in heaven's name, Kepler was talking about).

Kepler had good reasons for trying to please the Bear. He was only just managing to exist on a poverty wage as a teacher of mathematics and astronomy at the Protestant school in the provincial backwater of Graz, and was about to lose even that position because Styria had begun expelling Protestants. He must have seen in the Bear, who in 1591 had managed, despite Tycho's sniping, to become the *Imperial Mathematicus* to the Holy Roman Emperor Rudolph II, a possible patron. Initially, he did not even get a reply.

In the following year Kepler published his own first work, the Mysterium Cosmographicum, containing his 'Platonic solids' theory, and started

---

[9]Letter reproduced in facsimile in Rosen (1946). I am enormously indebted to Mrs. C. Donahue for her help in translating Kepler's almost incomprehensible Mediaeval Latin, but take full responsibility for any errors in the rendering of his cosmology.

to send out copies. To the Bear he sent two, asking, incredibly, for one to be forwarded to Tycho (who was still in Denmark and so only marginally closer to Prague than to Graz). It would be an exaggeration to say that the book was a best-seller, but it attracted enough interest for the Bear to see its author as a potentially useful, if unwitting, ally in his campaign against Tycho. His next book, *Nicolai Raimari Ursi Dithmarsi de Astronomicis Hypothesibus*, printed in Prague in 1597, included a copy of Kepler's letter.

In publishing *Astronomicus Hypothesibus*, the Bear had (not for the first time) overstepped the mark. One historian, who has been more sympathetic to him than most, described the book as *savage and scurrilous even by the ferocious standards of sixteenth century polemic* (Jardine 1984). It not only attacked Tycho's astronomy, but it mocked his nose and made obscene suggestions about his wife and daughter. It was so abusive that the Archbishop of Prague, who acted also as the Imperial censor, refused to licence its publication. The Bear published anyway, with a banner headline on the title page glorying in the fact that this was being done without permission. This, it turned out, was a very bad idea indeed.

Kepler, when he eventually heard about the publication of his letter, was horrified. He later claimed that, although he could not (quite reasonably) remember exactly what he had written, he remembered enough to know that what the Bear had published was an edited version, from which favourable references to Tycho had been removed. If he had ever had hopes of the Bear as a patron, he had long since abandoned them, and he had in any case realised that the only way that he would ever be able to test his beautiful theory would be by using Tycho's observations. Fearing, with good reason, that the copy of the *Mysterium* sent via the Bear would not have reached Tycho, he had already sent a second one, accompanied by another of his embarrassing letters. By an extraordinary coincidence the package reached Tycho on the same day in March 1598 that he first saw a copy of *De Astronomicis*.

At this stage in his life, Tycho had much to think about. His old patron, Frederick II, had died and the new king, while initially favourably disposed towards him, was far less ready to comply with his financial demands. His treatment of his tenants on Hven had become a national scandal, and he had enemies at court. Faced with these problems, he had left Denmark in a huff a year earlier, and any hopes he might have had of an early (or any) return had been dashed by an ill-tempered exchange of letters with the king. One of his admirers had generously provided him with a castle at Wandesbeck, near Hamburg as a new home, but this, while comfortable, was temporary, and he was desperate for a new sponsor who would fund a new observatory. And, although he had only just turned fifty, his much abused body was beginning to show definite signs of wear.

In the midst of all this, a distant and silent Bear would have been no more than a minor irritant, but the publication of *De Astronomicis* changed all that. He became, and remained for the rest of Tycho's life, a major obsession, but his downfall was plotted more skilfully and less impetuously than might have been expected. Kepler was one of the beneficiaries. Tycho's reply to his letter was written almost as between scientific equals, and concerned chiefly the *Mysterium*. Only in a postscript was he gently chided for his involvement with the Bear.

This letter was not, however, the whole of Tycho's strategy for involving Kepler on his side. He also wrote, much more forthrightly, to Kepler's old Tübingen tutor Maestlin, assuming, rightly, that his complaints would be passed on. In all probability it was not until the arrival of a letter from Maestlin that Kepler heard about the Bear's new book, and his place in it. Faced with this disaster, he despatched another of his famous letters, full of grovelling apologies, to a Tycho whose fortunes had taken a turn for the better. Thanks to *De Astronomicis*, the Bear was in disgrace and had temporarily fled Bohemia. Tycho had replaced him as *Imperial Mathematicus* and a new observatory was being built in a castle at Benatky, near Prague. No-one who had known the old Tycho would have given Kepler's latest letter much chance of success, but the new Tycho was thinking ahead. His first response was brief, but friendly enough, and a second, and much more positive, letter followed shortly afterwards.

Koestler, ever the fan, believed that the reason Tycho responded as he did was that he had '*immediately realized young Kepler's exceptional gifts*' on reading the *Mysterium*, but this is unlikely. Although the book had attracted some attention, it was full of errors and improbabilities, and it also revealed Kepler's acceptance of the heliocentric hypothesis, which was anathema to Tycho. Moreover, in trying in his second letter to excuse his praise for the Bear's *Fundamentum*, Kepler had written that a 'doctor' on his way back from Italy had shown him the book when he stopped in Graz but had only allowed him three days to read it. He added, potentially fatally, that the geometric and trigonometric theorems and proofs that he found there had been new to him, and had been the reason for his enthusiasm for the Bear, but that he had subsequently found the same material in Regiomontanus and Euclid. Thus, in trying to excuse himself to Tycho, Kepler had shown himself to be ignorant of some things that even a provincial mathematics tutor should have known.

Whatever the truth, by the time the second letter reached Graz, Kepler, jobless and almost at his wits' end, had already left.[10] A Protestant in a province that was becoming more and more stridently Catholic, he was being forced out, and on the first day of the new century he had taken a gamble and had left to see Tycho in Bohemia. It is a measure of his desperation that he made the journey through the Alps unprompted and in the middle of winter. It took him two weeks to reach Prague (where he accidentally met, and 'upbraided', the Bear, revealing his own identity only as they parted) (Rosen 1946), and it was not until the middle of February that he finally reached Benatky and Tycho.

There can scarcely ever have been two less likely collaborators. Tycho was an aristocrat, arrogant, irascible, extroverted and enjoying, at least in his early life, the rudest of health. He was raised in luxury within a vast and supportive family network. Kepler was the poor and sickly son of a determinedly downwardly mobile and dysfunctional *petit-bourgeois* Swabian family. Tycho had been brought up in the house of an admiral, Kepler's father had been a mercenary foot soldier, despised in his Protestant home town for fighting in the Netherlands on the Catholic side. His aunt may have been burned as a witch and his mother only just escaped the same fate. Kepler had, throughout his life, an almost painful need to be liked, which to Tycho would have been incomprehensible. Tycho was an unenthusiastic Catholic, Kepler a convinced Lutheran whose life would have been much easier had he taken one of the many opportunities given to him to change his faith. But had he done so, he might never have met Tycho.

One reason for rejecting Koestler's interpretation is that, as far as assistants were concerned, what Tycho needed scientifically after leaving Hven were not volatile blue-sky thinkers but automata capable of carrying out hundreds of routine calculations without making too many mistakes. What he wanted from Kepler was quite different. He wanted testimony against the Bear in the lawsuit he had begun alleging plagiarism and defamation, and for most of the time that they were together, Kepler was put to work on this. In March 1600 he produced a two-page deposition for the law courts entitled '*Quarrel between Tycho and Ursus over Hypothesis*', but his letters show that he resented having to do so (Thoren 1990, p. 459).

After only two months, and following a quarrel that was basically over pay and conditions, a disillusioned Kepler left Benatky for Prague. Tycho

---

[10]The four letters, exchanged during the period from December 1597 to December 1599, are reproduced as Letters 82, 92, 112 and 145 in Caspar (1945).

then showed just how badly Kepler was needed by going personally to Prague to persuade him to return, and it may have been at this time that he was offered responsibility for Mars. The work had been previously given to Tycho's senior assistant, Longomontanus, who had made the trek from Denmark with him, but it would have suited Tycho to make the change because Mars was to him a lower priority and he really wanted Longomontanus to work on the Moon. Nevertheless, and to Kepler's dismay, once back in Benatky he found himself still being pressed to write polemics against the Bear. However, his options were by that stage very limited, and on 1 June 1600 he left for Graz to collect his family.

While he was away, two important things happened. Longomontanus went back to Denmark and the Bear died in Prague. Tycho, in the ascendant, had been unrelenting, and had sent to the deathbed

> … two doctors of jurisprudence together with a public notary to ask whether he was willing to retract that malicious publication, chockfull of insults. At the same time I prepared the main items of the insults to be read to him…. But the defendant died.

He later noted that

> … death had struck that wild beast with special kindness and saved him from a thoroughly deserved punishment

since had he

> … lived a while longer, he would have been sentenced, as I learned from the commissioners, to be branded in infamy, and beheaded or quartered according to Bohemian Law.[11]

They evidently took plagiarism pretty seriously in Bohemia at the start of the 17th Century—although the Bear's main fault had presumably been his impertinence, as a commoner, in daring to defame a member of the nobility.

Kepler returned to Prague in October, accompanied by his family and ill. He may have feared that, with Ursus dead, his usefulness to Tycho would be at an end, but if he did, he was mistaken. Denied his day in court, Tycho was determined to restore his own reputation by destroying the Bear's, and

---

[11]All three quotations from Thoren (1990).

for that Kepler was even more important. In letters to Maestlin, Kepler complained that during the whole of that winter he was employed only in writing against the Bear. His *Apologia Tychonis contra Ursum* (*Defense of Tycho against the Bear*), an expansion of the earlier pamphlet, was written then but not published until 1858, following its discovery amongst his personal papers.[12] In April 1661, with Tycho's approval, he returned to Graz to try (unsuccessfully) to extract some money from the estate of his recently deceased father-in-law.

All in all, Kepler's second trip back to Graz must have been a welcome relief, even though he failed in his main objective. During the four months that he spent away, his health improved and he was able to do some serious scientific work on optics (published in 1604 as *Astronomiae Pars Optica*). He must have been ready for almost anything when he returned to Prague in August, which was just as well because he was almost immediately taken by Tycho to meet the Emperor (Thoren 1990, p. 460). That this meeting would be a turning point not only for Kepler but for the science of astronomy, was unintended. For Tycho it was about persuading Rudolph to pay Kepler a salary, and in order to achieve that, he must have praised his young assistant to the skies. It was agreed that Tycho's observations were to be processed and published as the Rudolphine Tables, and that Kepler would be assigned to the work as Tycho's main assistant.

The unintended consequence came two months later. On 13 October 1601 Tycho accompanied the Imperial Councillor to a formal dinner in Prague and during it felt a great need to urinate. Rather than breach etiquette by leaving the table before the Councillor, he remained seated, and something inside him burst. He died five days later, having suffered periods of great pain and other periods when the pain retreated and he ate ravenously. Within a few days of the funeral, and presumably because Tycho's praise was still fresh in his mind, Rudolph appointed Kepler as his new *Imperial Mathematicus*.

Kepler Alone

It is in dealing with Kepler that Koestler's strengths as a biographer come to the fore. He took a liking to this strange character, and excused actions and obsequiously embarrassing correspondence that he would have roundly con-

---

[12]The genesis and contents of this pamphlet were discussed in Rosen (1946). It is translated in Jardine (1984).

Fig. 3.9 Tycho Brahe (left) and Johannes Kepler (right). The portrait of Tycho is taken from the frontispiece to Astronomica Instatae Mechanica, and seems less of a caricature than most others, while retaining the famous and impressive moustaches. Oddly, it suggests a more thoughtful and introspective character than the rather frightening Kepler of Jakob von Heyde's copper engraving, made in about 1620, when Kepler was just reaching the end of his time as Imperial Mathematician

demned in Copernicus or Galileo. As a sympathetic portrait of an extraordinary personality, the *Sleepwalkers* is hard to better, and worth reading for that alone. It is also sometimes, as in its description of Kepler's search for a second wife, very funny. That search, which turned into an erratic odyssey between eleven possibilities, ended with his choice of Susannah, the least socially acceptable but quite possibly, for him, the most suitable. It perhaps says less about Kepler and more about Koestler that he interpreted her near-absence from Kepler's subsequent diaries as the sign of a happy marriage.

Kepler had a miserable start in life and a miserable end, but for twelve years in its middle, and through a series of coincidences and accidents, he was almost unbelievably lucky.[13] Koestler (1959, p. 300) pointed out that had Tycho stayed in Denmark for what remained of his life it was unlikely

---

[13]So much so that some have claimed that Kepler made his own luck by murdering Tycho. This seems completely incompatible with his character, painfully revealed in his own diaries, even though possibly consistent with the rather sinister portrait in Fig. 3.9b. It would also have been a colossal gamble, since he could not have known that the erratic and unreliable Rudolph would act as he did.

that Kepler could have met the cost of a visit to him, and that it was the fact that they were both exiles that made their meeting possible. In an uncharacteristic excursion into metaphysics he speculated that the pair might have been brought together not by chance or providence but by the existence of some hidden law of gravity in History. After all, gravity, he said, was a word used to describe an unknown force acting at a distance.

This ignores the fact that Tycho might have lived considerably longer had he remained in Denmark, but it is also true that he might then not have needed Kepler as an assistant. It was Ursus the Bear, not gravity, that brought this ill-assorted couple together, and without his intervention Isaac Newton would not have had the information that led to him to a mathematical formulation of the Law of Universal Gravitation. Somebody would have done it eventually, but it might have taken much longer.

Had things been only slightly different, Tycho's death could have been a disaster for Kepler. Rudolph may not even have known that he existed before their meeting in August, and might have forgotten him again a few weeks later, so the timing was crucial. Even so, his appointment as *Imperial Mathematicus* gave him no rights to any of the Tycho's observations, which were legally the property of Tycho's heirs, and one of these (his son-in-law, the Dutchman Frans Tengnagel) fancied himself as an astronomer. Acting with quite uncharacteristic decisiveness, Kepler turned this particular crisis into an opportunity, and committed one of the most important crimes in the history of science. He stole the data.

Theft may be too strong a description, for Kepler never attempted to hide what he had done, either from the real owners or his wide circle of correspondents—in fact he seemed to glory in it. A year after Tycho died, he was writing to David Fabricius, a Lutheran pastor and astronomer who had worked briefly for Tycho, to the effect that …

> it is true that Tengnagel had good reason to suspect me. I had the observations and declined to hand them over …. (Koestler 1959, p350)

and three years later he wrote to Christopher Heydon, an English country squire who somehow managed to combine almost fanatical Protestant faith with a firm belief in astrology, that

> I have to admit that when Tycho died, I seized on the absence or laziness of the heirs and took possession of the observations … . (Koestler 1959, p350)

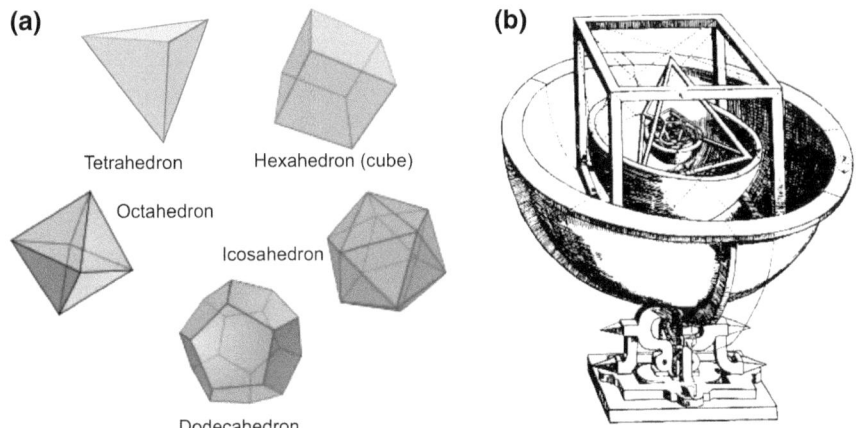

Fig. 3.10 **a** The five regular Pythagorean or Platonic solids. In each solid, the faces are identical regular polygons. No other such solids are possible. **b** Model commissioned by Kepler to illustrate his idea that the orbits of the six planets then known could be defined by inscribing them around or within the five solids (Kepler 1596)

It was just as well that he did. Tengnagel did get his hands on Tycho's instruments and tried to sell them. He was unsuccessful, because he priced them far too high, and in the course of a few years they simply disintegrated. The observations would probably have suffered a similar fate.

As long as Rudolph lived, Kepler had physical and financial security, and because of his position, recognition as Europe's leading astronomer. Money worries were not entirely absent. His agreed salary was only a sixth of Tycho's and he had to fight the burgeoning Austro-Hungarian bureaucracy for every promised penny,[14] but he was at last able to concentrate on what he did best. He took Tycho's raw data and processed them, always with the aim of confirming the great idea that had come to him whilst he was still a lowly teacher in Graz, that he had first set out in his unfortunate letter to the Bear and that he later expanded in the *Mysterium*. He was determined to show that the orbits of the six planets fitted within and around the five regular solids (Fig. 3.10).

The idea was his answer to the question, often posed by the astronomers of the time, of why there were just six planets. His solution was to link this incorrect belief to the fact that there were known to be (and this has not changed) only five regular solids. Even if his idea had been compatible with

---

[14]It is no coincidence that Kafka, the great chronicler of bureaucratic nightmares, was from Prague.

Tycho's observations (which it was not), it would still have been destroyed by the eventual discovery of Uranus and then Neptune but, as Koestler, who tracked the odyssey in detail, pointed out, it was only by testing the idea to destruction that Kepler arrived at the two laws he published (with a glowing acknowledgement to Tycho) in 1609 in his *Astronomia Nova* and the third law, published nine years later in the *Harmonices Mundi*. These laws stated that

1. *The orbit of a planet is an ellipse with the Sun at one of the two foci.*
2. *Any line joining a planet to the Sun sweeps out equal areas during equal times.*
3. *The square of the orbital period of a planet is proportional to the cube of the semi-major axis of its orbit.*

None of these laws involve the regular solids, and the first law, with its elliptical rather than circular orbits, is incompatible with spheres inscribed around or within those solids. Despite this, Kepler never quite abandoned his idea, introducing instead the concept of zones or spherical shells within which the ellipses could be contained. That did not matter. What did matter was that the three laws accounted for all the existing planetary observations, including those of Mars, which is the planet whose orbit departs most from a circle and so was the one that presented the greatest difficulties.

By the time that *Harmonices Mundi* had been published, Kepler's relatively tranquil days as Imperial Mathematician were over and his habitual bad luck had reasserted itself. Rudolph was effectively deposed in 1611, and died in 1612, and there was no place for a Protestant under the new regime. It was also in 1611 that his first wife and his favourite child died. The first of these losses may even have been welcome, for few husbands can ever have written less fondly of their wives, but his evident dislike of her did not prevent the fairly regular arrival of offspring.

For the remainder of his life Kepler moved erratically around Europe, propelled by the fluctuating fortunes of Catholics and Protestants. He went first to Linz, where he held a position little different from the one he had occupied in Graz so many years before, but which did allow him the leisure to continue with his life's work. Indeed, the post had been created for him by some of his supporters for that very purpose (Koestler 1959). It was while in Linz that he conducted his epic search for a new bride and he was still there in 1618 when the mass insanity known as the Thirty Years War began with the ejection of some Imperial envoys from a high window in the Hrad in Prague (they landed in a dungheap and survived).

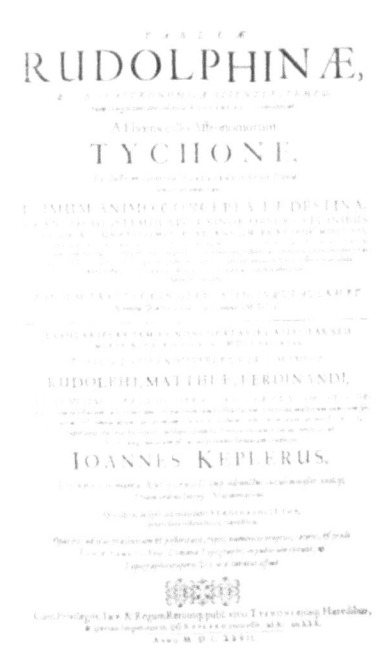

Fig. 3.11 The Rudolphine Tables. It was typical of Kepler that he gave pride of place on the title page to Tycho, despite having spent almost as many years in processing the data as Tycho did in making the original observations or having them made. Few of his contemporaries would have been as generous. A map of Hven, which Kepler never visited, is prominent on the plinth of the rotunda

He was also distracted during his time in Linz by his need to defend his mother, still in Swabia, against accusations of witchcraft. From Koestler's description of the saga, which began in 1615 and continued until the old woman's death in 1622, it might be felt that, if witches really had existed, then Katharina Kepler would have been one. Rublack (2017) provides a much more sympathetic portrait of the woman, who is now commemorated by a statue in her home town of Eltingen-Leonberg, but it was almost certainly only her son's still retained title as *Imperial Mathematicus* that saved her from the stake. The miracle is that despite all his troubles, he was able to produce first the *Harmonices Mundi* (in 1619) and then, at long last in 1627, the *Tabulae Rudolphinae* (Fig. 3.11).

In publishing the tables, Kepler honourably and unnecessarily paid debts to two people no longer able to help him. Rudolph had died, deposed and mad, fifteen years earlier but the tables were dedicated to him. They were also a memorial to Tycho, acknowledged on the title page as the senior author. The work may not have been Kepler's greatest scientific achievement, but it was one of his most impressive practical ones. The mere mechanics of extracting the money needed for its publication from a reluctant Imperial treasury and getting the printing done in the middle of the wars and peasant uprisings that were engulfing Linz, cost him years of his life, and destroyed what remained of his health.

During the last stages of the production of the *Tabulae*, Linz followed the example of Graz in its treatment of Protestants, and Kepler was forced to be once more a wanderer. He never found another permanent home, and even returned briefly to Prague. In an extraordinary shift of loyalties, he became astrologer to the Catholic general Wallenstein, but the relationship was short lived. He died in poverty in Regensburg, in eastern Bavaria, whilst searching for another post, leaving behind him an almost completed copy of *The Somnium*, arguably the world's first science fiction novel.[15] It comes as a shock to realise that the man whose work destroyed for ever the Ptolemaic theory had been dead for three years when Galileo stood trial in Rome. Ironically, his grave was destroyed by the armies of Gustavus Vasa, king of a Sweden that by that time had swallowed up Tycho's homeland of Skåne and the island of Hven.

References

Caspar M (ed) (1945) Johannes Kepler: Gesammelte Werke. vol 13, Briefe 1590–1599. C. H. Beck, Munich

Czartoryski P (ed) (1985) Nicholas Copernicus collected works: III minor works. Translation and commentary by Edward Rosen with assistance of Erna Hilfstein. Macmillan, London

Gingerich O (2004) The Book nobody read. Walker & Company, New York

Jardine N (1984) The birth of history and philosophy of science. Cambridge University Press, Cambridge

Jardine L (1999) Ingenious pursuits. Little, Brown & Co, London

Kepler J (1596) Mysterium Cosmographicum. George Gruppenbach, Tübingen

Koestler A (1959) The sleepwalkers. Hutchinson, London

Oster M (ed) (2002) Science in Europe, 1500–1800: a primary sources reader. Palgrave, London

Rosen E (1946) Kepler's defence of Tycho against Ursus. Popular Astron 54:405–411

---

[15]It imagined a trip to the Moon and was published posthumously in 1634.

Rublack U (2017) The astronomer and the Witch. Oxford University Press, Oxford
Sobel D (2011) A more perfect heaven: How Copernicus revolutionized the Cosmos. Walker, New York
Thoren V (1990) The lord of Uraniborg. Cambridge University Press, Cambridge

4

The Synthesis

It took just over a century for Copernicus, Tycho Brahe and Kepler to between them destroy Ptolemy's geocentric universe of spheres and circles and replace it with a heliocentric universe of elliptical orbits. During a part of this period, Galileo, the contemporary of Tycho and Kepler, introduced the idea of acceleration, and established by experiment the relationship between the time of fall of a body and the time of swing of a pendulum. Mersenne and Riccioli took those ideas and, with Herculean efforts, made the first usable estimates of the length of a seconds pendulum and therefore of 'g'. But although Kepler and Galileo corresponded they never collaborated, and any chance of their working together disappeared with their disagreement on the origins of tides. It took new minds to connect 'little g' with 'Big G'. It is a sad fact that the two men who were most influential in forging this new unity wasted so much of their lives in tearing each other apart.

The Royal Society

In the late 1640s the English Civil War was coming to its messy conclusion, but small groups of like-minded individuals were still able to meet to discuss the new, experiment-based, ways of studying the natural world that we now call science. Some of those involved were for King and some were for Parliament but for those who favoured the Royalist cause it was safest, after Charles lost his head in 1649, to be interested in things other than politics. In Oxford, a particularly active group was meeting at Wadham College

© Springer International Publishing AG, part of Springer Nature 2018
J. Milsom, *The Hunt for Earth Gravity*,
https://doi.org/10.1007/978-3-319-74959-4_4

under the auspices of the Master, John Wilkins. Wilkins had taken the Parliamentary side (he would not have been Master of a college had he not done so) and had cemented his position by marrying Cromwell's youngest sister, Robina, but the group included many who took the other view.

Soon after the restoration of the new King Charles in 1660, twelve members of the group met at Gresham College in London (Fig. 4.1) for a lecture by its Professor of Astronomy, Christopher Wren. They included Robert Boyle, Sir Robert Moray, a soldier of fortune who had been present when Charles was crowned in Edinburgh immediately after his father's execution, and also Wilkins, now in his turn in need of support and protection. Together they agreed to found *a Colledge for the Promoting of Physico-Mathematicall Experimentall Learning*, and adopted for its motto the words *Nullius in Verba*, often translated as 'take nobody's word for it'. Moray put his high standing at court to good use by securing the King's approval and encouragement for the 'Colledge', and by 1661 it had been transformed into *The Royal Society of London for Improving Natural Knowledge*. In 1663 it received a second Royal Charter under that name, and a library and

Gresham College *as it appeared before it was taken Down to Build an Excise Office*

Fig. 4.1 Gresham College, London, the first home of the Royal Society and home also, for almost forty years, to Robert Hooke. The building almost miraculously escaped destruction in the Great Fire of 1666 and was then requisitioned by the Corporation of the City of London for use as its headquarters during the planning and execution of the rebuilding programme (Gresham College Archive)

museum for specimens of scientific interest was established at Gresham College. Henry Oldenburg became its first Secretary.

The Society was organised around weekly meetings at which experiments were demonstrated and what would now be called scientific topics were discussed. After the Great Fire of 1666 the meetings moved for some years to Arundel House, the London home of the Dukes of Norfolk, and it was not until 1710, under the Presidency of Isaac Newton, that the Society acquired its own home, two houses in Crane Court, off the Strand. The 1662 Charter allowed it to publish, and the first issue of its *Philosophical Transactions*, now recognised as the world's oldest scientific journal in continuous production, appeared in 1665. The first two books published were *Sylva* (subtitled *A Discourse of Forest-Trees and the Propagation of Timber in His Majesty's Dominions*) by John Evelyn and *Micrographia* by Robert Hooke.

From the beginning, Fellows had to be elected, but the criteria were vague and the vast majority were not professional scientists. The astonishing frequency of the meetings suggests that many of them had little else to do. In 1731 a new rule specified that every candidature had to be accompanied by a certificate signed by those who supported it, and these survive and give some insight into the reasons why Fellows were elected and also the relationships between them. For the years before they were introduced, there is only speculation, but it is clear that the Society in its early days was very much a gentleman's club.

The First Skirmishes

Kepler described the ways in which the planets behaved but not why they did so. In fact, he very much disapproved of their failure to follow perfectly circular orbits. The first person to attempt a physical explanation was René Descartes who, being unable to accept the almost mystical idea of forces propagating in empty space, suggested that the planets were propelled by vortices in an all-pervading *aether*. This theory was wrong in almost every respect but its basic idea, of a space entirely filled with a 'something' in continual motion, might imaginatively be seen as anticipating the modern conceits of string theory and dark matter. Christiaan Huygens, who first introduced the concept of centrifugal force, might have arrived at a better solution had he not deferred too much to Descartes, a close family friend. Foreshadowing some of the methods of what came to be known as calculus, he did work out the equation of motion of a pendulum, and so provided an explanation for Galileo's ratio between times of swing and times of fall (see Chap. 14 Coda 3). This led him to 'little g', but for Huygens, as for Descartes, 'Big G' was a step too far.

The man generally credited with having put Descartes right was Isaac Newton, but for many years even he believed in vortices and an invisible *aether*. He might never have produced the synthesis that finally brought 'Big G' and 'little g' together had it not been for his decades-long conflict with Robert Hooke, the author of *Micrographia*. The hostilities were concentrated in three main campaigns. The first, and relatively brief, Optics War, had nothing to do with gravity and ended in an uneasy truce, brokered by their friends. The second, the War of the Six Letters, began with Hooke occupying a strong strategic position and ended with a tactical withdrawal by Newton to his semi-isolation in the Fens. In the third, the Principia War, Hooke was totally defeated. It may not be an exaggeration to say that, after it was over, he died of his wounds.

The Optics War began when, in 1671, Newton submitted to the Society a telescope, built by himself and to his own design, that magnified using reflections from a spherical mirror rather than refraction by spherical lenses. 'Reflectors' have two major advantages over refractors. They are inherently shorter, for the same magnification, and the images do not suffer from the chromatic aberration that occurs when refraction splits white light into its component colours. This telescope and his appointment in 1669 as Lucasian Professor of Mathematics in Cambridge were reasons enough for Newton to be elected a Fellow of the Society, on 11 January, 1671/2.[1] A few weeks later he submitted a paper summarising his experimental and theoretical work on optics, in which he set out his idea of light as made up of tiny particles that moved in straight lines except where deflected by reflection or refraction.

In the early 1670s almost nothing happened in London 'Philosophy' without Hooke being involved. He was an innovative designer of experiments, a skilled craftsman when it came to preparing laboratory apparatus, an unusually gifted draughtsman renowned for his drawings of details visible only under a microscope and the possessor of a fertile brain from which ideas spun off in all directions. He had progressed from early employment as Robert Boyle's lab assistant to the important position of Curator of Experiments at the Royal Society and two years later had, despite being only just a gentleman (his father had been a minor curate on the Isle of Wight),

---

[1]It was not until 1752 that England moved officially from a year beginning on 23 March (Lady Day) to the present system, but 1 January had been commonly accepted as the 'real' start of the year for a very long time. To avoid ambiguity, it was common practice in the late 17th and early 18th centuries to show both possible years for dates before 23 March and this was followed in all the proceedings of the Royal Society.

been elected to full membership. He was also the surveyor appointed by the City of London Corporation for the (quite literally) monumental task of rebuilding the city after the fire, working closely with Christopher Wren, the surveyor appointed by the king. His '*Micrographia*', published in 1667 but actually authorised for publication by the Society three years before, included an entire section on his experiments with colours, and his theoretical explanation of their origin.[2] It was almost inevitable that the Society should ask him (together with Robert Boyle and the Bishop of Salisbury) for an opinion on Newton's work.

Unfortunately, Hooke's review, delivered only nine days later, was all about him. He had seen an opportunity to promote his own 'wave theory' of light, and in order to do so had dismissed Newton's 'particle' alternative out of hand. The History of the Society for 15 February 1671/2 notes that

> Mr. Hooke's considerations upon Mr. Newton's discourse on light and colours were read. Mr. Hooke was thanked for the pains taken in bringing in such ingenious reflections, and it was ordered that this paper should be registered, and a copy of it immediately sent to Mr. Newton: and that in the mean time the printing of Mr. Newton's discourse by itself might go on, if he did not contradict it; and that Mr. Hooke's paper might be printed afterwards, it not being thought fit to print them together, lest Mr. Newton should look upon it as a disrespect, in printing so sudden a refutation of a discourse of his, which had met with so much applause at the Society but a few days before.[3]

Clearly the Society, or at least its Secretary, who was no friend of Hooke's, was far from happy with the review, and with good reason. The antagonism between Newton and Hooke dates from this time but the quarrel was patched up in an exchange of letters remarkable for their flowery (and, one suspects, entirely insincere) expressions of regard. Newton's letter to Hooke included one of his most widely quoted remarks, to the effect that if he had seen further, it was *by standing on the shoulders of giants*.[4] It is not necessary to agree entirely with those who see in this a coded and slighting reference to Hooke's small stature to doubt whether the writer's true feelings were being

---

[2]The Micrographia also included remarkable drawings of animals and plants seen through the microscope, testifying to Hooke's extraordinary skills as a draughtsman. Two of the most famous are of a flea and a nettle. It seems somehow appropriate that Hooke should be remembered for depicting, in remarkable detail, things that bite and things that sting.

[3]Birch (1757). Hooke's review appears immediately after this note.

[4]Letter dated 5 February 1675/6. Although often cited as evidence of Newton's humility in acknowledging his debt to all his predecessors, the context and the circumstances make it clear that it was only Descartes and Hooke who were being acknowledged, and those only in the field of optics.

expressed, but the truce held. The two seldom met. Hooke was (as he was contracted and paid to be) a weekly presence at the London meetings, while Newton was isolated in Cambridge, which he seldom left. His communications with the Society during the next few years were mainly about improving the performance of his reflecting telescope, and gradually tailed away, and he only published his *Opticks* twenty years later, almost immediately after Hooke's death. It is hard to believe that any scientist who was confident of his own work would wait so long, and until his severest critic was dead, before publishing, and it may well be that he feared that there were flaws in his theories that a hostile reader might expose.

The stitched-together truce might well have ended the conflict, but in 1677, following the death of Henry Oldenburg, Hooke was appointed Secretary of the Society. He was probably the worst choice ever made for the post because, whatever his merits as a scientist, he was an appallingly bad archivist. Moreover, he seems to have spent much of his time searching the Society's files for documentary proof of his predecessor's supposed vendetta against him (Adams and Jardine 2006). He did, however, take seriously one part of his duties, which was to write to scientists who were not in London and not active in the Society, soliciting scientific contributions. It was in that capacity that, on 24 November 1679, he fired the first shots in the War of the Six Letters, by writing to Newton.

Had there been no history of conflict between the two men, Hooke's letter would have been unexceptional. Indeed, he attempted to address that history, saying, with an almost touching disregard for the realities of academic life, that *differences of opinion, if such there be (especially in philosophicall matters where interest hath little concern) me thinks should not be occasions of enmity*. He also solicited Newton's opinion on a new and rather bizarre cosmological theory then being put forward in Paris by Claude Mallemont, and suggested that Newton might like to take part in an astronomical measurement of the difference in latitude between London and Cambridge.

Newton's reply, written only a few days later (Koyré 1952), was very odd. He began by telling Hooke that he *had for some years last been endeavouring to bend myself from philosophy to other studies*, which could refer to his obsessions with biblical chronology, or to alchemy, or to mathematics, which was not then regarded as philosophical. Despite this disclaimer, and after having rightly dismissed Mallemont's ideas in a few short sentences, he went on to discuss, in considerable detail, the question of whether a weight dropped from a great height would land vertically beneath the dropping point, or some distance to the east (i.e. in the direction of the Earth's rotation) or some distance to the west. Opinion up to that time had been that the fall-

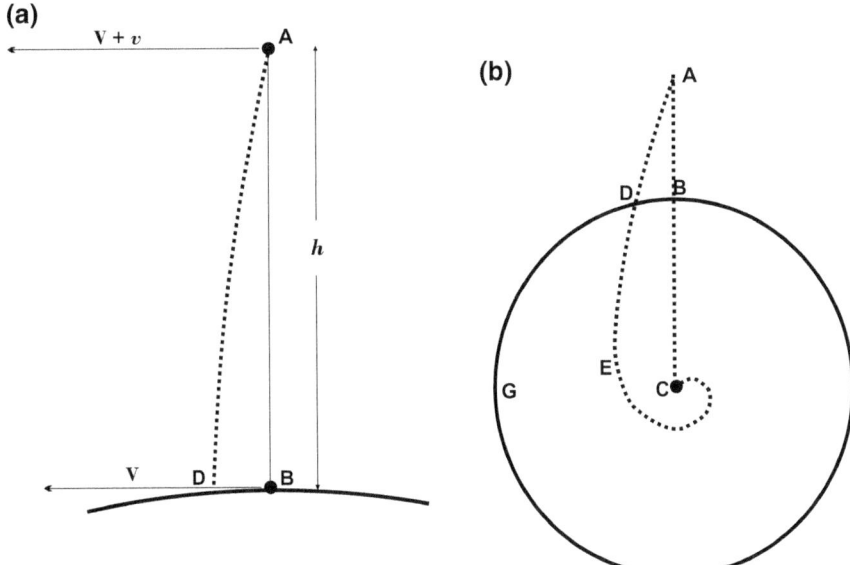

Fig. 4.2 **a** Idealised trajectory of a body falling from a height h to the surface of a rotating Earth, radius R. The point B on the Earth's surface is moving east at a velocity V which at the equator is equal to about 460 m/sec. The point A, h metres vertically above B, is moving at the slightly higher velocity V + v, where $(V + v)/V = (R + h)/R$. If air resistance is neglected, this higher velocity implies that a body falling from A will reach the ground at a point offset slightly to the east of B. **b** Redrawing of Newton's sketch in his letter to Hooke of 28 November 1679, in which he quite unnecessarily extended the trajectory of the falling body below the surface of the Earth

ing weight would lag behind the Earth, and so fall to the west, but Newton devoted almost half of his long letter to arguing that it would fall to the east (Fig. 4.2a). In doing so he, quite unnecessarily, included in his discussion the (impossible) path that it would follow if it continued its trajectory unimpeded below the surface of the Earth, and included a sketch to show what he meant (Fig. 4.2b).

Had Hooke been a tactful man (and not even his most enthusiastic admirers would ever claim that he was), he would have limited himself to thanking Newton for his ideas, let him know that they had been discussed by the Society and tell him that the members were keen to try a practical test. Sadly, he could not resist arguing about the trajectory the weight would follow below the surface of the Earth, were such a thing possible. On 9 December he sent Newton a letter pointing to what he considered to be errors, and two days later he made this criticism public by airing it at a meeting of the Society.

To Newton, this must have looked like a re-run of the War of 1671/2. His answer, written on 13 December, while providing an alternative diagram, was terse and uncompromising. He ended by saying that

> ... the thing being of no great moment I rather beg your pardon for having troubled you thus far with this second scribble wherein if you meet with any thing inept or erroneous I hope you will pardon the former and the latter I submit and leave you to your correction ....

Hooke, however, could not take the hint. There is something terrier-like in the way that he worried away at the problem, regardless of the likely consequence. His reply of 6 January 1679/80[5] was to become a crucial element in the later conflict, since in the very first sentence he assumed the inverse square law, stating that *my supposition is that the attraction always is in duplicate proportion to the distance from the center reciprocal.* He also took the opportunity to backtrack, rather as had been done by Newton in his letter, saying that

> What I mentioned in my last concerning the descent within the body of the earth was but on the supposal of such an attraction, not that I really believe there is such an attraction to the very center of the earth, but on the contrary I rather conceive that the more the body approaches the center the less it will be urged by the attraction.

He also mentioned that he had made *three tryalls* of the trajectory of a body dropped from a great height, following Newton's suggestion, and that in each case it fell to the southeast of the vertical by at least a quarter inch. On 17 January, he wrote again to Newton reporting similar results from another two experiments and effectively asking Newton to do the math to establish what the trajectory really would be if the inverse square law applied. Newton, however, was not interested, and was also probably by that time very conscious that everything he wrote to Hooke on the subject would be read out (as Hooke was bound to do, given his position) at a meeting of the Royal Society. It was not until 3 December 1680 that he made any sort of a response, and then merely to say that

> For the trials you made .... I am indebted to you thanks which I thought to have returned by word of mouth, but not having yet had the opportunity must be content to do it by letter.

---

[5]Reproduced in the Appendix to Rouse Ball (1893).

It is interesting, and surprising, that neither man seems to have attempted to calculate what the results of the trials should have been, even though this would have been well within the state of knowledge and mathematics at the time. The size of the Earth was already known, and this, with the length of the day, allowed the velocity of a point at or above its surface to be calculated. Also, 'g' was known with sufficient accuracy. Had they done the calculations, they would have realised that the effect was too small to be measured by any of the means they had to hand.[6]

This letter also showed that Hooke's interest in gravity, on display since as early as 1666, remained unabated. He included the note that:

> Mr. Halley, when he returned from St Helena, told me that his pendulum at the top of the hill went slower than at the bottom … I presently told him that he had solved me a query I had long desired answered …. To know whether the gravity did actually decrease at a greater height from the center.

The Path to the *Principia*

Again there was a pause in hostilities, for about six years, but 1684 saw the start of the events that led to both the publication of the *Principia* and the final conflict. The story is told that Halley, Hooke and Wren met early in the year after a session of the Royal Society and speculated on the path a planet would follow when acted upon by a force that obeyed the inverse square law. What actually prompted this discussion is not known. There is no mention of the problem in the *History of the Royal Society* covering that period, and it seems odd that the question should have been posed in that form, rather than by taking the widely accepted elliptical form of planetary orbits as the starting point.[7]

Whatever its beginning, the discussion ended with Hooke claiming to have proved that the path would be an ellipse. He was no great mathematician, which was why he had tried to persuade Newton to solve the problem

---

[6]If the trial were to be made at the equator, where the effect would be greatest, the offset for a 5 m drop would be only about a third of a millimetre. A longer drop would produce a greater offset but not proportionately, because of the object's acceleration, and the practical difficulties would be increased. The situation is more complicated at the latitude of London, and the theoretical offset would be even smaller.

[7]The first person to have proposed in writing an inverse-square law seems to have been the French astronomer, Boulliau (1645). Boulliau was later elected a foreign member of the Royal Society, and Hooke, Wren and Halley may all have known of his work.

four years earlier, so this was almost certainly a guess. It was not a very wild guess, because it was seventy years since Kepler had first told the world that planetary orbits were elliptical. Wren then offered a prize of *a book worth forty shillings* to the first of the other two to produce a solution to the problem. Neither he nor Halley was able to do so, and neither, it turned out, was Hooke—or at least he never offered any proof that he could.

Shortly after the meeting, Halley's father, his main source of financial support, disappeared, and his body was not found until five weeks later. Distressingly, it turned out that he had been murdered and, also distressingly, that he had died intestate. Halley became heavily involved in the resulting legal problems and it was probably in the course of dealing with these that he travelled to East Anglia and decided to call on Isaac Newton (Fig. 4.3).

The tale is then taken up by Abraham De Moivre, a French Huguenot mathematician who, while making a precarious living as a refugee in London, was befriended by both Newton and Halley. In his account, set down years after the event (he did not arrive in England until 1690), he said that Newton told him that

> … in 1684, Dr. Halley came to visit him at Cambridge, after they had been some time together, the Dr. asked him what he thought the Curve would be that would be described by the Planets supposing the force of attraction toward the Sun to be reciprocal to the square of their distance from it. Sr Isaac replied immediately that it would be an Ellipsis, the Doctor struck with joy & amazement asked him how he knew it, why saith he I have calculated it, whereupon Dr. Halley asked him for his calculations without any further delay, Sr Isaac looked among his papers but could not find it, but he promised him to renew it, & then sent it him.[8]

Newton's claim to have already obtained the proof is as suspect as Hooke's. It was three months before he sent it to Halley, in the 9-page 'De Motu' which, ultimately and enormously expanded, became the *Principia Mathematica*. The *Principia* itself followed, two years later. The manuscript of the first volume was presented to the Society on 28 April 1686 by Nathanial Vincent, the former King's chaplain, on Newton's behalf, and on 2 June the Council ordered it to be printed, with the very unfair proviso that *the business of looking after it, and printing it at his own charge* should fall on Halley. It was not, however, that little piece of penny-pinching that upset Hooke.

---

[8]Letter from Abraham De Moivre, 1727, quoted in Whiteside (1991).

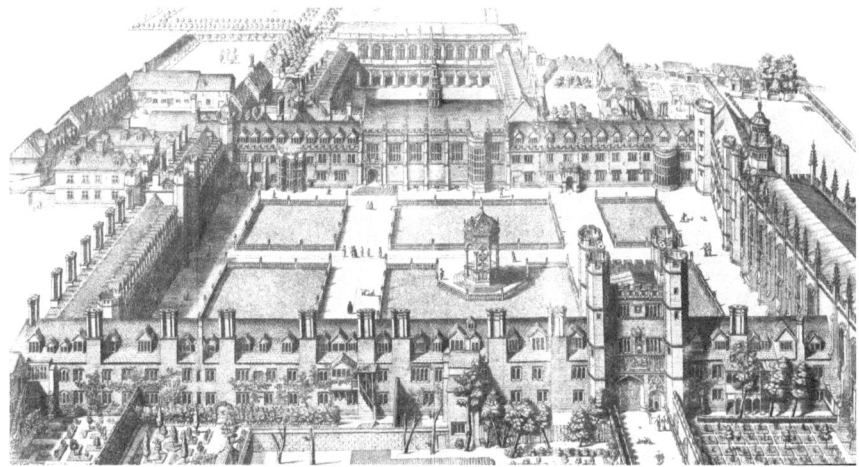

Fig. 4.3 Trinity College Cambridge, as it would have appeared when Isaac Newton was Lucasian Professor of Mathematics. In 1684 Halley would have found him in his rooms on the first floor, just to the right of the gatehouse. Engraving from Loggan (1690)

The Principia War

On January 1685/6, Halley had been elected clerk to the Royal Society, despite being married with children and thus failing Stipulation No. 5 of the council's criteria. It was in that capacity that, as soon as the Society had decided to publish the *Principia*, he wrote to Newton informing him of the decision and adding that

> There is one more thing that I ought to inform you of, viz that Mr. Hooke had some pretensions to the invention of the rule for the decrease of gravity being reciprocally as the squares of the distances from the center. He says you had the notion from him, though he owns the demonstration of the curves generated thereby to be wholly your own. How much of this is so, you know best, so likewise what you have to do in this matter. Only Mr. Hooke seems to expect you should make some mention of him in the preface, which 'tis possible you may see reason to prefix. I must beg your pardon that 'tis I that send you this ungrateful account; but I thought it my duty to let you know it, so that you might act accordingly, being in myself fully satisfied that nothing but the greatest candour imaginable is to be expected from a person who has of all men the least need to borrow reputation.

This letter was recorded in the Society's archives, which make no other mention of Hooke's objections. At first sight it might seem astonishing that Hooke could even have found the offending section in the three hundred closely argued pages of the original Latin text of the *Principia*, but he would have been forewarned. A week earlier Halley had read to the Society a paper of his own entitled *A Discourse Concerning Gravity, and its Properties*[9] which prepared the ground. It began with a critique of Descartes' vortices and went on to summarise the theory of *our worthy Country-man Mr. Isaac Newton (who has an incomparable Treatise of Motion almost ready for the Press)*. The inverse-square law was then stated, far more clearly than it ever was by Newton himself.

Of all the people involved, Halley is the most sympathetic. He seems never to have tried to take more credit than was his due, for anything, and in some cases settled for very much less. Without his urging, and willingness to provide the money, the *Principia* might never have been published, and he later did his best to reconcile Newton and Hooke. He was, however, financially reliant on the successful publication, at a time when his own financial position was uncertain because of his father's intestacy, and it is not surprising that, in correspondence at least, he tended to take Newton's side. Once the first part of the book had been delivered, his main concern was to get his hands on the remainder.

Newton's reply gave little prospect of a peaceful resolution.

> Sir, In order to let you know the case between Mr. Hooke and me, I give you an account of what passed between us in our letters, so far as I could remember; for 'tis long since they were writ, and I do not know that I have seen them since. I am almost confident by circumstances, that Sir Chr. Wren knew the duplicate proportion when I gave him a visit;[10] and then Mr. Hooke (by his book Cometa written afterwards) will prove the last of us three that knew it
> ....

The letter continued along these lines, citing earlier documents but with neither dates nor sufficient information for them to be identified. It was followed on 20 June by a second letter, with a postscript clearly prompted by a second-hand report that had reached Cambridge after the main part of the letter had been completed. The postscript alone ran to three pages and began

---

[9]The paper was published in the *Philosophical Transactions* on 1 January 1686. The 'reading' was thus a mere formality and might not have been done in full, since the text would already have been in the hands of the Fellows.

[10]Halley's letter on p. 442 of Turnbull (1960) suggests that Wren did not confirm this story.

… I am told by one, who had it from another lately present at one of your meetings, how that Mr. Hooke should there make a great stir, pretending that I had all from him, and desiring they would see that he had justice done him.

Evidently, the story of Hooke's behaviour had lost nothing in the telling, because Halley, when he replied on 29 June, offered a more moderate version of the events. As he put it:

As to the manner of Mr. Hooke's claiming this discovery, I fear it has been represented in worse colours than it ought; for he made neither publick application to the Society for justice, nor pretended that you had it all from him.

Hooke's objections had been less to the *Principia* itself and more to the praise heaped upon it for its extreme originality in its use of the inverse-square law, but Halley also told Newton that the opinion of the members of the Society present had been:

… that nothing thereof appearing in print, nor on the books of the Society, you [i.e. Newton] ought to be considered as the inventor. And if in truth he knew it before you, he ought not to blame any but himself for having taken no more care to secure a discovery, which he puts so much value on. What application he has made in private, I know not; but I am sure that the Society have a very great satisfaction, in the honour you do them, by the dedication of so worthy a treatise.

Sir, I must now again beg you, not to let your resentments run so high, as to deprive us of your third book ….

Halley was at this time desperate for that third volume. He thought that he had the second already in his hands, but suffered a serious reverse when even this was withdrawn by Newton on the grounds that it was too easy to read. However, the pleas on Hooke's behalf did have some effect, because on 14 July Newton wrote to approve the incorporation of woodcuts into the publication and, while restating his differences with Hooke, ended by saying that:

And now having sincerely told you the case between Mr. Hooke and me, I hope I shall be free for the future from the prejudice of his letters. I have considered how best to compose the present dispute, and I think it may be done by the inclosed scholium to the fourth proposition …. "The inverse law of gravity holds in all the celestial motions, as was discovered also independently by my countrymen Wren, Hooke and Halley".

This went no distance at all towards mollifying Hooke, nor does it seem that Newton was very happy with the concession, removing Hooke's name from all subsequent drafts. The printing went ahead. The revised second book went to the printers in March 1687 and the third in July. The entire work was dedicated to the Royal Society and the first volume was prefaced by Halley with a set of Latin hexameters lauding its author. It sold rapidly, and not only in England, and soon became hard to obtain. Halley presumably got his money back, and he may even have made a profit.

The dispute sputtered on. It is hard to gauge the effect that it had on the other Fellows, but probably most just did their best to ignore it. There is very little evidence of support, or even sympathy, for Hooke, whose eternal combativeness had probably tried most people's patience. There might have been rather few people active in the Society whose ideas this 'universal claimant' had not by that time claimed as originally his own. Whatever truth there may have been in such claims (and Hooke had been extraordinarily productive, over an enormous range of topics), they would not have won him many friends. John Aubrey was almost the only significant contemporary to take Hooke's side, and he was probably lodging with Hooke at Gresham when he wrote the relevant entry in his *Brief Lives*.

The Fellows would, moreover, have had their minds on matters far from philosophical. 1686 may be one of the most important dates in the history of science, but 1688 was a turning point in the history of Great Britain, being the year in which the country's last Catholic monarch was ejected. The upheaval affected both Newton and Hooke, but in very different ways. In Newton's case, the publication of the *Principia* had transformed him from a respected but little known academic into a celebrity, and that in turn had thrust him, possibly unwillingly, into a leading role in Cambridge University's opposition to appointments imposed upon them by King James. Had the king remained on the throne, Newton would doubtless have suffered for this, but after the king had fled Newton was appointed to the convention that approved the enthronement of William and Mary.

Newton was not the only one of Hooke's enemies to become powerful under the new regime. William brought with him from the Netherlands an entire retinue of trusted advisers and servants, and amongst these were members of the powerful Huygens family. Christiaan Huygens, with whom Hooke had already contested priority for the design of pendulum clocks and watches, was not himself a politician, but his relatives certainly were. There was going to be no future for Hooke in any sphere that depended on royal patronage, and he ceased to try very hard to argue his case against Newton. Many of his presentations to the Society during the following years concerned fossils, and particularly the spectacular ammonites (which he identified as nautiloids, a very near miss) that are found in abundance along the 'Jurassic coast' of Dorset. The presence of marine fossils well above sea level in both Britain and on the continent led him to innovative speculations about past Earth movements.[11] Remembered until recently for almost nothing except his Law governing the extensions of springs (one of his least impressive achievements), he would be more fittingly honoured as one of the founders of Palaeontology, and hence of Geology.

Things might have gone even worse for him had Newton not, in 1692, suffered a mental breakdown, sometimes attributed to clinical depression that may have been triggered by an infatuation with a handsome young Italian scientist, Nicholas Fabio, or by despair over the failure of his alchemical experiments, or by poisoning from the mercury that he used in those experiments. Hooke would probably have liked to think that the Principia War was also at least partly responsible, but that seems unlikely because by then he had suffered the fate most dreaded by ambitious scientists; he had simply become irrelevant. From 1690 onwards his health worsened rapidly, due in part, as with Newton, to unwise experimentation (in his case with self-medication) (Jardine 2003). His most serious error was to die first, in the spring of 1703, leaving Newton in control of his legacy. He is only just beginning to emerge from the results of his rival's efforts to erase him from the history of science.

The Man About Town

After recovering from his mental collapse, Newton quit science, and Cambridge, and sought more profitable employment elsewhere. Imitating Copernicus, he became interested in money and was appointed first as Warden and then as Master of the Royal Mint, at a time when a fifth of the coins in England were said to be counterfeit. Counterfeiting was a major crime, regarded as high treason and therefore punishable by hanging, drawing and quartering, but it was very difficult to obtain convictions. Evidence was hard to come by, and the barbarity of the sentence (even though generally commuted to 'mere' hanging) made juries reluctant to bring in guilty

---

[11]On 6 December 1686 he read a paper to the Society on shells *wherein he gave several material instances to prove, that there have been very great changes in the earth's surface, as of rows of oistershells found in a cliff in the Alps* … (Birch 1758).

verdicts. It was, after all, one thing to apply that sort of punishment to the people who had actually removed the head of a reigning monarch, but the people who pressed copies of it on to unauthorised bits of metal, while clearly very naughty, were not, to most juries, in the same league.

Newton, however, took to the chase with enthusiasm.[12] He created a network of informers and was quite prepared to threaten them with the gallows if they did not say the right things in court. Most had themselves been arrested for counterfeiting, and the threat was all the more effective because it was sometimes carried out. Using these methods, he was credited (if that is the right word) with at least a hundred arrests and a score or more of executions. Even if allowances are made for different times and different attitudes, this makes uncomfortable reading, but there is no need to go as far as the anonymous internet commentator who claimed, as evidence of Newton's inhumanity, that he snubbed one of his victims by failing to turn up to see the poor fellow die. It is hard to imagine that this cast much of an additional cloud over the experience.

Newton's move to London, where interaction with other 'philosophers' could be a daily event, also prompted a return to science, although his major late publication, the English-language *Opticks*, relied largely on work done twenty years earlier.[13] In effect, he 'coasted' on the reputation he had established with the *Principia*. With something of that magnitude to his credit, who could blame him. His election as President of the Royal Society in 1704 gave him a splendid opportunity to pursue not only his vendetta with his dead enemy, Hooke, but those with living ones such as Leibniz, with whom he disputed priority in the invention of calculus. Leibniz rashly submitted his case to the adjudication of the Society, which unsurprisingly decided in its President's favour. The jury had been fixed. De Moivre was just one of the known friends of Newton who was invited to sit on the panel that made the decision, but was still not rewarded with the proper job that he so earnestly sought.

Newton never married, but there is some suggestion that towards the end of his life he became a quite avuncular uncle. His late mellowing might have

---

[12]This part of Newton's life is entertainingly described in Levenson (2009).

[13]The nucleus of this book was contained in the 'letter' sent by Newton to the Royal Society on 6 February 1671/2, shortly after his election as a member, *concerning his discovery of the nature of light, refractions and colours*. It was printed in the *Philosophical Transactions* 6 (80), p. 3075, and it was Hooke's dismissive comments that were the original cause of the enmity between them. It is said that Newton deliberately withheld publication until Hooke was dead, which, if true, suggests he might have been less confident in the truth of his theories than he seemed. A man confident of his own work would have relished the opportunity to challenge a rival.

had something to do with financial security, since the Master of the Mint could become very rich. This was even legal, since he was entitled to a percentage of the value of all the coins minted (a privilege that has today been passed from the Bank to the bankers), but it seems that his reading of the works of Copernicus had been too limited. Had he studied the economic ones, and taken them to heart, he might have lost less heavily than he did in the collapse of the South Sea Bubble, but even after that disaster he was financially comfortable when he died in 1727.

The Missing Portrait

It is darkly rumoured that, once he was in a position to do so, Newton took the feud with Hooke to the extreme of destroying the Society's portrait of him, or at least allowing it to disappear. There is no proof of this, but it is remarkable that no likenesses survive of a man who had been so prominent in London society for so long. Hooke was also poorly served by his literary executor, Richard Waller, who collected together and published his unpublished works, but did it through the Royal Society and dedicated the book to the Society's President, Isaac Newton. It is to be hoped that Hooke never knew, during his lifetime, that after his death his collected works would end up with a dedication to his bitterest enemy. The book also contains, as an introduction, one of the most remarkable obituaries ever penned. It is not every eulogist who describes his subject as *despicable, being very crooked*, and follows that by adding that *his temper was melancholy, mistrustful and jealous*. *Nil nisi bonum* was evidently not the creed by which Waller lived.

Waller's preface is very detailed in its accounts of some of Hooke's experiments, and in the details of a long conflict with Hevelius, but contains a remarkable omission. There is no mention of any events between the end of 1682 and the beginning of 1687, when the death of Hooke's niece is noted. All that is said to cover the years when the *Principia* was being published is that from 1682 onwards *he* (Hooke) *began to be more reserved than he had been formerly*. Neither gravity nor Newton is mentioned. Waller was supposedly a friend, but one might well feel that with such people for friends, it was really not necessary for Hooke to devote so much of his time to making enemies. The disloyalty has sometimes been explained on the grounds that he was describing Hooke in his later years, when illness and disappointment had taken their toll, but that is a poor excuse. Waller was elected to the Society in 1681 and became its secretary in 1687. He would have known Hooke in his prime, energetic and surrounded by admirers. Perhaps he felt that during those years he had chosen the wrong side, and that the time had come to ingratiate himself with the new scientific regime.

A Question of Priority

Did Hooke have a case? Or was he simply, as Newton would probably have said, an untalented mountebank who had so many contradictory ideas that some were almost bound to be right?

It seems very likely that Newton in the years leading up to Halley's visit had been interested mainly in religion, alchemy and developing the method of *fluxions* (his personal, user-hostile, version of calculus), that he had not given much thought to either dynamics or cosmology and that, rather passively, he accepted Descartes' vortex theory. That Halley's visit prompted him to apply 'fluxion' analysis to the mathematics of elliptical orbits, and that his motive for doing so was, at least in part, to humiliate Hooke, with whom he had already clashed twice. That it was while he was doing this, and not under an apple tree twenty years earlier, that he had his great revelation and realised that he had in his hands the means to fundamentally reshape mankind's view of the universe. And that, once he had begun, he was able to think of almost nothing else (including food and sleep) until the work was finished two years later. The extent to which he drew on Hooke's work, only he could know. We know what he said, but we have also to recognise that he was not the most trustworthy of witnesses. The real answer must depend, to a considerable degree, on when he abandoned his belief in the vortex theory.[14]

One thing is certain. Hooke's interest in gravity predated any known interest on Newton's part. As early as 21 March 1665/6 he had read a paper to the Society in which he described experiments in Westminster Abbey and 'St Pauls tower' (Fig. 4.4. It was semi-derelict at the time and was soon to be destroyed in the Great Fire) and also in deep wells at Banstead, near Epsom, '*to find the difference of the weight, if any, between a body placed on the surface of the earth, or at a considerable distance from it, either upwards or downwards*' (Birch 1756; pp. 70–72). In each place he weighed large masses, raising and lowering them to different heights and depths on light wires or strings, and concluded that '*If … there be any such inequality of gravity, we must have some ways of trial more accurate, than this of scales*'. Anticipating many future

---

[14]A much fuller account of Newton's flirtation with Vortex Theory has been given in Kollestrom (1999).

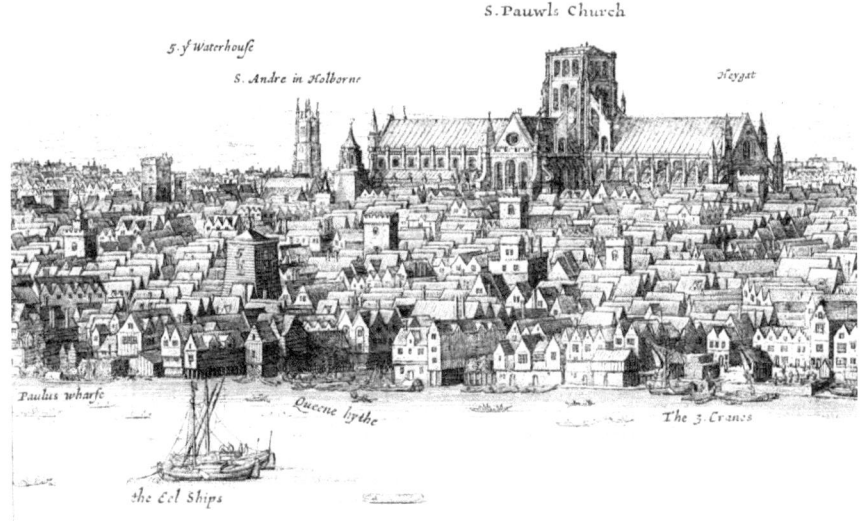

Fig. 4.4 Old St Paul's, before the Great Fire of 1666. The tower had lost its impressive spire to fire almost a hundred years before and was, by the time of Robert Hooke's gravity experiments, in a very dangerous condition. Engraving by Wenceslas Holler in Dugdale (1658)

developments, he went on to propose the use of a *'swing clock'* to make further tests, recognising that although the effect might be *'so small as not to be sensible at … an hundred vibrations, yet in many thousands of them, it will not be difficult to find…'*[15]

It is also clear that Newton's claim that Hooke believed that *the duplicate proportion … reached down from hence to the centre of the earth* is not supported by the evidence, because he had already stated, correctly, in his 1665/6 paper that … *a body at a considerable depth, below the surface of the Earth, should lose somewhat of its gravitation … by the attraction of the parts of the Earth placed above it*. Where, in the letters of 1671/2, Hooke appeared to be ignoring that insight, he was working with the very unrealistic model proposed by Newton. On the other hand, Newton's use, in the same context, of a constant gravitational force from the surface of the Earth to its centre is evidence for his adherence at that time to Descartes vortex theory. Had he been, as he was to claim later, already in possession of the inverse

---

[15]It is not necessary to suppose that Hooke knew of Huygen's analysis of pendulum motion, which at this time had probably been completed but not published. He would only have needed to know of Galileo's work showing a constant ratio between a time of fall and a time of swing to realise that a pendulum could be used to measure gravity.

square law, it is inconceivable that he would have proposed such a model. His scientific soul would have rebelled at the very idea.

1666 was also the year in which, as he was to claim many years later, Newton saw an apple fall from a tree and begun to speculate that the force that made it do so might also hold the Moon in its orbit.[16] This was the story that he told to his friend William Stukely and to John Conduitt, his assistant at the Royal Mint and the husband of his niece,[17] but there is no independent evidence that it ever happened. Even if it had, it would have been after Hooke's paper on his gravity experiments had been read in London, since there would have been no apples on the tree in March. Interestingly but irrelevantly, Newton was at his family home in Woolsthorpe at the time because he was taking refuge from the plague, which was also the reason why, a year earlier, Hooke left London for Epsom and there become aware of the deep borehole at Banstead.

After Banstead, Hooke occupied himself an extraordinary number of other things, but in 1674 he returned to gravity and published, through the Royal Society, a short paper entitled '*Attempt to prove the Motion of the Earth*', which ended with

... three Suppositions. First, That all Coelestial Bodies whatsoever, have an attraction or gravitating power towards their own Centers, whereby they attract not only their own parts, and keep them from flying from them, as we may observe the Earth to do, but that they do also attract all other Coelestial Bodies that are within the sphere of their activity and consequently that not only the Sun and Moon have an influence upon the body and motion of the Earth, and the Earth upon them, but that Mercury also Venus, Mars, Jupiter and Saturn by their attractive powers, have considerable influence upon every one of their motions also.

The second supposition is this, That all bodies whatsoever that are put into direct and simple motion, will so continue to move forward in a straight line, till they are by some other effectual powers deflected and bent into a Motion, describing a Circle, Ellipsis, or some other compounded Curve Line.

---

[16]*Why not as high as the Moon said he to himself & if so that must influence her motion & perhaps retain her in her orbit* Preserved as Keynes Ms 130.04 at King's College Cambridge. Accessible on-line at www.newtonproject.sussex.ac.uk/view/texts/normalized/THEM00167.

[17]John Conduitt, in a memoir written at about the time of Newton's death, recorded that ... *In the year 1666 ... whilst he was musing in a garden it came into his thought that the same power of gravity (which made an apple fall from the tree to the ground) was not limited to a certain distance from the earth but must extend much farther than was usually thought.*

The third supposition is, That these attractive powers are so much more powerful in operating, by how much the nearer the body wrought upon is to their own Centers. Now what these several degrees are I have not yet experimentally verified; but it is a notion, which if fully prosecuted as it ought to be, will mightily assist the Astronomer to reduce all Coelestial Motions to a certain rule, which I doubt will never be done true without it.[18]

Here he is anticipating not only the ideas behind Newton's Law of Gravity but also his First Law of Motion (which had, in any case, first been stated by Descartes). Newton at the time was provably still thinking in terms of an all-pervading ether, since his *Properties of Light, discoursed of in my several Papers*, read to the Society in December 1675 and largely concerned with refuting Hooke's attack on his work, contained the speculation that

> ... the gravitating attraction of the earth be caused by the continual condensation of some other such like aetherial spirit, not of the main body of phlegmatic aether, but of something very thinly and subtilely diffused through it, perhaps of an unctious, or gummy tenacious and springy nature. (Birch 1757)

It is thus almost impossible to accept as true what he wrote to Halley in 1686

> Between ten and eleven years ago there was an hypothesis of mine registered in your books, wherein I hinted a cause of gravity towards the earth, sun and planets, with the dependence of the celestial motions thereon; in which the proportion of the decrease of gravity from the superficies of the planet (though for brevity's sake not there expressed) can be no other than reciprocally duplicate of the distance from the centre. And I hope I shall not be urged to declare, in print, that I understood not the obvious mathematical condition of my own hypothesis. (Turnbull 1960)

No modern scientist could expect to get away with a claim for priority based on such tenuous grounds, but Newton went further, claiming priority not only over Hooke for the inverse square law but over Kepler for the elliptical form of the planetary orbits.

---

[18]The paper is devoted mainly to experiments, completed and proposed. The section quoted occurs on the last two pages. The division into paragraphs was not in the original, but does make it more digestible.

But, grant I received it afterwards from Mr. Hooke, yet have I as great a right to it as to the ellipse. For as Kepler knew the orb to be not circular but oval, and guessed it to be elliptical, so Mr. Hooke, without knowing what I have found out since his letters to me, can know no more, but that the proportion was duplicate quam proximè at great distances from the centre, and only guessed it to be so accurately, and guessed amiss in extending that proportion down to the very centre, whereas Kepler guessed right at the ellipse ....

... And so, in stating this business, I do pretend to have done as much for the proportion as for the ellipsis, and to have as much right to the one from Mr. Hooke and all men, as to the other from Kepler; and therefore on this account also he must at least moderate his pretences.

Here is a real, and unappealing, insight into Newton's mind. For him Kepler's ellipses could be no more than guesses, because they were based purely on observation, and not on mathematically-supported theory. That Tycho's observations had been meticulously analysed to such an extent that Kepler had been able to go beyond the elliptical shapes of the orbits to the realisation that the Sun must be at one of the focii, and not at the centre, was of no account.

Also as part of his defence, Newton wrote to Halley on 20 June 1686 a long letter in which he summarised the six letters of 1679/80 saying

That in my answer to his first letter I refused his correspondence, told him I had laid philosophy aside, sent him only the experiment of projectiles (rather shortly hinted at than carefully described); could scarce be persuaded to answer his second letter; did not answer the third.

Even a cursory examination of the actual letters shows that the truth is here being dealt with very economically indeed.

It might well have been their shared interests in optics that ultimately made both men receptive of the inverse square law. They would both have been well aware that in a three-dimensional universe the intensity of illumination due to a single source inevitably decreases as the square of distance from it, because the area to be illuminated increases as the square of distance. Once the Sun had been recognised as the source of the force that kept the planets in their orbits, the possibility of an analogy would have been almost bound to be considered, and a link would soon have been made with the force that made things fall to Earth.

In the end, what Hooke and Newton were amongst the first to raise was a question that has bedevilled science ever since, the question of what, exactly, constitutes priority. Galileo's ideas may have been triggered by the inanimate swing of a chandelier, but in many cases the trigger has been an all too animate chance remark, perhaps made by someone who never would, or never could, take the idea any further. Should they then have some of the credit, when the new ground was broken? The question is unanswerable and perhaps, except for those directly involved and whose continued career or livelihood may depend on it, unimportant. Only Newton could know (and might quite possibly have been unwilling to admit even to himself) the extent to which Hooke's interventions prompted him to abandon the semi-mystical concept of vortices in favour of the much more mathematically quantifiable inverse-square law. Hooke, on the other hand, might have been better advised to enjoy the uncontroversial acclaim he enjoyed for the buildings that he created and which, unlike the Law of Gravity, would never have existed in the form that they took had he not been there to make them so.

Throughout the controversy, it is mainly Newton's voice that we hear. Hooke's is barely audible, but it seems that the claims he made were actually rather modest. All that he was seeking was some acknowledgement of a role in the development of Newton's ideas. There, he almost certainly had some right on his side, but Newton was having none of it.

> Now is this not very fine? Mathematicians, that find out, settle and do all the business, must content themselves with being nothing but dry calculators and drudges; and another, that does nothing but pretend and grasp at all things, must carry away all invention ....[19]

Most, but not all, commentators have agreed with Newton, and it was only he, and certainly not Hooke, who was, at that instant of time and in that place, capable of providing the mathematical proofs without which all the talk of ellipses and '*Reciprocall*' square laws was mere speculation. No-one else in England could have done what he did. Christiaan Huygens might have done it before him had he not been so much in thrall to Descartes, and Gottfried Leibniz might have applied his own version of calculus to planetary orbits, had not Newton forestalled him. Without Hooke's prompting, however, Newton might never have interested himself in the subject. The view of Clairaut, who said that '*the example of Hooke serves to demonstrate the distance that exists between an idea perceived and an idea proved*' (Clairaut 1759) is one way of looking at the controversy, but modern science, per-

---

[19]Postscript to letter to Halley of 20 June 1686, in Turnbull (1960).

haps because computers now do the routine number-crunching, often values inspiration more highly than proof. The picture of Hooke's misshapen figure zig-zagging its way through life and spraying out ideas in all directions, with Newton plodding along behind doing the maths, is not a widely recognised one, but it is defensible.

'Big G', 'Little g'

What Newton wrote in the *Principia* can be summarised as:

The gravitational force between two point masses is proportional to the product of those masses and inversely proportional to the square of the distance between them.

$$\text{or, as an equation,} \quad F = G.m_1.m_2/r^2$$

where m_1 and m_2 are masses, r is the distance between them and G (or 'Big G') is the universal constant of gravitation, which remains to this day the least accurately measured of all the important physical constants.

But Newton didn't actually say this, or anything so simple. For one thing, since he wanted to reach an international audience, he wrote in Latin (Newton 1687). He was not going to get the recognition he craved using the Low German dialect of a cold, foggy offshore island, remote from the European intellectual mainstream. It was a time when only Italian and the French scientists could get away with writing in anything other than Latin, although this was beginning to change as the Royal Society made its presence felt.

In any case, saying things simply was not his way. He liked to take things step by tortuous step. In the *Principia* he began by considering, in what now seems tedious, but at the time may have been very necessary, detail the forces that were needed to make a particle follow each of the four possible conic-section paths (circles, ellipses, hyperbolas and parabolas), and only then did he generalise this to a universal force. He then took the vital step of calculating the gravitational force that would be exerted by a uniform sphere, and showed it to be the same as the force that would exist were the whole mass concentrated at its centre. It was that calculation that finally cemented the link between the motions of objects falling to Earth and planets orbiting the sun.

So it was that, from a starting point in a Europe where the geocentric world view was almost universally accepted, Copernicus, Tycho Brahe, Kepler and Newton, with the help of the Bear and Robert Hooke, presented

'Big G' to the world. The path was then clear for practically-minded people who were prepared to travel to do their science to begin to study the changes in 'little g'. Judging by what he had to say in his first draft for the second volume of the *Principia*, this was not something that Newton himself would have considered worth doing. This, his most accessible work, remained unpublished during his lifetime, specifically to prevent it being read by people who had not *first made themselves masters of the principles established in the preceding books.*[20] Eventually rediscovered and printed in 1728 in an unauthorised English translation, it contained the thought that

> Perhaps it may be objected, that according to this philosophy, all bodies should mutually attract each other, contrary to the evidence of experiments in terrestrial bodies. But I answer, that the experiments with terrestrial bodies come to no account. For the attraction of homogeneous spheres near their surfaces are as their diameters. Whence a sphere of one foot in diameter, and of a like nature to the earth, would attract a small body placed near its surface with a force 20,000,000 times less than the earth would do if placed near its surface; but so small a force could produce no sensible effect. … Nay, whole mountains will not be sufficient to produce any sensible effect. … it is only in the great bodies of the planets that these forces are to be perceived.[21]

From this point onwards, the history of gravity measurement becomes the history of the increasingly successful efforts of Newton's successors to prove him wrong, in this respect at least.

References

Adams R, Jardine L (2006) The return of the Hooke folio. Notes Rec R Soc 60:235–239

Birch T (1756) The history of the Royal Society, vol II. A. Millar. London

Birch T (1757) The history of the Royal Society, vol III. A. Millar. London

Birch T (1758) The history of the Royal Society, vol IV. A. Millar. London

Boulliau I (1645) Astronomia Philolaica. Piget, Paris

Clairaut A (1759) Explication abregée du systême du monde, et explication des principaux phénomenes astronomiques tirée des Principes de M. Newton. Paris

---

[20] Preface to *Principia, Volume III.*

[21] Newton (1731). Newton's preoccupation with motion, rather than force, is evident in this paragraph, and it is ironic that it is to the unit of force that that his name has been attached.

Dugdale W (1658) History of St Paul's. Thomas Warren. London

Jardine L (2003) The curious life of Robert Hooke. Harper Collins, London

Kollestrom N (1999) How Newton failed to discover the law of gravity. Ann Sci 56:331–356

Koyré A (1952) An unpublished letter of Robert Hooke to Isaac Newton. Isis 43:322

Levenson T (2009) Newton and the counterfeiter: the unknown detective career of the world's greatest scientist. Faber & Faber, London

Loggan D (1690) Cantabrigia Illustrata. Cambridge

Newton IS (1687) Philosophiae naturalis principia mathematica. Royal Society, London. English edition: Motte A (1729) The mathematical principles of natural philosophy, London

Newton IS (1731) A treatise of the system of the world (unauthorised and anonymous translation of unpublished MS). Fayram, London

Rouse Ball WW (1893) Essay on Newton's principia. Macmillan, London

Turnbull HW (ed) (1960) The correspondence of Isaac Newton, vol 2. Cambridge University Press, New York

Whiteside DT (1991) The prehistory of the Principia, from 1684 to 1686. Notes Rec R Soc London 45:27

5

The Figure of the Earth

Chapter 2 ended with a gravity map of Eastern Papua, but before it could be drawn the measured values of 'g' had to be manipulated in ways that are now standard but which took two hundred years to become so. The effects of changes in latitude were removed by subtracting the gravity field of an ideal Earth from 'g', and the effects of the differing distances of the observation points from the centre of the Earth were then removed by a 'free-air' correction, so called because it ignores the effects of the rocks above sea level. The next stage was to account for the effects of those rocks, and this was done, rather roughly, by subtracting from each measurement the gravity effect of a flat plate with a constant density and a thickness equal to the height of the measurement point. This is now known as the Bouguer plate, and the correction as the Bouguer correction. But who was Bouguer, and does he deserve to be commemorated in this way?

The Pendulum Clock

If you want to lose weight, head for the equator. If you weigh 150 lb in London, you will weigh about twelve ounces less in Singapore. Sadly, your mass will stay exactly the same.

Newton discussed gravity very specifically in terms of the Sun, Moon and the planets, including the Earth and, as the quotation at the end of Chap. 4 showed, he thought that the effects of all other objects would always be too small to measure. He was wrong about that, but he was right when he suggested

© Springer International Publishing AG, part of Springer Nature 2018
J. Milsom, *The Hunt for Earth Gravity*,
https://doi.org/10.1007/978-3-319-74959-4_5

that the Earth could not be a perfect sphere but would have been given an equatorial bulge by centrifugal force. The distortion is small, but it is partly because of it that 'g' depends on latitude, being greater at the poles, where the sea level surface is 6353 km from the centre of the Earth, than at the equator, where it is 31 km further away. It was, however, only when people began to measure time accurately in different parts of the world that this effect was noticed.

Despite all Galileo's work on pendulums, he never built a pendulum clock. Nor was his mathematics equal to the task of establishing the theory behind the proportionality between 'g' and the period of a pendulum, and he never realised that it was not only air resistance that made wide swings take slightly longer than small swings. All these things had to wait until Christiaan Huygens (Fig. 5.1) became interested in the subject and, starting from his investigations into evolutes and the cycloid, became the first person to obtain a mathematical expression relating the length of a pendulum to its period (see Chap. 14, Coda 3).[1]

Huygens was born four years before Galileo faced the Curia in Rome, and built his first clock just fifteen years after Galileo's death, but he lived in a very different world, populated by people such as Halley, Hooke and Newton. The night sky was no longer studied mainly in the hope of predicting the future, or to argue obscure theological points. The new elite of practical merchants and empire builders believed not in fortune-telling but in fortune making, and to do so they needed not only to travel but to know where they were. Astronomy became a branch of navigation, and the delicate and sickly Huygens was happy to put his talents at the service of his more robust contemporaries. His battlefields were the courts of law, and he sought patents on his clocks as devices for measuring longitude even before they had been tested. There is no doubt that he was a genius, but a difficult, hypercritical and pernickety one, which may have hindered rather than helped him in his search for profit. Certainly, there are unanswered questions concerning the sea trials of his instruments.

By the time Huygens was born, the Protestant United Netherlands had been battling the Catholic powers of Spain and France for fifty years. His family had been prominent in the struggles, but he was no patriot. He is famously quoted as having said that '*the World is my Fatherland, Science is my*

---

[1]During a single period, the pendulum bob passes through every point (except the two extremes) twice, moving in opposite directions. The most accurate measurements of time and position are made when the bob is vertically below the support and moving at its greatest speed. This happens twice in every period, and it is for this reason that what came to be known as the 'seconds' pendulum was defined as having a half-period, rather than a full period, of one second.

Fig. 5.1 Christiaan Huygens. The pastel-on-paper original by Bernard Vaillant in the Huygensmuseum Hofwijck is traditionally dated to 1686, when Huygens would have been 57, but seems to show a much younger man

Religion'. Reversing the trajectory followed by Descartes, a close friend of his family who became his mentor, he left the Netherlands and settled in France and, once there, relied on the Académie des Sciences (and so, effectively, on the French state) for the support he needed. In 1670 two of his clocks were placed on a French frigate bound for North America, under the care of a young assistant astronomer called Jean Richer.

The experiment was not a success. One of the clocks stopped during a storm only a few hours after the ship had left harbour, and Richer failed to follow his instructions to restart it. The second clock stopped a day later, and Richer made no attempt to restart that one either. For the rest of the journey he simply ignored them, and one broke free of its mountings and was destroyed when it crashed to the deck. Richer's own report on the journey has not survived, but we know what Huygens said about it, and he was not happy. From then on he refused to use French ships for his experiments and in February 1672 he wrote to Henry Oldenburg, the secretary of the Royal Society in London, saying

I think that I will myself have to go on some small voyage to ensure the success of this invention, as I see that it depends very much on the commitment of those entrusted with it, with which I have not so far been very satisfied.[2]

Desperate measures indeed, but there is no record that he ever carried out his threat. We know from other, quite unrelated, events that, like many a more recent designer of field equipment, he was prone to blame the users, and never his own designs, when things went wrong. This may have already been widely known, because no-one in Paris took much notice of his complaints. Richer, the man who, in his eyes, had failed him so miserably, had taken insufficient care, had not applied a little oil when needed, and had not restarted the clocks when they stopped, seems to have suffered no sanctions when he returned to France. Far from being in disgrace, he was sent, only a year later, to make astronomical observations *useful for navigation* in Cayenne, the capital of French Guyana. So it was that the first clear evidence that changes in latitude produced changes in 'g' was provided by the very man whose efforts (or lack of them) had so enraged Huygens.

Measuring 'g' was no part of Richer's remit, and he made meticulous measurements of many things while in Cayenne, but the one for which he is remembered is the one that he was never expected to make. For his astronomical work he needed an accurate clock, and the one he took with him was pendulum-based. Even the earliest such clocks were built so that the pendulum lengths could be altered very slightly to ensure that they kept perfect time, and on his return he wrote that:

One of the most important observations that I made was of the length of the seconds pendulum, which was found to be shorter in Cayenne than in Paris: for the same measure that had been marked there on an iron bar, recording the length found necessary for a seconds pendulum was, when taken to France and compared with the length in Paris, found to be different by one and a quarter ligne, that of Cayenne being shorter than that of Paris, which is three feet and eight and three fifths lignes. This observation was repeated during ten whole months, during which not a week passed without it being made several times, with the greatest of care.[3]

---

[2]Huygens to Oldenburg, 13 February 1672 (Bosscha 1897). Original in French.

[3]Richer (1679), p. 66. The length of the seconds-pendulum is, in round numbers, about 994 mm in Paris, about 991 mm at sea level at the equator and at sea level at the poles (if this can be reached) about 996 mm (Richer's one and a quarter ligne difference is equivalent to about 2.8 mm). The corresponding half periods for a pendulum of constant length are, respectively, about one and a half thousandths of a second more and about one thousandth of a second less than at Paris. These are small differences, but they add up. In a day at the equator, a pendulum clock that kept perfect time in Paris would lose more than two minutes.

This does not sound like the man described by Huygens.[4] Similar experiments had been carried out by other observers (including, as already noted, by Halley on St Helena), but Richer's seem to have been the best. In the *Principia*, Newton noted that his

> … diligence and care seems to have been wanting to the other observers. If this gentleman's observations are to be depended on, the earth is higher under the equator than at the poles, and that by an excess of about 17 miles; as appeared above by the theory. (Newton 1687; p. 412)

The correct answer is close to 19 miles.

The Shape of the Earth

By the beginning of the 18th Century even kings and emperors had realised that maps were important and were funding their production, although the results were not always to their liking. In France, improved measurements of longitude had shifted the Atlantic coastline eastwards and reduced the area of the country by about a tenth, causing Louis XIV to complain that his mapmakers had lost him more territory than his generals had gained. Despite this, he and his successors agreed to fund the Académie des Sciences to send expeditions overseas on potentially hazardous journeys with the obscure aim of determining the shape of the Earth. In the same year as Richer's expedition to Cayenne a scientific team was despatched to Tycho Brahe's island of Hven to measure the difference in longitude between Paris and the Uraniborg, because only with this knowledge would French sea-captains be able to use the new telescope-based navigation instruments to exploit to the full Kepler's painstakingly-compiled Rudolphine Tables.

The visitors were shocked by what they found. Their leader, Jean Picard, wrote that

---

[4]History seems to have repeated itself. In December 1662 Alexander Bruce, Earl of Kincardine, made a series of modifications to two of Huygens' pendulum clocks and attempted to test them on a packet boat sailing from the Hague to England. The crossing was rough, one of the clocks fell from its mountings and the other failed to work properly. According to his own account, Bruce himself was too ill to do anything. Richer's similar failure may quite possibly have been due to simple sea-sickness. He was, after all, an astronomer, not a sailor.

We eventually arrived at the enclosure, where we found signs enough to tell us that we really were at Uraniborg. The outline of the building could still be traced by the remains of the foundations that I found in a number of places. But as well as being angry that I was having to search for Uraniborg when I was actually at Uraniborg, I was disgusted to see this famous place, which is spoken of wherever astronomers meet, filled with the carcases of dead animals .... (Picard 1680)

In less than a hundred years, thanks to Danish and then Swedish neglect, and the desire of Tycho's former tenants to profit from a providential supply of free building materials, the observatory had all but disappeared. Only the underground Stjerneborg survived, to be excavated as the basis for the modern museum.

Sixty years after Picard's eventual success at Uraniborg the work begun by Louis XIV was being continued under his successor, and the Académie was making ambitious plans to find out how the Earth's radius varied with latitude. This was of limited importance to contemporary map makers, but national pride was also involved. Newton, an Englishman, had postulated an Earth flattened at the poles, whereas the followers of Descartes, a Frenchman, had claimed the opposite. The jury was still out, because the decrease observed by Richer might be blamed on centrifugal force alone, and any Frenchman who favoured Newton could find himself in trouble, as Victor Hugo discovered. For a conclusive test, measurements had to be made as close to the equator as possible and as far away from it as possible, and sites were selected in Swedish Lapland and what is now Ecuador, then part of the Spanish colony of 'Perou'.

The French kings had little interest in pure science, and the support of Louis XV for the South American expedition may not even have been due to royalty's very definite interest in the more useful science of navigation.[5] Ever since the days of the Conquistadors, Spain had tightly controlled access to her American colonies. Visitors from other countries were discouraged, sometimes lethally, from travelling around, and all information, particularly if in the form of maps and charts, was closely guarded. Even the death of the last of Spain's Hapsburg kings, which led to the War of the Spanish Succession and the eventual placing of a Bourbon on the vacant throne, had not made French citizens welcome in Spanish America. To Louis the expedition must have seemed a heaven-sent opportunity to obtain detailed information on the territories that were the main, and almost the only, source of Spain's wealth. Perhaps without realising it, the scientists were to be his

---

[5]The political history of French science during this period is summarised in Saunders (1984).

spies. Had only science been the object, the work could have been done, with many fewer political problems, in French Guyana.

The idea behind the projects was a simple one. The first step, in both Lapland and South America, would be to find the difference in latitude between two points on the same line of longitude (the same meridian) by reference to the fixed stars. This was comparatively easy, although with the instruments available a difference of at least one degree was desirable. The difficult part was to measure the distance on the ground between the two points. For this, the lengths of baselines had to be measured with obsessive accuracy and the distances between the end points of the arcs had then to be established by triangulation using theodolites. This sort of work had already been done within France itself, but the latitude differences were too small for the results to be decisive. They seemed to favour Descartes rather than Newton but the Académie decided that confirmation over greater latitude differences was needed and the two teams were despatched.[6]

An Expedition to Peru

Making the measurements needed had been hard enough in France, and neither Lapland nor Peru could be described as hospitable. Once committed, the teams were going to be away from home for a very long time; the survivors from South America did not return for ten years. Of the eight who went, one died, one was murdered and another got married.[7] English speaking readers are fortunate in having two recent books that describe what happened. One (Ferreiro 2011) deals with the story as a whole, the other (Whitaker 2004) concentrates on a marriage that led to one of the greatest epics of travel ever recorded. The journey was made by the Spanish-Peruvian wife of one of the Frenchmen who, when separated from her husband by an outbreak of hostilities between France and Spain, decided to walk across the continent to rejoin him in French Guyana. Her story is brilliantly told by Whitaker. Here it is enough to note that almost everyone who went with her, or tried to help her, died on the way.

---

[6]Ferreiro (2011) provides a vivid description of the factions, personalities and manoeuvrings within the Académie that led to the despatch of the two expeditions. At the time that Bouguer left France for South America the Lapland expedition may have been no more than the germ of an idea in the mind of Maupertuis, its eventual leader. It did not leave France until a year later.

[7]Given the hazards of 18th Century life, a 75% survival rate over ten years was actually very good. It is quite possible that more of the group would have died had they all remained in France.

The leaders of the South American expedition (Fig. 5.2) were Pierre Bouguer, a mathematician who was none too fit and already, at 37, rather old for that sort of thing, Charles-Marie de La Condamine, aged 34 but in much better health, and Louis Godin, who was only 31 but already a famous scientist. It was Godin who had originally suggested the project, and whose prestige had been sufficient to ensure that it was funded. He was the nominal leader. Along with him came his 21-year old cousin Jean, who, with another assistant, would make the first forays into the field to find a strip of land over which triangulation would be possible. It was Jean who would marry in Quito. The five other members of the team were a 'watchmaker', to look after the scientific instruments and repair them when necessary, a botanist, a surgeon, a draughtsman and an engineer. The work was done almost entirely within what is now Ecuador, but at the time even the name did not exist. The area was simply part of the Spanish Viceroyalty of Peru.

The expedition left France in May 1735 and long before it reached its destination the three principals had begun to quarrel. The mutual toleration maintained during the month-long crossing of the Atlantic did not survive an enforced stay in Santo Domingo, where they had to wait for a Spanish vessel before being allowed any further into Spain's American empire. The ostensible cause of the rift was Godin's extravagance, prompted by his too enthusiastic appreciation of the Dominican beauties, but it would surely have happened anyway. Tempers fray during long periods of isolation on

Fig. 5.2 Left: Charles-Marie de la Condamine. Pastel on paper portrait by Maurice Quentin de la Tour, now in the Frick Collection, Pittsburgh. Right: Pierre Bouguer, by Jean-Baptiste Perronneau, now in the Louvre, Paris. Accounts of the expedition suggest that the portraits accurately reflect the characters of the two men

fieldwork, and it is more remarkable that Bouguer and La Condamine remained on generally good terms, with only occasional differences, throughout the whole expedition, than that others fell out. Sadly, the friendship did not survive the eventual return to France, where they became undying enemies thanks to an argument over publication. Which is, of course, also typical of modern science.

One consequence of the quarrels in Santo Domingo was that, having safely reached the coast of the Province of Quito, the team travelled inland in three separate groups. Bouguer and La Condamine decided to leave the ship at Manta, its first stopping point, rather than spend another two weeks sailing on to their planned destination of Guayaquil with uncongenial companions. They could justifiably do so because part of the agreement with the Spanish government had been that measurements of longitude would be made along the coast, and Manta, being within a degree of the equator, was a good place to start. However, it was also likely that they were happy to get away from Godin and do some independent work, and also to see if it would be possible to measure the degree of latitude in the coastal lowlands instead of in the highlands. They took ashore with them a complete set of instruments, including a pendulum with which they made the first measurement of 'g' at the equator (Richer had been about 5° north of it). When Bouguer eventually returned to Paris he used these results, and those from similar measurements in Haiti and Panama, to show that the changes in 'g' with latitude recorded by Richer and others could not be due to centrifugal force alone. The mathematical journey that he made to prove this was almost as tortuous and difficult as the physical journeys that he and his colleagues were to make in carrying out their primary task (Chap. 14, Coda 4). It would have been simpler had he taken Newton's *Principia*, written with unaccustomed clarity on this particular topic, for a guide, but he was either unaware of the possibility or rejected it.

The country between Guayaquil and Manta is dominated by low barren hills that seem, in the dry season, to be almost ideally suited for the measurements the Frenchmen had come to South America to make. In the 1730s, however, the forests that covered those hills had not yet been stripped away, and the climate was wetter than it is today. When they arrived, at the height of the wet season, they found inland travel almost impossible because of heavy rains that had made the roads and tracks impassable (and the mosquitoes especially ferocious). Even today the rainfall increases and the vegetation thickens a hundred kilometres to the north of Guayaquil, and Godin, who over-ruled the others and insisted that the work should be done in the long highland valley running between Quito and Cuenca, as originally planned, was probably, and almost unprecedentedly, right.

Bouguer and La Condamine stayed together for about a month, mapping and measuring, but Bouguer's health deteriorated and he was forced to make for Guayaquil. He was to be the last of the three to arrive in Quito. La Condamine continued exploring, heading further north. His attempts to find a direct route from Manta to Quito failed, because the forests that then stretched from the coast to the Andes proved to be impenetrable. It rained every afternoon, and his guides abandoned him. He lost his supplies, including his supplies of gunpowder, and for eight days was reduced to living on bananas and *quelques fruits sauvages*. The swamps sucked him down and the heat and humidity dragged him down even further. The leeches and mosquitoes conspired to drain him of blood, and he suffered an attack of fever. In Papua we felt hard done by if we had to spend more than a couple of days in such conditions before a helicopter arrived to take us away. La Condamine had weeks of it.

For all three parties the journey inland was horrible. Godin, who started first, was the first to complete it. The Spaniards had inherited an exceptionally good system of roads from the Incas, but had not bothered to maintain them. The frequent crossings of deep ravines were made on bridges woven out of jungle vines, and were terrifying, and in the high Andes it was very, very cold, even during the day. La Condamine, who eventually took the road from Esmeraldas in the north rather than from Guayaquil in the south, had the worst of it. Once in the mountains he had to leave his scientific instruments and most of his clothes behind at various points as pledges for debts incurred. He arrived in Quito, then the capital of the Peru viceroyalty, with just the rags he stood up in, and in no fit state to call upon the governor. He also distanced himself for some time from his colleagues in Godin's party, who were already there, but with them he had good reason for being upset. He had depended on them to ensure that the personal belongings he had left on the ship were brought up from the coast. No-one had bothered.

It is at this point that we see La Condamine at his worst and at his best. First he upset the Spanish governor, on whose co-operation everything depended, by not even telling him of his arrival. Then he compounded his offence by sending a fairly brusque reply to an official letter demanding his presence. But finally, when he did appear, he charmed the man to such an extent that they remained firm friends for the next ten years. Without this friendship, it is doubtful whether anything at all would have been achieved.

The Frenchmen were also very fortunate in another way. Jorge Juan and Antonio de Ulloa, the two Spanish naval officers assigned to them to make sure that they behaved themselves, were enthusiastic participants in the work rather than the obstructive bureaucrats they might well have been. They too wrote an account of the expedition (Juan and de Ulloa 1765), and anyone who wishes to follow the entire story as seen through the eyes of the participants should read the Spaniards as well as the French. At the end of the expedition the pair travelled separately back to Spain and de Ulloa was captured by a British warship. Science over-ruled politics to such an extent that, while a prisoner in London, he was elected a member of the Royal Society and freely attended its meetings. Happier days!

Eventually Bouguer also arrived in Quito, but the work went slowly. As Frenchmen in Spanish America, the scientists were regarded with suspicion, and not just by the governor. In Cuenca, this suspicion, plus extraordinarily bad behaviour on his part, led to the death of Senièrgues, the expedition's surgeon. Ferreiro devotes an entire chapter to this event, and one is left feeling more sorry for the assassins, who ended up spending long years in prison, than for the victim. The site of the murder is today a sleepy square a short walk from the centre of Cuenca, with a beautiful small museum of modern art at one end, but in the 18th Century it was the Plaza del Torros, where the bullfights were held. The citizens of Cuenca might possibly have excused Senièrgues for flaunting his new mistress in public, in defiance of all local codes of propriety, but his actions led to the bullfight being cancelled, and that was unforgivable. Fortunately for the success of the expedition, he was probably its least important member. Eighteenth Century surgeons were as likely to kill as cure, and this one had nearly got them all killed.

Elsewhere, the general atmosphere of suspicion had its comic side. Juan wrote that:

The other adventure I shall mention, happened to myself in particular, and not with simple and ignorant Indian peasants, but with one of the principal inhabitants of Cuenqa. ... As I was cheerfully descending the mountain, ... I happened to be overtaken by a gentleman of Cuenqa, who was going to take a view of his lands in that jurisdiction, and had observed me coming from our tent. He was, it seems, acquainted with my name, though he had never seen me; but observing me dressed in the garb of the Mestizos, and the lowest class of people, the only habit in which we could perform our operations, he took me for one of the servants, and began to examine me; and I was determined not to undeceive him till he had finished. Among other things, he told me, that neither he nor anybody else would believe, that the ascertaining the figure and magnitude of the earth, as we pretended, could ever induce us to lead such a dismal and uncouth life; that, however we might deny it, we had doubtless discovered many rich minerals on those lofty deserts; adding, that persons in his circumstances were not to be satisfied with fine words. Here I laboured to remove the prejudices he entertained against our operations; but all I could say only tended to confirm him in his notion; and, at parting, he added, that, doubtless, by our profound knowledge in the magic art, we might make much greater discoveries than those who were ignorant of it. These opinions were blended with others equally absurd and ridiculous; but I found it impossible to undeceive him, and accordingly left him to enjoy his own notions. (Juan and de Ulloa 1765; pp. 226–227)

Almost all geologists and geophysicists who do research in the field have met with similar disbelief, and with similar critiques of their dress codes.

Everyone got ill. Couplet died on the slopes of Cayambe when the work had barely begun. Juan describes the event as follows

On his arrival, however, his distemper rose to such a height, that he had only two days to prepare for his passage into eternity; but we had the satisfaction to see he performed his part with exemplary devotion. (Juan and de Ulloa 1765; p. 213)

Some people, it might be thought, come by their satisfactions altogether too easily.

Instruments were damaged, the weather was terrible and the terrain in some places proved impassable. Somehow, the job was done, but by then the team had split into two groups, working independently. This was partly a matter of practicality, but also because Godin, on the one hand, and Bouguer and La Condamine on the other, were barely on speaking terms. Extraordinarily, when the time came for them to measure their separate final baselines, neither team was in error by more than a few feet. It was a triumph for French science, even though it proved that Newton, the Englishman, and not Descartes, the Frenchman, had been right all along.

The Blow Falls

The triangulations that took up most of the expedition's time were only rather distantly linked to 'g', but Bouguer and La Condamine, in addition to their pendulum observations at different latitudes, made two other sets of measurements that were much more directly relevant. Both projects depended on the fact, which had played no part in the choice of the place where the equatorial degree of latitude would be measured, that they were in the high Andes. Measuring the gravity effects of mountains was not what the expedition was for, and Ferreira devotes just two of his three hundred pages to it. He does, however, make an important point about the timing.

On 9 September 1738, about two years after the work had begun, the news reached Quito that the team that had gone to Lapland had completed their measurements along the eastern bank of the Tornionjoki (in modern Finland) in just a little over a year and were back in Paris. They had benefited from the full and enthusiastic cooperation of the Swedes, including the logistical support of the Swedish army. The Swedish physicist Celsius, now remembered as the originator of the centigrade temperature scale, had accompanied them, and may even have been the person who suggested the project. Working in easier country, they had also sensibly chosen to measure an arc only a quarter of the length decreed by Godin. The baseline measurements were left until winter, when they could be made on the almost level surface of the frozen river. The expedition's leader, Maupertuis, wrote an account in which he made as much as he could of the difficulties and privations encountered (Maupertuis 1738),[8] but these were insignificant compared to those in South America (Fig. 5.3). It seems that the greatest hazard that he personally faced was the infatuation of a local girl, to whom he wrote excruciatingly bad poetry and who, together with her sister, followed him back to Paris. Things didn't turn out well for either sister, and the Finns have made their story into an opera.[9]

Maupertuis was known to be a supporter of Newton, and when he returned to Paris with evidence that the Earth was flattened at the poles it was treated with a certain amount of suspicion, but even so the status of the Peru expedition had been drastically diminished. The results were still going to be useful, but they were not going to make any scientific reputations, and it was reputation that Bouguer had come to Peru to earn. He decided that if he couldn't do it by measuring latitude, then other ways would have to be found. It had already been established (by Edmond Halley amongst others) that 'g' decreased with height, but no-one had yet measured by how much. Finding the rate of change over the range of heights available in Ecuador would be scientifically well worthwhile, and if some good solid scientific conclusions could be drawn from the results, that would be even better. La

---

[8]Confusingly, the titles of both this report and Bouguer's are often abbreviated as '*La Figure de la Terre*'.
[9]Also entitled La Figure de la Terre. Libretto by Jaakko Nousiainen, with music composed by Miika Hyytiäinen.

Fig. 5.3 Maupertuis in Lapland, from Maupertuis (1738). He is obviously having far too much fun for his tales of hardship to be taken very seriously

Condamine was persuaded that this was a good idea, and he played his part in most of the actual experiments, but it is clear that Bouguer was the prime mover.

Measuring Gravity

Bouguer was more generous than Larrie Ferreiro in describing the gravity work, spreading his discussion over more than seventy pages. He began by describing the pendulum he used (Fig. 5.4), and the corrections he thought should be made to the results, and only when he had done that did he start to talk about the implications.

Richer in Cayenne had used a pendulum clock, and Bouguer took with him a very accurate version manufactured by George Graham in London. Graham was one of the most important clock makers of his generation and made two important innovations in design. The first was the so-called 'dead-beat' escapement that provided a better way of delivering the energy needed to keep the clock going, the other was an ingenious form of temperature compensation. His pendulum shafts were made of steel, but were terminated in a stirrup holding a glass tube containing mercury. If temperature increased, the shaft lengthened, but the mercury expanded even more, and upwards. With the right sizes, shapes and volumes of all the components, it was possible to arrange for the centre of oscillation of the entire system to remain in very much the same place.

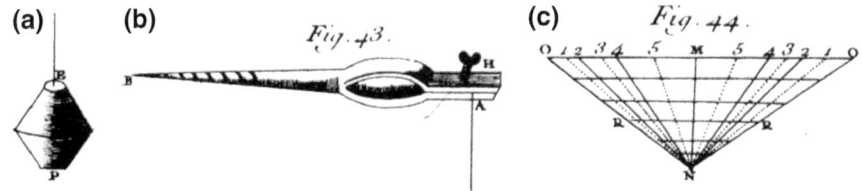

Fig. 5.4 Bouguer's pendulum: extracts from Figs. 43 and 44 of 'La Figure de la Terre' **a** The lower end of the fibre and the copper weight **b** The clamp and the top of the fibre **c** The scale used for estimating fractional swings

In Bouguer's experiments the clock had only an auxiliary role. At each site it was adjusted to tell the right time by reference to stellar transits observed with a zenith sector. This could take a long time and required clear skies (and, as can be seen in Fig. 5.5, the sector was not a comfortable thing to use) but was standard procedure. The adjusted clock was then used to time the oscillations of a very simple pendulum, made by Bouguer himself, that beat seconds approximately but not exactly. It consisted of a weight made of two truncated copper cones bonded together (Fig. 5.4a), with holes drilled so that it could be suspended with either cone uppermost. He made his 'shaft' from fibres extracted from the leaves of a local variety of sisal, and secured it at its upper end with a rather rudimentary clamp (Fig. 5.4b). During set-up and before the jaw was fully engaged, the fibre was looped around it so that its length could be altered, crudely but effectively, by inserting thin spacers. These adjustments were made until flat surfaces cut into an iron ruler coincided exactly with the bottom of the jaw and the top of the weight, and only when this had been achieved was the clamp was tightened. One advantage of this rather complicated procedure was that the instrument was assembled anew for each experiment and could be disman-tled for travel between sites.

For each measurement the weight was pulled about two inches sideways and then allowed to swing free, with the swing amplitude reducing to about half its original value in ten to twenty minutes (the longer times at high alti-tude where air resistance was smaller). Only a few thousand swings, at the very most, would have been observable but Bouguer made an important innovation that to some extent compensated for this. Behind the pendulum he mounted a scale that allowed him to estimate fractions of a swing over a wide range of swing amplitudes (Fig. 5.4c). He seems to have been especially proud of his use of both eye and ear in making his measurements, using his eyes to observe the movement of the pendulum while his ear told him when

Fig. 5.5 An observatory hut in the Andes, from de La Condamine (1751). The second observer is uncomfortably using a zenith sector to observe the passage of stars across the field of view. The figure in the doorway is perhaps one of the 'servants and Indians', trying to get in out of the cold

the clock ticked. The ear, however, was the less important, because the way he set up the experiment allowed him to see simultaneously the clock pendulum and the fibre shaft of the *pendule d'expérience*.

Down on the coast and in Quito all this would have been relatively straightforward. But in a temporary hut in the thin air on top of freezing cold volcano? Jorge Juan described what it was like:

The strange manner of living which we were reduced to, may not, perhaps, prove unentertaining to the reader, and therefore I shall, as a specimen of it, give a succinct account of what we suffered on Pichincha. For this desert, both with regard to the operations we performed there, and its inconveniences, differing very little from others, an idea may be very easily formed of the fatigues, hardships, and dangers, to which we were continually exposed….

We generally kept within our hut. Indeed we were obliged to do this, both on account of the intenseness of the cold, the violence of the wind, and our being continually involved in so thick a fog, that an object at six or eight paces was hardly discernible. When the fog cleared up, the clouds, by their gravity, moved nearer to the surface of the earth, and on all sides surrounded the mountain to a vast distance, representing the sea, with our rock like an island in the centre of it. When this happened, we heard the horrid noises of the tempests, which then discharged themselves on Quito and the neighbouring country. We saw the lightnings issue from the cloud, and heard the thunders roll far beneath us; and whilst the lower parts were involved in tempests of thunder and rain, we enjoyed a delightful serenity; the wind was abated, the sky clear, and the enlivening rays of the sun moderated the severity of the cold.

But our circumstances were very different when the clouds rose; their thickness rendered respiration difficult, the snow and hail fell continually, and the wind returned with all its violence so that it was impossible entirely to overcome the fears of being, together with our hut, blown down the precipice on whose edge it was built, or of being buried under it by the daily accumulations of ice and snow. The wind was often so violent in these regions, that its velocity dazzled the sight; whilst our fears were increased by the dreadful concussions of the precipice by the fall of enormous fragments of rocks. These crashes were the more alarming, as no other noises are heard in these deserts. And, during the night, our rest, which we so greatly wanted, was frequently disturbed by such sudden sounds....

Though our hut was small, and crowded with inhabitants, besides the heat of the lamps, yet the intenseness of the cold was such, that every one of us was obliged to have a chafing dish of coals. These precautions would have rendered the rigour of the climate supportable, had not the imminent danger of perishing by being blown down the precipice roused us, every time it snowed, to encounter the severity of the outward air, and sally out with shovels, to free the roof of our hut from the masses of snow which were gathering on it. Nor would it, without this precaution, have been able to support the weight. We were not, indeed, without servants and Indians; but they were so benumbed with the cold, that it was with great difficulty we could get them out of a small tent, where they kept a continual fire. (Juan and de Ulloa 1765; pp. 216–217)

How they got any results worth having under those conditions must remain a mystery. As must the fact that the 'servants and Indians' did not rise up and slaughter them in righteous anger and then occupy the hut. To cap it all, after all that they had been through to get them, the Pichincha results were almost ignored and only the results obtained in Quito and at the coast, which were thought to be more reliable, were used.

The Effect of Height

Bouguer set about calculating the effect of height on 'g' in two stages, as is still done today. His reasoning is not always easy to follow. Accounts written by 18th Century scientists tend to be, to modern eyes, rambling and discursive. The mathematics are often buried in the text, and the units of length and weight are unfamiliar. Decimals are rarely used and results are expressed as fractions or ratios. It takes imagination and empathy to understand the thinking of people who, while very much cleverer than most of us, knew a lot less. In Chap. 14 (Coda 4) an attempt is made to retrace Bouguer's route to his final conclusions but even if this is done successfully there is a surprise in store for any geophysicist who reads *La Figure de la Terre*. Bouguer never mentioned the 'uniform flat plate extending on all sides to infinity' that today bears his name.

What he did do was ground-breaking, and it took the geophysical world almost two centuries to make full use of it. He went systematically through all the factors that he thought might influence 'g', and then used his pendulum measurements to make deductions about the composition of the Earth. His deductions were wrong, by a very large margin, but that was because of mass distributions of which he was unaware and whose very existence would not be suspected for more than a hundred years.

Having first considered at length the effect of latitude, Bouguer calculated the effects that he would have observed at his Andean locations had there been only air between his measuring points and sea level. For this he needed only the basic inverse-square law and Newton's proof that the gravity pull of a sphere of uniform density was the same as if its entire mass were concentrated at its centre. Using these principles he showed that, within the limits of what he could measure, the decrease in 'g' would be proportional to height. Correcting for this 'free-air effect' is an essential step in the conversion or *reduction*[10] of measurements of 'g' to quantities that are geologically meaningful, but because rock masses above sea-level are ignored it is very far from being all that is needed. Maps of 'free-air gravity', made by correcting only for this and for latitude, are today used almost exclusively for the results of marine surveys, where the observations are made at sea level and the correction is effectively zero.

Bouguer's obvious next task was to calculate the effect of the topographic masses, but to do it he took a rather curious route. He devoted three pages of *La Figure de la Terre* to calculating the effect of a spherical shell occupying the space between the observation point and sea level and assigning to it a density different from that of the 'whole Earth'. Then, after having put considerable effort into the exercise, and having arrived at his equivalent of the

---

[10]It is just possible that this use of the term 'reduction', which is now standard in processing gravity data and which seems slightly odd in English, goes all the way back to Bouguer's French, but this would be a difficult thing to prove.

modern expression $4\pi G\text{đ}h$ for the gravity at the outer surface of an ideally thin spherical shell, density 'đ' and thickness h, he abandoned this approach because

> … the Peruvian cordillera, however large, will not be able to produce the same effect as the imagined spherical shell.

He then decided to approximate the mountain mass by an infinitely long ridge with a triangular cross-section, with an apex angle of 90° and the observation point at its apex. He claimed, correctly but without providing any proof, that the gravity effect observed would be only a quarter of that of a spherical shell, but discarded this model in its turn, saying that the topography of the Andes implied an apex angle closer to 170° than 90°. This, he said, again correctly, would have an effect almost twice that of the 90° ridge. It is a curious fact that the gravity effect of an infinite ridge at its apex is directly proportional to the ridge angle (see Chap. 14, Coda 4). In modern notation this implies a field of $\pi G\text{đ}h$ for the 90° ridge and $2\pi G\text{đ}h$ for the infinite flat plate, where 'h' represents, in the first expression, the apex height and in the second expression the plate thickness. We cannot know how Bouguer got to his equivalent expression because he did not show his working, but he did it.

The Density of the Andes

Bouguer may well have been the first person to realise that the rate at which 'g' decreases with height must depend both on the average density of the whole Earth and on the density of the rocks between the observation point and sea level. From that idea it was only a short step to using his pendulum observations to estimate the ratio between the two and, because he, like Galileo, was working entirely with ratios, he needed to know neither 'Big G' nor the total mass of the Earth to do so. Having found that the seconds pendulum was shorter in Quito than at sea level by one part in 1331 and knowing that the ratio of the height of Quito to the radius of the Earth was 1/2217, he calculated that the ratio of the average density of the rocks of the Andes to the average density of the Earth was 850/3993, from which he deduced that the Peruvian cordillera, despite the metallic minerals it contains, has not even a quarter of the density of the interior of the Earth. (Bouguer 1749; p. 362)

This is not even approximately correct. The average density of the rocks of the Andes is between 2.6 and 2.7 gm/cc and the currently accepted value for the average density of the whole Earth is 5.515 gm/cc, so the true ratio is much closer to a half than a quarter. Over the years a variety of explanations have been offered for this discrepancy, the most usual being that Bouguer's measurements were simply not sufficiently accurate. More sympathetic commentators have suggested that the nature of the terrain, which departs noticeably from his simple model, was responsible, but for a long time the lack of accurate maps prevented this idea from being tested. It was 21st Century access to high-precision global terrain models (Digital Terrain Models or DTMs) compiled from measurements made by space-shuttle radar that allowed Smallwood (2010) to show that the effects of irregular terrain would have been quite small, at least in Quito, and that this solution must also be rejected. Managing somehow to fit the study into his busy work schedule as a senior oil company geologist/geophysicist, he also provided what must surely be the correct answer, in which the deep structure of the Andes plays a major role.

Beyond the coastal plains across which the members of the French expedition made their separate and tortuous journeys from Guayaquil, Ecuador rises steeply to the Andean cordillera, associated with an additional gravity effect which is discussed in Chap. 6 but which Bouguer could neither have known about nor even suspected, and which is quite enough to explain his 'error'. Modern gravity surveys have shown that even after allowing for the 'free-air' and topographic effects, which together amount to about 400 milligal, there remains a difference of about 200 milligal between 'g' at the coast and at 3000 m above sea level in the flat and fertile Quito valley. Smallwood, with all the facilities available to a modern geophysicist,[11] found that the original measurements were quite astonishingly accurate and that the pendulum lengths determined both on Pichincha and in Quito were well within the range of the possible errors in his own calculations due to uncertainty as to the precise locations of Bouguer's observation points. Unexpectedly, and probably because of a statistical fluke, when all the effects had been properly calculated the Pichincha measurements, made under

---

[11]The grids obtained using the measurements made during the Shuttle Radar Topographic Mission (SRTM) now provide an estimated surface height for every 1′ × 1′ rectangle over most of the Earth's land surface.

quite appalling conditions, were slightly more accurate than those made in Quito.

Although in the extract quoted Bouguer is speculating about the density of the rocks of the Andes, the remainder of his text makes it clear that what he had originally wanted to do was find the mean density of the Earth. His answer may have been wide of the mark, but there were still quite respectable scientists around at the time who thought that the Earth might be either hollow or filled with water, and he certainly made both of those ideas untenable.

Aftermaths

There was a sad postscript to the expedition. For what must have been a very trying ten years, Bouguer and La Condamine had tolerated each other's foibles, with no more than the sort of temporary squabbles that could scarcely have been avoided (the worst occurring when Bouguer discovered an error in La Condamine's calculations). They then fell out, and permanently, once they were back in France and under very different pressures. It seems to have been almost entirely Bouguer's fault, since he published his own account of their work without waiting for his colleague, who got back to France a few months later. This was considered, not unreasonably, to be dirty dealing, and the two fought it out for much of the rest of their lives. It was La Condamine who, in his lifetime, had the last laugh. He lived longer, in better health, and seems to have enjoyed his extra years to the full. He had certainly enjoyed his time in South America more than did the ascetic Bouguer, leaving behind him at least two illegitimate children (which did not prevent him from obtaining, once he was home, a Papal dispensation to marry his sister's daughter). When he died, he left his scientific papers to Maupertuis.

Today, however, it is Bouguer's name that is familiar to every geophysicist. Is this deserved, and when, posthumously, did he and the flat plate first come together? In Appendix 1 of the US Coast and Geodetic Survey Annual Report for 1894, G. R. Putnam noted that:

The reduction known as Bouguer's formula (sometimes improperly called Young's rule) has been very generally applied in reducing pendulum observations to the level of the sea. This formula is $dg = + 2gH(1-3d/4Đ)/r$, ... The first term takes account of the distance from the earth's center and the second term of the vertical attraction of the matter lying between the sea level and the station, on the supposition that the latter is located on an infinitely extended horizontal plain.

It is clear from this that even at this comparatively late date Bouguer's name was still being linked to the free-air correction as well as to the flat-plate correction to which it is now firmly, and exclusively, attached. The formulation used by Putnam is also close to Bouguer's original, with its emphasis on densities and, in this case, on sea-level gravity field rather than on 'Big G'. In 1912 Putnam's remarks were quoted approvingly by William Bowie, but in a way that showed that by that time the separation of the Bouguer correction from the free-air correction had begun (Bowie 1912).

The Bouguer plate has served geophysicists well ever since, but there has been some dissent. The authors of one important recent attempt at a global gravity map have used the spherical shell approximation and have laboriously computed the effects of the deviations from it around the entire globe (Balmino et al. 2011). It can be argued that their efforts were not justified by the results, but at least the work was done with a full understanding of what was involved. Other commentators have been less aware. In 1971, a University of Wisconsin physicist called John Karl dashed off a paper in which he brashly claimed that his geophysical brethren had been doing the wrong thing for more than fifty years, and were in error by a factor of $2\pi G d h$ (Karl 1971).

This was a time when the stories of the early 20th Century rejection of continental drift and the then very recent rejection of the first outlines of plate-tectonics were much in people's minds, and the editors of geophysical journals were hyper-sensitive to the risks of suppressing new ideas. It must surely have been for a reason of that sort that they decided to accept the contribution, rather than justify its rejection with a quick quotation from *La Figure de la Terre*. Bouguer's simple statement of the inadequacy of the spherical shell was far more succinct than the refutation that was published with Karl's paper and which evidently lacked the clarity necessary to dissuade him. Twelve years later he was back (Karl 1983), with the equally implausible claim that geophysicists had also, ever since Bouguer, been using the wrong free-air factor, because local variations in the vertical gradient of 'g' were not being taken into account. Remarkably, he had the confidence to do this despite admitting, in a reply to criticism, that "*I know nothing about geodesy*".[12]

---

[12]What Karl ignored, and Bouguer did not, was the principle of superposition, which says that the gravity field at any point can be obtained by adding together the individual contributions of all masses producing measurable effects at that point (with due regard for direction when the addition is done). The free-air correction is designed to remove the effect of the simple model of the ellipsoidal Earth of uniform density. The effects of topographic masses are removed separately and the effects of all other masses are the very things that can give the geophysicists some understanding of the geology.

1971: Etna

Pichincha, the mountain that Bouguer climbed for his experiments and used in his calculations, is a volcano. He was well aware of the fact, and actually worried about the effect that its internal structure might have on some of his observations. He would surely have been interested to know that it would one day be an almost trivial matter to make accurate measurements of the gravity effects of volcanoes and that by making repeated measurements it would be possible to measure their inflation as magma rises and eruptions become imminent. Detecting these changes would for him have been completely impossible, but with the gravity meters available in the second half of the 20th Century it could be done, and has been.

GPS instruments now measure inflation more accurately still, but did not exist in the 1970 s, and it was in the 1970 s that a project was set up at London's Royal School of Mines to monitor Mt Etna. In those days airline tickets were relatively much more expensive than they are now, but once bought they could be used for travel by almost any route that went in roughly the right direction, and journeys could be broken almost anywhere. Fortuitously, at just the time that the first measurements were needed on Etna, I was on my way from London to Fiji with a gravity meter, and was asked to visit Sicily *en route* and set up a series of very precise gravity stations that could be regularly reoccupied.

My base was the old Rifugio Sapienza, high up the mountain and long since buried under a lava flow. It was low season for tourism, and when I arrived the barman was serving one solitary drinker. It seemed churlish not to join him, especially as, unusually for those days, he spoke good English (with the inevitable American accent). Equally inevitably, he wanted to know why I was there. Happily, he seemed rather more ready to believe in pure science than the citizen of Cuenca encountered by Jorge Juan.

I didn't see him again, but the work went amazingly well. Impeded only at the weekend by entire families from nearby Catania making their excruciatingly slow way up to the mountain picnic sites in monstrously overloaded Fiat 850s, in four days I had measured all the stations that could possibly be needed by the most enthusiastic volcanologists, to acceptable accuracy. This was due, in no small measure, to the willing assistance I received wherever I went. Leaving the Sapienza, I commented to the barman on how helpful and friendly everyone had been.

Of course. They know you're a friend of the Capo.

The word had evidently gone out and, on an island where suspicion of strangers has deep roots, that was probably just as well.

1988: Grimsvötn

Iceland in summer was very different from the rainforests of New Guinea and Indonesia that I was used to. For one thing, it never got dark.

It would have been nice to have been able to think that I had been chosen as PhD supervisor by a student from far-away Iceland because of my stellar reputation in the geophysical world. Unfortunately, I knew that he had actually chosen UCL, and me by default, because he was married to one of Iceland's leading young artists, who wanted to study at the Slade. The Slade is part of UCL and really does have a stellar reputation, and so I got Magnus. He arrived knowing exactly what he wanted to do, and with no intention of letting any supervisor get in his way. Since he knew far more about almost everything connected with his chosen subject than I did, and wrote better English than any PhD student that I had ever supervised (with some minor difficulties with his 'v's and his 'w's), it seemed best just to sit back and let him get on with it.

The Vatnajökull ice cap covers about one tenth of Iceland. It rests on the volcanic rocks that are all that Iceland has to offer by way of geology, and several active volcanoes poke their way up through the ice. Grimsvötn, with its 1000 ft high caldera walls, is the giant amongst these, and the most active. The snow that collects in the caldera is continually being melted, from the bottom up, by volcanic heat, forming an ever-deepening lake that is mostly covered by a thin shelf of ice. As more snow falls, and the volume of meltwater increases, the ice shelf rises within the caldera, until the pressure is great enough, and the level is high enough, for the water to spill out under the ice and lift it up all the way to its southern edge. The result is a *jökulhlaup*, a sub-glacial flood, and the Grimsvötn versions regularly destroy the road that runs along the south coast. There is no way to stop this happening, but the government of Iceland and, presumably, the people living nearby would at least like to be able to predict when. Magnus had decided that he would get the answer, and one of his methods (he ended up trying almost everything) was to use measurements of 'g' to estimate the volume of molten rock in the magma chamber a few kilometres below the caldera floor. My own thought was that volcanoes were dangerous things, and glaciers were dangerous things, and a combination of the two was likely to be very dangerous indeed. However, since my student was going to do his fieldwork there, I was honour-bound to put in an appearance.

Australia has surf bums, who work only to earn enough money to go surfing. Iceland has glacier bums, and most of Magnus' friends seemed to fall into that category. Magnus was slightly different, because he actually had a full-time, if slightly insecure, job at the Science Institute. There were also proper tenured lecturers there who were in the same glacier-bum category. Getting together a field party was not difficult, and nor was borrowing the equipment. We got a satellite phone, as well as the gravity meter, from the Science Institute, snowmobiles and sledge-trailers from goodness knows where (Fig. 5.6), and Loran radio locators that would, in those days before GPS, let us know where we were. The Icelanders seemed to have a relaxed attitude to lending stuff out but, it turned out, an equally relaxed attitude to ensuring that it actually worked before they handed it over.

We began with three snowmobiles and two unpowered sledges loaded with large wooden boxes carrying generous supplies for four people for three weeks. When we reached the edge of the glacier the sky was blue and the sun was shining. It was also about ten o'clock in the evening, but that's the

Fig. 5.6 Icelandic workhorses. The nerve-rackingly unreliable snowmobiles. The Grimsvötn caldera wall is in the background left, with the Glaciological Society hut at its highest point

Icelandic summer for you. Getting on to the glacier would, I was told, be the tricky bit, and be best done in the early morning, when the melt-water pond that surrounded the ice during the summer would be slightly shallower. This turned out to be the very least of our problems, but it took most of the day to get on to the ice proper. The nice big comforting Toyota Land Cruisers that had brought us that far then left us to it.

Ice caps are not, near their edges, smooth. They are rough, they are lumpy, they are not good places for vehicles that run on rubber tracks. Travel became a boring routine. Track comes off. Offload, turn the vehicle over, loosen the power train, get the track on again, tighten everything up, turn the vehicle right-side up, load it up, get back on it and away. For a kilometre or so, if we were lucky, before it happened all over again. Human beings can only stand so much of that sort of thing, and the largest of the snowmobiles decided that it could stand even less. It decided, in fact, that it would go no further, and that was that. But at least it was once again a nice sunny day, or night. We camped.

The sun did, in fact, drop briefly out of sight, but there was still light enough to wander around and examine some of the crevasses, which seemed all too plentiful. About these, Magnus was reasonably comforting. *You know, most people who fall down crevasses, we usually get them out in the end.* Then he pointed to one of the round swallow holes down which melt-water was pouring in fine style. There seemed to be rather more of those. *Now, if anyone goes down one of those, we NEVER get them out.* I ceased to be comforted.

Two snowmobiles would still get us to the hut at the edge of the Grimsvötn caldera but not, certainly, with all our supplies. We loaded up the essentials and went on and, after a time, the going got better. Away from its edges, the glacier smoothed out. The tracks stayed on, and we covered the ice at a decent rate, heading into the great white mass ahead of us, which turned out to be fog. Or cloud. Or something else that cut visibility down to no more than fifty yards in any direction. From then on, we were completely dependent on the Loran. It might, I suggested to Magnus, be a good idea to cross-check ours with the Loran on the other snowmobile. *That one,* he said, *doesn't work. Never has.* I was even less comforted. We were on a glacier, in something approaching a white-out, heading for a 1000 ft vertical drop over a caldera wall, with just one suspect radio locator to steer by.

And, by some miracle, a great mass of bare rock appeared out of the gloom. *Edge of the caldera* shouted Magnus and turned sharply right so that we were now running parallel to the drop, but far too close to it for my liking. I suggested, rather feebly, that we might keep a little further away. *Too many crevasses over there,* was the answer.

So, We had a 1000 ft drop on one side, a field of crevasses on the other, we could barely see where we were going, and the wind was rising. I was not at all comforted.

We did reach the hut, but there was still a problem. It had been built on a patch of bare rock close to caldera wall. The rock was bare because it was warm, thanks to the volcano, and bare rock is not a surface for snowmobiles. We had to unload and then manhandle the supplies up to the hut. Not, under normal circumstances a problem, but by now the wind was blowing at hurricane force (I don't know what hurricane force actually is, but I do know that it was dangerous to stand up, which is hurricane enough for me). I can still recall vividly the business of getting those boxes up to the hut. Lie on your back. Grip the handle. Pull it towards you. Edge a foot or so up the slope. Pull. Edge. Pull. Edge …. I'm not sure how long it takes to cover 100 metres in that way, but there was one big consolation. When we had finished, the hut was warm. In fact, it had been warm when we arrived. It was warmed by steam coming up through a pipe connected directly to the volcano below.

It took two days for the wind to drop enough for us to begin work, and we passed some of the time by trying to contact Reykjavik on the satellite phone, but without success. Things were, we felt, a little marginal. We had left much of the food and fuel with the broken-down snowmobile, and one of the ones that had brought us this far had lost a windshield—to the wind. It wasn't until the weather had changed and we were actually doing some work that the phone came to life. It was Magnus' cousin, in Reykjavik, calling us.

You haven't been able to use the phone, have you?

No

You borrowed it from the Science Institute, didn't you?

Yes
They hadn't paid their bills. You'd been cut off

Now I do know that Bouguer and La Condamine worked in Ecuador in a vastly more hostile environment than anything we faced in Iceland, and that they had many more life-threatening things to worry about, but it is also true that being disconnected by their phone company had not been among them. However, at least we were finally able to call for back up, which, when it arrived, made glacier travel seem all too easy. With the variety of vehicles coming across the ice, it looked like one of the convoys in a Mad Max movie. Snowmobiles were obviously *passé*, although the one that had broken down had been located and repaired, and had been brought along. What people who are serious about getting around on Vatnajökull use are big powerful 4-wheel drive SUVs. As the going gets softer, more air is let out of the tyres. Traction is never lost.

Before we left, Grimsvötn had one last treat in store. I was sitting peacefully on the back of a snowmobile crossing the ice shelf (no one would trust me to drive one of those things) when there was a sudden acceleration, a bump and a bang. We slid gracefully to a stop on the far side of a wicked-looking crack in the ice. The last-minute acceleration had just managed to jump us across a crevasse. Even Magnus was upset.

> There have never been crevasses here before. You know, when I saw it, I thought - not only have I killed myself AND my supervisor, but I have destroyed Iceland's one and only Lacoste gravimeter.

Even then, Magnus took his geophysical responsibilities very seriously.

These days he has to. After he had finished with London he went back to Iceland and his rather insecure job at the Science Institute, still studying his beloved Grimsvötn. Three years later it erupted in spectacular fashion, grounding air travel throughout much of Western Europe and he was on every radio and TV channel, explaining to the world what was happening. After that they had to give him a permanent job.

Did we do anything useful? By putting together the gravity measurements and the magnetic measurements Magnus could begin to build digital models of the interior of the volcano. It is a sad fact that, for every map of 'g' and no matter how well it has been measured, there are infinite numbers of possible mass distributions. The only comfort is that there are even larger (?) infinities of distributions that will not produce that particular gravity field, and some of those might otherwise have been geologically acceptable. Even that is not the full story, because any plausible density model still has to be interpreted in geological terms, and there can be conflicting explanations for the presence of a low density, or high density, body in a particular area.

In the end, there were two main groups of models that seemed to 'work', and with these the paper could be written (Gudmundsson and Milsom 1997) and I could get back to the calm and safety of the rainforest. Magnus went on to use even more geophysical methods on the glacier, and further refined his picture of what was going on within the volcano, but with the eruption everything changed and it all had to be done again. The repeats of the measurements made in 1998 showed that data were needed over a much wider area, and those additional surveys were made (Gudmundsson and Högnadóttir 2007). When dealing with something like Grimsvötn, a geophysicist's work is never done.

References

Balmino G, Vales N, Bonvalot S, Briais A (2011) Spherical harmonic modelling to ultra-high degree of Bouguer and isostatic anomalies. J Geodesy. https://doi.org/10.1007/s00190-011-0533-4

Bosscha J (ed) (1897) Oeuvres complètes de Christiaan Huygens. Tome VII: Correspondance 1670–1675. Martinus Nijhoff, Den Haag

Bouguer P (1749) La Figure de la Terre. Jombert, Paris

Bowie W (1912) Effect of topography and isostatic compensation upon the intensity of gravity. US Coast and Geodetic Survey Special Publication 12

de La Condamine C-M (1751) Journal du voyage fait par ordre du Roi a l'équateur, servant d'introduction historique a la mesure des trois premiers degrés du méridien. Imprimerie Royale, Paris

Ferreiro L (2011) Measure of the Earth: the enlightenment expedition that reshaped our world. Basic Books, New York

Gudmundsson MT, Högnadóttir T (2007) Volcanic systems and calderas in the Vatnajökull region, central Iceland: Constraints on crustal structure from gravity data. J Geodynamics 43:153–169

Gudmundsson MT, Milsom J (1997) Gravity and magnetic studies of the subglacial Grımsvotn volcano, Iceland: implications for crustal and thermal structure. J Geophys Res 102:7691–7704

Juan J, de Ulloa A. (1765) Relación histórica del viaje á la América Meridional. English second edition: Adams J (1765) A voyage to South America. Alexander Ewing, Dublin

Karl JH (1971) The Bouguer correction for the spherical Earth. Geophysics 36:761–762

Karl JH (1983) The normal vertical gradient of gravity. Geophysics 48:1011–1013

Maupertuis PL (1738) La Figure de la Terre, determinée par les observations faites par ordre du Roi au Cercle Polaire. Imprimerie Royale, Paris

Newton IS (1687) Philosophiae naturalis principia mathematica. Royal Society, London. English edition: Motte A (1729) The mathematical principles of natural philosophy, London

Picard J (1680) Voyage d'Uranibourg ou Observations Astronomiques faites en Dannemarck. Imprimerie Royale, Paris

Richer J (1679) Observations astronomiques et physiques faites en l'isle de Caienne. Imprimerie Royale, Paris

Saunders ES (1984) Louis XIV: patron of science and technology. Purdue University Libraries Research Publications

Smallwood JR (2010) Bouguer redeemed: the successful 1737–1740 gravity experiments on Pichincha and Chimborazo. Earth Sci Hist 29:1–25

Whitaker R (2004) The mapmaker's wife. Basic Books, New York

6

The Attraction of Mountains

Once Bouguer and La Condamine had made the measurements they needed to work out the effects on 'g' of latitude and elevation, they decided to take things a step further and make a direct estimate of the ratio of the average density of the whole Earth to the density of the rocks near its surface. Theirs was the first of a series of attempts made over a period of some seventy years, all aimed at obtaining a value for the mass of the Earth by measuring the gravitational pull of mountains.

The Deflection of the Vertical

Bouguer and La Condamine made their attempt towards the end of 1738. According to La Condamine:

> It was a question of determining, by direct experiment in observing the same star from two different places, whether the proximity of a very large mountain could deflect the plumb line of a quadrant; in conformity with M. Newton's theory of universal gravitation. This idea was due to M. Bouguer. I merely took part in its execution …. (de La Condamine 1751)

It was generous of him to say this, because by the time he did so, he and Bouguer had become enemies.

The vertical is the direction in which a dropped weight will fall or a plumb line will hang. It is the direction of the local gravity field, and not necessarily the direction of the centre of the Earth. If there is a large mass

© Springer International Publishing AG, part of Springer Nature 2018
J. Milsom, *The Hunt for Earth Gravity*,
https://doi.org/10.1007/978-3-319-74959-4_6

away to one side, a plumb bob will be pulled towards it. Newton took the example of a hemispherical mountain with a radius of three miles and with the same density as the Earth and calculated that it would deflect a plumb line through an angle of less than two minutes (Newton 1731). He was very far from proposing that such an experiment be made. Rather, he was recording his doubts about the possibility of doing so. Bouguer, however, knew himself well able to measure changes in angle of much less than a minute since that had been a requirement of the project that had brought him to South America in the first place. He set about finding a suitable mountain.

The choice eventually fell on Chimborazo. Pichincha, the obvious candidate because so much work had been done on it (Bouguer himself had spent almost three months there), was rejected because its multiple summits made the gravity effect hard to calculate. Cotopaxi was considered but was known to be a volcano and it was feared that there might be internal cavities that would significantly reduce its, at first sight impressive, mass. Tongouragoura, another possibility, had erupted in 1640 and again in 1645, and access to it anywhere between the snowline and the cultivated land around its base was in any case very difficult. Chimborazo had not erupted in living memory (and has not since), its height had already been measured and, as can be seen from Quito, it has a very regular shape. It was only when the scientists were coming back down the mountain after completing their work that they saw signs of past volcanic activity and learned of a tradition of pre-Colombian eruptions (Bouguer 1749; p. 389).

In principle all that they had to do was observe suitable stars from some point where a large deflection was to be expected and from another point at exactly the same latitude that was far enough away for the effect of the mountain to be negligible. Placing both observatories at the same latitude should not have been a problem, since the true North-South direction could be determined by either sun or star sightings and the true East-West direction would be at right angles to this, but it did require a clear line of sight, and when Bouguer wrote of Chimborazo that "*je sçavois qu'il étoit un accès assez facile*" this was by the standards of 18th Century South America, where nothing was ever easy. Although several possible arrangements of observation points were considered, the one eventually adopted (Fig. 6.1) was also the one least likely to record a strong effect, and was chosen simply because it was the simplest logistically. Even so neither horses nor mules were able to reach the mountain observatory (at A) and the final approach had to be made on foot, over rough ground in terrible conditions. It also proved impossible to find a place for the second observatory that was sufficiently far from the mountain yet visible from it and also at exactly the same lati-

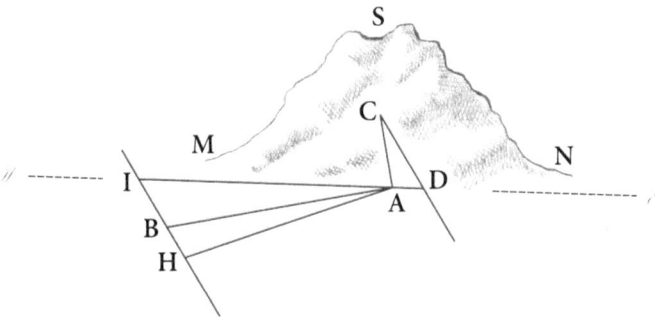

Fig. 6.1 Chimborazo. C is the location of the centre of mass of the mountain and CD is the meridian line through it. The line IAD is the parallel of latitude passing through the mountain observatory A. The remote observatory B was situated about 7 km to the west. H is a point from which observatory A was visible and which was used only for triangulation and BH is the directly measured triangulation baseline. The distance IB had to be determined so that the latitude difference between A and B could be calculated, and their height difference also had to be measured (Drawing: Kate Milsom, based on Fig. 47 of Bouguer (1749)

tude, and additional surveying was needed to find the difference in latitude between B, the point actually used, and I, its ideal location. Even B had no line of sight to A. For that, yet another point, H, had to be occupied.

Before setting out for Chimborazo, Bouguer had calculated that, thanks to the inverse square law, which emphasises the effects of nearby masses, the relatively tiny mass of the mountain might deflect his plumb line by as much as two minutes of arc at a suitably located observatory. This was a very rough estimate, and was based, among other things, on assuming that the mountain would have the same density as the Earth as a whole, but it was enough to satisfy him that the effect would be measurable. When it turned out to be only seven seconds of arc, he speculated again about the possible presence of large cavities within this newly-recognised volcano, but he may also have been quite satisfied with the result since it supported his earlier conclusion that the interior of the Earth was four to five times denser than the near-surface rocks. His report was, he felt, worth sending back immediately to the Académie, where it was read to an uninterested audience in 1739.

The true ratio of the densities is so different from the one calculated by Bouguer that the Frenchmen's experimental competence has again been questioned and, once again, it is John Smallwood who has ridden to their rescue (Smallwood 2010). Again using detailed shuttle-radar terrain models, this time to estimate the mass distribution represented by Chimborazo and to find the probable positions of the observatories, Smallwood concluded that the results that were recorded were within one or two seconds of arc of what they should have been. He was less impressed by Bouguer's mass calculations, as it was clear that part of the difference from his expected result was produced by an overestimate of the total volume of rock.

There was a third reason why Bouguer's estimate was very wrong, and it was one for which he could, indeed, be severely criticised. He assumed that the gravity effect of the mountain could be calculated by supposing that its entire mass was concentrated at its centre of mass. This is true only for uniform spherical bodies, and even with 18th Century understanding of gravity fields and their sources, he should have known that. On Chimborazo some of the mass that he included may actually have been to the south of the mountain observatory, and pulling the plumb bob the other way. However, since the 'centre of gravity' error is one that is still widely made, even in university common rooms, Bouguer should not be judged too harshly.

The Society Intervenes

The efforts to measure the gravitational attraction of the Andes did not go unnoticed in England. When in 1772 Nevil Maskelyne, then the Astronomer Royal, proposed that an attempt be made to do the same thing somewhere in the British Isles, one of his justifications was that *M. BOUGUER... expresses his wishes that a like experiment be made ... in France or England, where he thinks that some might be found of sufficient bulk for the purpose.*[1] He added that, although Bouguer's wishes had been expressed more than 30 years previously, *yet I believe no similar experiment has ever been made in Europe.*

Maskelyne has not had a good press lately. In *Longitude* (Sobel and Andrews 1998) Dava Sobel is very hard on him, describing him as the villain or, at least, the anti-hero, in John Harrison's long struggle to receive the prize that he had undoubtedly earned for his method of determining longitude at sea. Harrison relied on his ability to build marine chronometers based on balance wheels and hairsprings (and not pendulums) that

---

[1]Maskelyne (1775). Bouguer's original comment appears on pp. 389–390 of La Figure de la Terre. Maskelyne was one of those who believed that the Frenchmen's experimental technique had been at fault, pointing out that *their instrument [quadrant of 2½ feet radius] was too small and imperfect for the purpose; and ... they ... were subject to great inconveniences, being sheltered from the wind and weather by nothing but a common tent, and placed so high up the mountain as the boundary where the snow begins to lie unmelted all the year round.*

would keep accurate Greenwich time during weeks or even months at sea. Maskelyne had submitted his own proposal, based on measuring the distance of the Moon from the Earth, but then accepted a place on the Board of Longitude, the body responsible for awarding the prize. He therefore had a clear, and somehow very modern, conflict of interest, and he did not deal with it very well. It is, however, almost certainly not true that his opposition to Harrison's claim was motivated by greed, since a German astronomer called Tobias Mayer had a prior claim to the lunar distance method. His widow received £3000 for it at the same time that the main prize was awarded to Harrison. The surveyor Charles Mason later improved on Mayer's tables, and received £1317.

As described by Sobel, Maskelyne was a man in the unsociable Newton mould, although she did admit that he succumbed sufficiently to the ways of the flesh to get married, at the age of 52. The former Miss Sophie Rose, twenty years his junior, gave birth to their only child, a daughter, only ten months later. A very different opinion of the man was held by Derek Howse, sometime head of the Department of Navigation and Astronomy at the National Maritime Museum in Greenwich, who spent his spare time for some twenty years delving into the museum's archives. The Maskelyne he found there was obviously an obsessive, preserving every bill, both official and personal, that he ever paid, but these testify to, amongst other things, a level of alcohol consumption not usually associated with hermits. His devotion to good living is clear in the rather unappealing formal portraits of him that were painted late in life, but he was capable of enduring discomfort when he felt it to be necessary. He did sail to St. Helena and back, in an unsuccessful attempt to observe the transit of Venus across the face of the sun, and then spent months on a cold, wet and windy Scottish hillside. On this second occasion he succeeded in doing what he had set out to do.

At first Maskelyne had hoped to find somewhere in the north of England that could be used for the experiment, with Helvellyn and Skiddaw in the Lake District and Ingleborough or one of its neighbours in Yorkshire as possibilities. However, he obviously suspected that it might be necessary to go further north because when, in 1773, the Royal Society despatched Charles Mason to find a suitable place, his instructions were *to make a tour of the Highlands of Scotland* although with the additional task of *taking note of the principal hills in England which lay on his route.* Such a trip would be a daunting prospect even today, and at that time, with the Highlands still undergoing 'pacification' after the Jacobite uprising of 1745, it must have seemed a very tall order indeed. The requirements were quite stringent. Not only did the selected mountain have to be large enough for its effect to be measurable, it had to be very regular in cross-section for this to be calculable in the days before computers. It also had to be elongated E-W, not because the gravity effect depended on its orientation, but because the technique that was to be used certainly did.

Mason returned to London having found Schiehallion, *a remarkable hill ... of sufficient height, tolerably detached from other hills, and considerably larger from East to West than from North to South* (Fig. 6.2). That he succeeded in so difficult a task suggests that he might not have been travelling completely blind. There is an apocryphal story that Maskelyne had been describing his needs in his London club one evening and one of the military members, recently returned from harrying the Scots, told him that he knew of just the place. The story seems plausible. Derek Howse, who obviously saw in Maskelyne a fellow spirit, described him as '*a likeable, helpful, clubbable, friendly man*'.[2] The military man who provided the information might well have been William Roy, who between 1749 and 1755, was responsible, as a civilian, for producing the sensationally accurate maps of the Military Survey of Scotland and who later visited Maskelyne on Schiehallion to measure the heights of the 'observatories' and the summit, using a barometer.

In 1772 the main Longitude Prize was finally awarded to Harrison but the ill-feeling generated by the delays and arguments persisted for years afterwards, and for Maskelyne the prospect of a few months of isolation in the Scottish Highlands may well have had its attractions. Whether or not that was one of his motives for leaving London, on the last day of June 1775 he arrived at Schiehallion, apparently undeterred by (or perhaps blissfully unaware of) the English translation of its Gaelic name as '*constant storms*'.[3] Charles Mason, having actually seen the mountain, and being only recently back from the relative luxury of the American South, where he and Jeremiah Dixon had established the Mason-Dixon line, declined to take any further part in the project. He was not even tempted when the Society offered to double his salary, to a guinea a day.

Maskelyne did not intend to work alone. He sent ahead of him Reuben Burrow (or Burrough), a former assistant from the Royal Observatory, with the task of supervising the shipping of the bulky and delicate instruments and preparing a site for an observatory on the south side of the mountain. Even to reach that stage had not been easy, but Maskelyne had been lucky. In 1769 he had been funded directly by George III for the expedition to

---

[2]Bennet (1998). The writer added that '*the parallels (to Howse) are obvious*'.

[3]Which seems more appropriate than the commonly cited alternative as 'Fairy Hill of the Caledonians'.

Fig. 6.2 Schiehallion. The ridge is being viewed end on, showing its almost perfect triangular shape. *Photo* Neil Robertson, https://travelswithakilt.com

St. Helena, and there was still a reasonable amount of money left from the grant. Like many later researchers, he thought it only fair that he should be allowed to use this remnant to support his next project but, unlike many of his successors, he went back to source to get this approved. He might have had no option, because he also needed the king's permission to use instruments from the Observatory, and to take time off from his normal duties. Considering that the results from St. Helena had been disappointing (clouds rolled in at the crucial moment and no useful observations were made), it was rather decent of the king to approve the new project, especially since the receipts for the just under £400 that had been spent included a wine merchant bill for £56. The essential supplies included 16 gallons of Madeira and 65 gallons of 'Lisbon and port'. If I had been the king, I would have asked for my money back

Maskelyne on Schiehallion

The method Maskelyne used was very simple in principle, but not easy to put into practice. It was essentially a development of that used by Bouguer and relied on the fact that, thanks to centuries of star studies, first in the interests of astrology and then in the interests of navigation, the movements of the fixed stars were known and predictable to extraordinary levels of accuracy. Notably, the time on any given night when any selected star would cross the meridian was entirely predictable, as was its elevation angle at that time at any given latitude. Reversing this idea, if the meridional elevation angle was measured, the latitude could be calculated. This could be done at points on both sides of the mountain to obtain the difference in latitude between them, but all angles would be measured relative to verticals as defined by a plumb line. If the plumb bob was being affected by the mass of the mountain, then the latitude difference obtained by this method would differ from that calculated from a direct measurement of the distance between the two points. The difference would be a measure of the ratio of the horizontal attraction of the mountain to the vertical attraction of the bulk of the Earth, and from it the ratio of the average density of the rocks making up the mountain to the average density of the whole Earth could be calculated. This was what Maskelyne intended to do, and he was fortunate in having in George III a monarch who was interested in astronomy and could understand the point. It would have been much more difficult had he been dealing with George's hunting-obsessed predecessor or his dilettante son.

The preparations on site were necessarily extensive. An observatory consisting of a circular wall five foot in diameter and capped by a movable conical roof was built to house the astronomical instruments. The transit instrument was housed in a square tent nearby and a temporary hut was built for the scientists to live in. Maskelyne describes this as a 'bothie', the local name for a shepherds' hut. While small, these are usually very solid stone structures, but later events suggest that Maskelyne's was built mainly of wood. It would not have been much fun to have been living there, or in its counterpart on the north side of the hill, during the three-and-a-half months in which the mountain fully lived up to its evil reputation. Because of almost continuous cloud and fog (generally, on Schiehallion, the same thing) it was two weeks before it was possible to complete even the first and most basic part of the operation, which was to draw the meridian line on the observatory floor. The first useful measurements were made on 20 July, and work at the South Observatory was not completed until 15 August. During this time Maskelyne developed a method for verifying the position of his instruments with respect to the meridian which, had he 'thought of it at first, would have saved me much trouble'. It consisted of observing the transits of two stars with notably different elevations and comparing the apparent difference in their horizontal separation with the known values. If these agreed, then he could be reasonably confident that his instrument was properly aligned.

With the method perfected, things should have gone more smoothly on the north side of the mountain, but even though the distance between the two observatories was small (the measured separation of the parallels of latitude passing through them was only 1237 m), it took ten days to transfer all the equipment. Everything had to be carried on men's shoulders over the ridge *and some of the packages were very weighty*. However, they need not have hurried, because it was not until 4 September that there was a night clear enough for Maskelyne to position the meridian line using his new method. It was as well that he took the opportunity because it was not until 15 September that he saw enough of the sun to establish the same line by more conventional techniques. By then he had managed to make observations on just four nights, and he was only able to work on two more nights before he left the mountain on 20 October. During all that time, in appalling weather, the team of surveyors had been making the measurements needed to determine the relative locations and elevations of the two observatories. Understandably, by the time they had finished, everyone felt that they were entitled to a small celebration. The party was held in the northern bothie, and everyone who had helped was invited.

The events of that night are still the stuff of legend in Kinloch Rannoch and the surrounding area, and it is a story that shows Maskelyne in a much more sympathetic light than any cast by Sobel. As Derek Howse told it (Howse 1989), Maskelyne asked the 'Donnaeha Ruadh', the red-haired Duncan, who had been cooking for and generally looking after the scientists and surveyors, and who had livened up the evenings with his fiddle playing, to go down into Kinloch Rannoch and buy the necessary supplies, including a keg of whisky. A good time was evidently had by all, so much so that the bothie burnt down, and Duncan's fiddle burned with it. Duncan, understandably, was devastated, but Maskelyne consoled him, saying *'Never mind, Duncan, when I get back to London I will seek you out a fiddle and send it to you'.*

An easy promise to make, and one that might well have been forgotten when the scientists were safely home, but Maskelyne was as good as his word. A few months later, the replacement fiddle arrived in Kinloch Rannoch, and the delighted Duncan composed a song in its honour, 'A Bhan Lunnaineach Bhuide' (the yellow London lady). In translation, one of the verses goes

> It is Mr. Maskelyne, the hero
> Who did not leave me long a widower
> He sent me my choice treasure
> That will leave me thankful while I live.

Duncan might not, perhaps, have realised just how heroic Maskelyne had been. The original fiddle remained for some years in the hands of his descendants, but in 1840 a local copy was substituted by a crooked Edinburgh repairer. The label, however, was preserved. It read *'Antonio Stradivarius, 1729'*.

None of this sort of local colour was included in Maskelyne's reports to the Royal Society which, apart from recording the names of his numerous visitors, kept strictly to the science.

Hence, by calculation of the two triangles formed by the two carns and the two stations of the observatory, the distance between the parallels of latitude passing through the two stations comes out at 4364,4 feet, which, according to M. BOUGUER'S table of the length of a degree in this latitude of 56° 40', at the rate of 101.64 English feet to one second, answers to an arc of the meridian of 42,94.

This was 11.16 arc-seconds less than the difference calculated from the astronomical observations, a result that cried out for some very detailed further analysis. However, when Maskelyne returned to London, he was, like any normally busy bureaucrat who has made himself unavailable for several months, confronted by a desk piled high with correspondence and unpaid bills. Worse still, instead of dying down, the arguments at the Board of Longitude had become increasingly bitter. He therefore hastily wrote himself out of the script, saying merely that:

The attraction of the hill, computed in a rough manner, on supposition of its density being equal to the mean density of the earth, and the force of attraction being inversely as the squares of the distances, comes out as about double this. Whence it should follow, that the density of the hill is about half the mean density of the earth. But this point cannot properly be settled till the figure and dimensions of the hill have been calculated from the survey, and thence the attraction of the hill, found from the calculation of several separate parts of it, into which it is to be divided, which will be the work of much time and labour;

Very sensibly, from his point of view, he then left the actual expenditure of the *much time and labour* to the mathematician Charles Hutton, although he did go so far as to note that the results certainly disproved the suggestions of 'some naturalists' that the earth was 'only a great hollow shell of matter'. Hutton's final conclusion was that the mean density of the Earth must be about four and a half times the density of water (in modern units, about 4500 kg m^{-3}) and at the end of his paper he challenged future workers to identify areas where they felt his analysis could be improved. With the tools he had available, it is very doubtful if it could have been, but in 2007 it was again John Smallwood who took up the challenge, using a modern terrain model, subsurface effects estimated from the latest gravity maps, and modern estimates of rock density. Combining all these with

Maskelyne's eleven arc-second discrepancy gave him a mean Earth density of 5480 ± 250 kg m^{-3}. The accepted modern value is of 5515 kg m^{-3}. Whether hero, anti-hero or villain, Maskelyne was certainly a very, very good observational scientist. He was also clearly an expert at keeping in with the people who could be useful to him, concluding his account with these words:

> But whatever experiments of this kind be made hereafter, let it always be gratefully remembered, that the world is indebted for the first satisfactory one to the learned zeal of the Royal Society, supported by the munificence of GEORGE THE THIRD.

Nonetheless, and clubbable or not, there are suggestions, over and above his treatment of Harrison, that Maskelyne might have been less scrupulous in his dealings with those he thought of as potential rivals. His swift publication of an approximate answer secured him much of the scientific glory, while Hutton, who did the hard and boring work of the detailed calculations, was left feeling that his contribution had been undervalued. Reuben Burrow, who had done the topographic surveying on Schiehallion, similarly felt that he had been poorly credited, and although he is known to have been a difficult and vengeful man, described by one of his contemporaries as a '*paranoid genius*' (Smallwood 2007), there does seem to be a pattern in Maskelyne's behaviour. Everything that is known of him suggests a highly competent observer but an unoriginal thinker who made his major contributions by putting into practice other people's ideas, and scientists of that type sometimes find it difficult to cope with challenging subordinates. That does not stop them being, as Maskelyne evidently was, kindly and supportive to people who do not represent any sort of threat.

In 1992 I took a class of students to Schiehallion.[4] The weather was much kinder to us than it had been to Maskelyne and we had no difficulty in finding the site of the observatory on the north side of the mountain since it was very obviously cut back into the steep slope. We thought that we had also found the site on the south side, but the 'shelf' there was much less clearly artificial and, in those days before GPS, we could not be sure. It was a sobering thought that it had been from the villagers on that side of the mountain

---

[4]We had hoped to make enough measurements with our gravity meter, made in Canada by the now vanished Sodin company, to throw some light on the terrain effects that would have influenced Maskelyne's result. Sadly, we ran out of time, and the students' planned publication, *Sodin on Schiehallion*, was never completed.

that Maskelyne had obtained his labour force and his food supplies while working at the South Observatory. Now, in bleak testimony to the thoroughness of the Highland clearances, the valley is deserted.

Hutton in London

It was not until 1778, after *close and unwearied application for a considerable time*, that Hutton managed to complete his calculations, and publish them, but when he finally did so (Hutton 1778) he filled almost a hundred pages of the *Philosophical Transactions*. Editors were more generous to authors in those days, and Maskelyne's standing in the scientific world must have helped. There might also have been some sympathy in the Society for the '*long and tedious*' business of calculation, although that must have been tempered by the almost equally tedious business of reading it all. Most of the report was devoted to the calculations that defined the topography rather than to the interesting new science. There was a separate sketch of each of the seventy-two triangles that underpinned the trigonometric survey, as well as a truly excellent map of the area. The paper began on page 689, but the calculation of the gravitational pull of the mountain was not even started until page 748.

What followed should be fascinating to anyone who has been involved in a modern land-gravity survey. In these it is total gravity, defined by the magnitude of 'g', and not by its horizontal gradient, which matters, but both 'g' and its gradient include non-geological effects produced by the hills and valleys that surround each gravity station. If large, these effects must be removed before geological conclusions can be drawn and this has traditionally been done by making corrections based on the differences between the real topography and the flat upper surface of the 'Bouguer plate'. As Hutton himself put it:

> In a computation of this kind, we need only calculate the attraction of the matter above the plane or horizon of each observatory, and the attraction of so much matter as is wanting to fill up the vacuity below that plane lying between it and the surface of the lower part of the hill.

Or, as we would put it today, terrain corrections have to account for the gravity effects of the hills that rise above the level of the gravity station and the valleys below, while the effects of the bulk of the topography between that level and sea level are dealt with by the Bouguer correction. The coun-

(a)   **(b)**

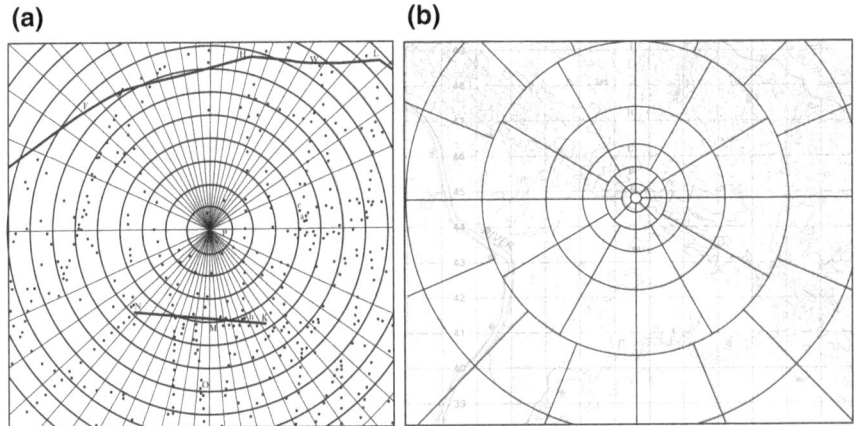

Fig. 6.3 **a** Hutton's terrain correction chart for the northern observatory, redrawn and modified from his 1778 paper to remove a rather confusing meridian line through the observatory. The lower thick line links the cairns established at either end of the main Schiehallion ridge and the upper thick line is a part of the traverse around the ridge that established the relative positions of the two observatories. **b** 'Hammer chart' for gravity survey terrain corrections, overlaid on a map of topographic contours

ter-intuitive corollary, which was recognised by Bouguer himself, is that the corrections are always positive; their sign does not depend on whether the deviations from a flat plane are above or below the gravity station.

In putting these ideas into practice, Hutton made two important innovations. In Fig. 6.3a the dots identify the points at which he had measured the heights, and his problem was to turn this information into an estimate of the gravity effect. His first idea was to divide the area up into compartments, to each of which he could assign a height based on the point or points within it, and he chose to do this using the graticule shown. He then realised that, '*numerous as the points were*', there would be many compartments that contained none. He said that

This circumstance at first gave me much trouble and dissatisfaction, till I fell upon the following method by which the defect was in great measure supplied, and by which I was enabled to proceed in the estimation of the altitudes both with much expedition and considerable accuracy. This method was the connecting together by a faint line all the points which were of the same relative altitude .... And as every base or little space had several of them passing through it, I was thereby able to determine the altitude belonging to each space with much ease and accuracy.

What Hutton is describing here is certainly one of the first, and quite possibly actually the first, use of contour lines to display and make use of topographic information.

Fast-forward a hundred and sixty years: in the days before computers became routinely available, terrain corrections for gravity surveys were being made by hand using charts and tables, the most widely used being those published by Sigmund Hammer (1939) and known ever since as Hammer charts and Hammer tables. The charts, printed on to transparent overlays at the appropriate map scales (Fig. 6.3b), were divided into compartments defined by circular zones and radial lines, for each of which the average elevation had to be estimated. For a description of this quite recent construction it is hard to improve on Hutton.

> Of all the methods of dividing the plan into a great number of small parts, I have found that to be the most convenient for computation, in which it is first divided into a number of rings by concentric circles and these again divided into a sufficient number of parts by radii drawn from a common centre ....

There are obvious differences between the two charts. Hutton was concerned with the N-S components of the horizontal gravity effect of the mountain, and he spaced his radial lines more closely when they were close to the N-S lines through the observatories than when they were at wide angles to them. He might also, without significant loss of accuracy, have allowed the widths of the zones to increase with distance from the observatory but either did not think of doing this or chose not to. Hammer, being interested in vertical 'g', gave his charts circular symmetry and progressively increased the zone widths to take advantage of the fact that coarser approximations could be used for more remote topography. Nonetheless, the basic concept is the same in both cases.

1970 The Woodlark Basin

Not every gravity survey needs terrain corrections, and they were avoided whenever possible in the days before computers were on every desk and digital terrain models could be downloaded from the internet. The rough magnitudes of the effects that could be expected from different sorts of terrain soon became well-known, and were, in many areas, recognised as being negligible. In the plains of the American mid-West or the Great Artesian Basin of Australia they could be, and almost always were, ignored. It is only in the most detailed surveys in such areas that the coverage must be so complete that 'g' has to be measured in some places where terrain effects are significant. New Guinea, however, boasts some of the most ferocious terrain in the world, and corrections must be made at every reading point for the pull of its mountains.

In the 1960s the labour-intensive business of calculating terrain-effects by hand was more suited to the (in those days) leisurely time scale of university research than to a government department that had every year to justify the money it paid its staff. I had never previously considered a return to university, but after completing the Eastern Papua surveys it began to seem a good idea. With impeccable timing, John Bruckshaw, Professor of Exploration Geophysics at Imperial College's Royal School of Mines, visited Australia in 1967 and his many former diploma students who had ended up working there were invited to a reception. I told him about what I had been doing, he was interested, the International Upper Mantle Project was just beginning, focussing on just the sort of rocks that formed the Papuan ophiolite, and by mid-1968 I was back in the UK. Uniquely among my fellow students, I arrived with my project defined, my fieldwork completed and my data partially processed. I had time to follow up on anything that seemed interesting, even if it was not quite necessary for the thesis.

Those were exciting times for geologists and geophysicists. The paper that brought plate tectonics to the world had been published by Fred Vine and Drummond Matthews only five years before and, although not everyone was yet a believer, the momentum was building. The geological paradigm was changing before our eyes, and the change was being greeted with especial enthusiasm in the southern hemisphere, where the evidence in favour of some form of continental drift had been so strong that its geologists had ignored the physicists and had always assumed that continents moved. To them, the ways in which the remains of similar land plants and animals were distributed across the widely separated southern continents could not be explained in any other way. All that was needed was a mechanism, and that is what sea-floor spreading was providing.[5]

I made my own, very minor, contribution to it all only because I was making terrain corrections. Today this is easy, with everyday access to computers that are thousands of times more powerful than Imperial College's CDC3600 mainframe into which I fed box after box of hand-punched cards defining the topography of Eastern Papua. The terrain that those cards represented had been laboriously written down on kilometre grids drawn on

---

[5]See Chap. 10 for a fuller account.

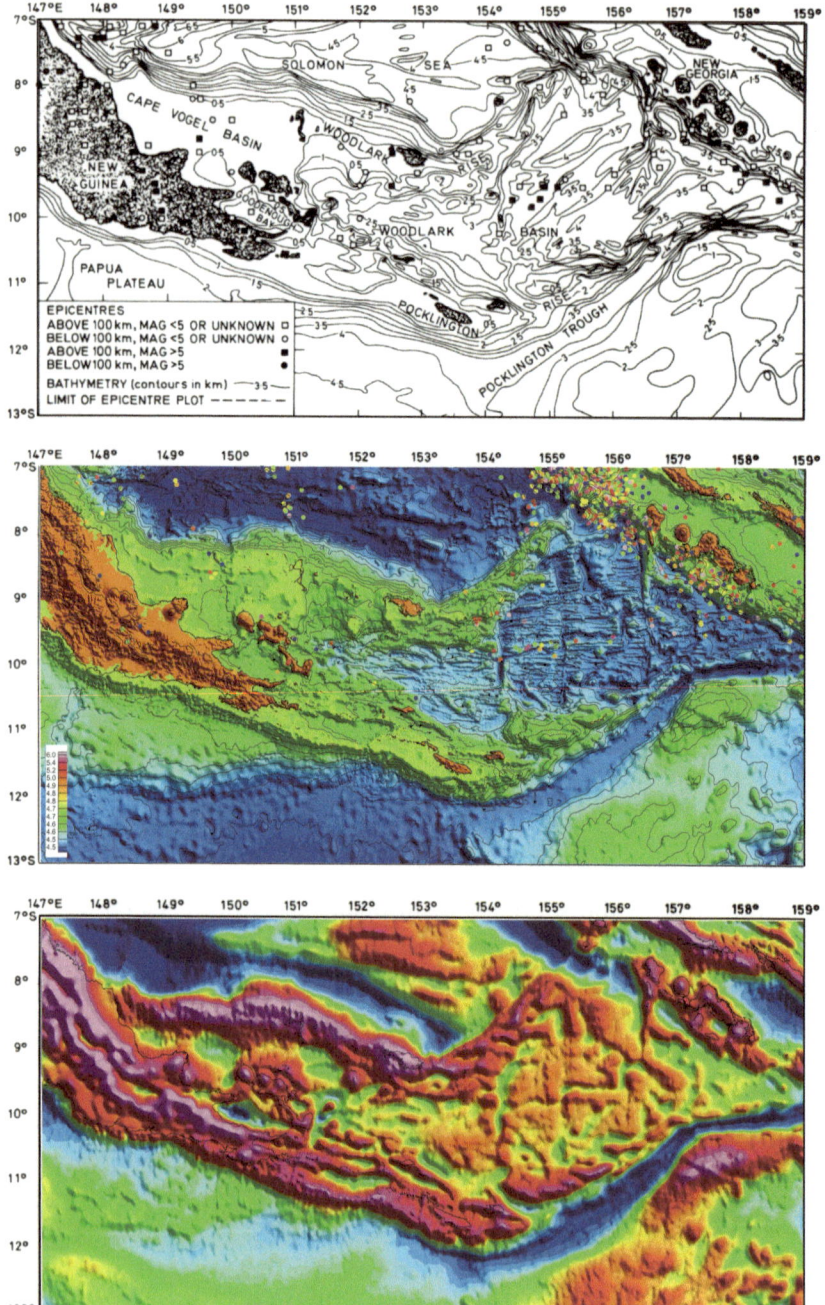

◀ **Fig. 6.4** The Woodlark Basin. Top: The 1970 version. Small, shallow earthquakes show where new sea floor is being created in the centre of the basin. The trough running from north to south between 154°E and 155°E is a fracture zone separating parts of the basin with different spreading rates. Centre: The 2010 version. The contour interval is 500 metres in both versions. Lower. Latitude-corrected (free-air) gravity, showing the strongly rectilinear trends in the eastern basin

tracing paper and overlain on topographic maps. A kilometre grid was really much too coarse for the corrections that had to be made, but the maps were not very good either.

There was another problem. The subsea topography that might also affect the gravity field was not shown on ordinary maps. For shallow waters there were Admiralty charts, but published soundings in deep water were few and far between. As a forlorn hope, I contacted the Australian Navy's Hydrographic Office, not really believing that that they might have anything better, and received an astonishing reply. A detailed echo-sounding survey of parts of the Woodlark Basin, one of the main gaps in my data, had just been completed and, with rare generosity, the hydrographers were happy for me to take the raw information and process it, and even to publish it at the sort of scale I needed. Even by hand it took only a few days to make a bathymetric map of the entire basin.

As far as terrain corrections were concerned, it was time ill-spent. When calculated, the effects proved to be far too small to make any significant difference to the very large, geologically-created gradients that dominated the gravity field. However, the work had been done and the map (Fig. 6.4: Top) went into the thesis to prove it. Its place there had to be justified, and it was only when I tried to do this that I realised that what the Navy had mapped had to be a very small and young basin extending from Eastern Papua to the Solomon Islands, within which oceanic crust was being created in exactly the way proposed by the new theory. What was even more remarkable, at the extreme eastern end of the rift the newly formed crust was almost immediately descending back into the Earth's interior at the deep trench along the southwestern flank of the Solomons. This was an idea that was publishable, and in the end it was published (Milsom 1970).

Other workers came later to the Woodlark and measured magnetic fields as well as water depths, and used that information to estimate the age of the sea floor and the rate at which it was being produced. Ultimately, the basin was selected as the site of Leg 180 of the Ocean Drilling Project, designed to provide a fuller understanding of the change from continental rifting, at the extreme eastern end of the Papuan Peninsula, to sea-floor spreading. It is one of the very few places in the world where this transition can be seen happening at the present day. My own role in all this was tiny, but it can stand as an example of the way in which science has advanced in many far more significant instances.

Firstly, the time was right. Sam Carey, then Professor of Geology at the University of Tasmania, had had a similar idea 30 years before, but sea-floor spreading had not been thought of, and no-one took any notice. By 1970, if I had not noticed what was going on, someone else would have, and very soon. Secondly, it was accidental. I was not even thinking about plate tectonics when I made the map, but it was impossible to ignore the deep N-S trough cutting across an area where every other sea-floor feature was trending east-west. The only reasonable explanation for that was that it was one of the fracture zones that were just then being recognised as essential parts of the sea-floor spreading story. It is easy to believe that Galileo noticed a swinging lamp and then went on to study pendulums because minds work by being triggered. If the story of Newton and the apple is not believed, it is not because it is, in its own right, improbable, but because the timing is wrong and the possible motives for the telling are only too clear.

There is a third point. The Australian Navy had concentrated on the western part of the basin, and the contours in the eastern part were based largely on conventional soundings. Logically, where there was doubt (and there was plenty of that), I drew them with trends more or less parallel to the basin's northern and southern margins. The more recent data have shown this was completely wrong, and that the dominant trends are either north-south or east-west. It is not only the bathymetric map (Fig. 6.4: Centre) but still more obviously the map below it, which is of latitude-corrected 'g', that shows this. Nature is always able to surprise us.

The Himalaya

It might have been thought that Bouguer, La Condamine and their successors, and Maskelyne and his successors, had between them discovered all there was to know about the gravity effects of mountains. Yet after corrections have been made for all the effects they had considered, the region of lowest gravity in the Bouguer gravity map of eastern Papua in Fig. 2.4 coincides rather precisely with the highest mountains. Evidently something else is going on, but it was not until almost a hundred years after Maskelyne had returned from Schiehallion that this was suspected.

The internet abounds in conspiracy theories, and at least one of them involves Maskelyne in a supposed plot by the modern scientific establishment to conceal the fundamental truths of physics. As far as Maskelyne's work was concerned, the blogger who made these claims advanced two main reasons for thinking that the accepted story was untrue. The first of these lay in his assumption that the observatories were some two miles from the foot of the mountain. Had they been, then the effects would indeed have been too small to be measured, but anyone who had taken the trouble to read the original papers would have known that the observatories were half-way up the mountain, as Maskelyne had known that they had to be. Typically, the author of the blog had not made the simplest of checks before writing.

The second argument was more interesting, although even less scientific, and was by analogy. How, it was asked, could Maskelyne have found such a large effect at Schiehallion when Bouguer in the Andes had measured a much smaller effect, and one that was much less than he had expected, despite the much larger masses involved? Once again, of course, the importance of the locations of the observatories was being ignored, but there was another factor. One of the things that contributed to the Maskelyne's success was not that Schiehallion was large enough, but that it was small enough.

In the 1800s British surveyors were busy providing Queen Victoria with accurate maps of her empire in India. This required detailed and painstaking survey of the entire sub-continent, and the name of the Surveyor General, George Everest, is forever associated with the project. His surveyors knew very well that because of the attraction of mountains their vertical reference, the plumb line, was not quite as reliable as they needed it to be, but they were not going to be lured into making corrections based on suspect theory or error-prone calculations. Instead, and using astronomical techniques very similar to those used by Maskelyne, they measured the deviation of the vertical throughout India and found that it did indeed increase as they approached the Himalayas. It was, naturally, in the pages of the *Philosophical Transactions* that these results were discussed and in the fine tradition of early modern science it was a clergyman, John Pratt, the Archdeacon of Calcutta, who led the discussion, observing that:

> It is now well known that the attraction of the Himalaya Mountains, and of the elevated regions lying beyond them, has a sensible influence upon the plumb-line in North India. .... It has been found by triangulation that the difference of latitude between the two extreme stations of the northern division of the arc .... is 523' 42.294", whereas astronomical observations show a difference of 523' 37.058", which is 5.236" less than the former.
>
> That the geodetic operations are not in fault appears from this; that two bases, about seven miles long, at the extremities of the arc having been meas-

ured with the utmost care, and also the length of the northern base having been computed from the measured length of the southern one, through a chain of triangles stretching along the whole arc, about 370 miles in extent, the difference between the measured and the computed lengths of the northern base was only 0.6 of a foot, an error which would produce, even if wholly lying in the meridian, a difference of latitude no greater than 0.006″. The difference 5.236″ must therefore be attributed to some other cause than error in the geodetic operations ….

A very probable cause is the attraction of the superficial matter which lies in such abundance on the north of the Indian arc. … Whether this cause will account for the error in the difference of latitude in quantity, as well as in direction, remains to be considered, and is the question I propose to discuss in the present paper.

This was real science, not unsubstantiated blogging. The data had been collected, and the theory was going to be tested.

And the theory was wrong. Pratt calculated with enormous effort what the deviation should be, and then stated that:

The conclusion, then, to which I come is, that there is no way of reconciling the difference between the error in latitude deduced in Colonel EVEREST's work and the amount I have assigned to deflection of the plumb-line arising from attraction – and which, after careful re-examination, I am decidedly of the opinion is not far from the truth, either in defect or in excess – but by supposing, that the ellipticity which Colonel EVEREST uses in his calculations, although correct as a mean for the whole quadrant, is too large for the Indian arc. This hypothesis appears to account for the difference most satisfactorily. The whole subject, however, deserves careful examination; as no anomaly should, if possible, remain unexplained in a work conducted with such care, labour, and ability, as the measurement of the Indian arc has exhibited.

This first contribution[6] was written by Pratt in Capetown in July 1854, when he was on his way either to or from his day-job in Calcutta. The sea voyage must have given him plenty of time to develop his ideas but, as with most of today's scientific publications, there was more going on than appeared in the pages of the journal. The paper was followed immediately by one from George Airy, the Astronomer Royal, who must have already been told about the discrepancy. While confessing that he had at first been astounded by it, Airy went on to say that, really, no-one should have been surprised. He

---

[6]The quotation forms the final paragraph of Pratt (1855).

pointed out that the strength of Earth materials was nowhere near enough to support the mass of the Himalayas, and that if they were not supported by strength, they must be supported by weakness. They must be floating on a fluid layer, in the same way as a visible iceberg is supported by the buoyancy forces acting on the nine-tenths that is below sea level (Airy 1855).

It took Pratt four years to reply (Pratt 1859). He then agreed with Airy that it must be buoyancy, not strength, that supported the mountains, but that was as far as his agreement went. He thought it improbable that the solid crust would be less dense than any fluid on which it could be supposed to rest, or that it would be at its thinnest beneath the deep oceans. To these intuitive arguments he added one that seemed to him conclusive. Airy's ideas, he pointed out, were incompatible with the estimate made by "*Mr Hopkins of Cambridge*" of an Earth's crust 800–1000 miles thick.

Pratt's deference to the opinions of a single individual might seem extraordinary, but there is reason behind his comment. Both he and Airy were graduates of Cambridge University, at a time when the only route to an honours degree, then a passport to a comfortable life in the academic world or the Church of England, was success in the Mathematical Tripos.[7] The highest classification in these examinations, equivalent to a modern first class degree, was that of 'Wrangler' (a good arguer) and Wranglers were ranked in order. In 1823 Airy graduated at the head of his year as Senior Wrangler, and ten years later Pratt graduated as Third Wrangler. Both men thus had impressive mathematical qualifications, but the uncomfortable fact was that mathematical teaching at Cambridge in the early to mid-19th Century was in a very poor state indeed. National pride was still preventing the advantages of Leibniz's calculus over Newton's more difficult method of 'fluxions' from being recognised, and geometry was still seen as the bedrock of all mathematics. Where similar propositions could be proven both geometrically and algebraically, the algebraic proofs were thought to be in some way inferior. Moreover, the teaching provided by the university was generally appalling. Few of the fellows and professors taught very much and some did not teach at all, and the vital tuition without which wrangler status could not be hoped for was provided by private tutors. By the time Pratt arrived in Cambridge, the most famous and successful of these, and the man that he chose to guide his own studies, was William Hopkins.

---

[7]The name 'Tripos' supposedly derives from the three-legged stool on which candidates had to sit while being orally examined, and not from its division into three parts (even though sometimes it was).

Hopkins had graduated in 1827 as 'only' Seventh Wrangler but was an inspiring teacher. His marriage shortly before graduation had made him ineligible for a university fellowship, but his decision to become a private tutor was eventually a profitable one. It also gave him time to follow a very wide range of interests, and one of greatest of these was geology. In 1851 he was elected president of the Geological Society and three years later, because of his geological work, he was made president of the British Association for the Advancement of Science. He remained, however, essentially, a mathematician, and it was as a mathematician that he attempted to determine the internal structure of the Earth from the small irregularities in its motion caused by the varying gravitational pulls of the Sun and the Moon. It was a brave attempt, but a flawed one. As one of his most successful pupils, William Thomson, was to write to the even more notable George Stokes concerning Hopkin's published work, *so far as the mathematical problems he attacks are concerned, they are all wrong.*[8] These were, however, fortunate errors, because without them Pratt might never have developed his own theory, in which he attributed the mass deficiency beneath the mountains to density changes in the thick and rigid crust proposed by Hopkins. To illustrate his idea, he divided that crust into vertical prisms with slightly different densities, all terminating at a constant depth that he called the isopiestic level. If adjacent prisms had different densities, their elevations above the isopiestic level, and hence above or below sea level, had to be different (Fig. 6.5).

Given a real world of real rocks, neither the Pratt nor the Airy mechanism for what came to be called isostasy could possibly be exactly true, and neither author claimed that it was. In particular, the 'pure' Airy principle must in practice be modified to take account of the finite strength of the crust and its ability to support small loads. Calculations based on the Airy-Heiskänen model, which incorporates a value for crustal rigidity, have shown that a small mountain such as Schiehallion can be largely supported whereas the Himalayas, or the Andes, cannot. That is why Schiehallion was a better place to weigh the Earth than Chimborazo.

---

[8]In a letter of 1862. This quotation, and most of the other information about Hopkins, is taken from Craik (2008).

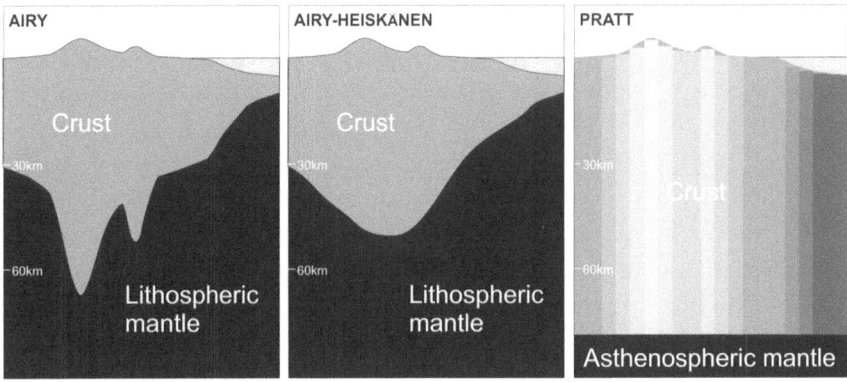

Fig. 6.5 Isostatic mechanisms. The same mountain mass, with heights up to 5 km, is supported by, from left to right, Airy point-for-point compensation, Airy-Heiskänen distributed compensation and Pratt variable density (with lighter shading to show lower density). The vertical exaggeration is roughly 50:1. Airy compensation implies that the topography of the base of the crust should be an approximately 6 times exaggerated mirror-image of the land topography but, because of the water layer, only a 4 times exaggeration of the seafloor topography. The base of the Pratt crust (the isopiestic level) is shown at 75 km, but the isostatic corrections work best if it is set at a little over 100 km

The Proofs

Although the Pratt and Airy models were very different, their gravity effects were too similar for a choice to be made between them, and there the matter rested for some fifty years. It was then not gravity but seismology that seemed to provide the definitive answer.

In 1909 a major earthquake shook the South Slav lands of Austria-Hungary and the arrival times of the main shocks at seismographs scattered throughout the area were analysed by a Croatian meteorologist named Mohorovičić. He concluded that the travel times of some of the waves could only be explained by supposing there to be an interface at an average depth of about 30 km at which their velocity increased abruptly. For almost all known types of rock this implied an increase in density. Since he also showed that this surface (the Mohorovičić Discontinuity, now inevitably known as the Moho) shallowed beneath the Adriatic and deepened beneath the Dinaric Alps, it seemed that Airy had been proven right, and Pratt wrong.

The final twist came fifty years later. It took the Second World War and its Cold War aftermath to persuade governments that the floors of the deep oceans were worth mapping (the better to hide their submarines). Previously it had been known that some parts of the oceans were much shallower than others, but the existence of a world-wide system of continuous ocean ridges was barely even suspected. Once this had been established it became one of the key pieces of evidence that went into the formulation of Plate Tectonics. In this theory the rift valleys at the crests of the ridges are the places at which new crust and upper mantle are being formed, and this material, which is very different from the material that forms the continents, slowly cools as it moves away from the central rift. Other things being equal, hot rocks are lighter than cold rocks, and the mid-ocean ridges are high because they are hot. The broad topography of the oceans is thus determined by isostasy according to a mechanism very similar to that proposed by Pratt. He, as well as Airy, had been right and for very large parts of the globe.

The Re-emergence of Finland

Weikko Heiskänen, who modified George Airy's isostatic model to allow for crustal rigidity, was a Finn. With such a name he could scarcely be anything else. The history of isostasy is peppered with similar names, and the Isostatic Institute of the International Association of Geodesy (IAG), established in 1936 and now incorporated into the Finnish Geospatial Research Institute, was located near Helsinki. All of which might seem a little odd, because isostasy was initially recognised by measuring the gravity effects of mountains and Finland is not a mountainous land. The glaciations of the last 70,000 years have scraped it almost flat, and beneath a thin and patchy cover of glacial soil the bedrock is everywhere at least 1700 million years old. Despite the almost constant crustal thickness this implies, many of the great names in 20th Century isostatic studies have been distinctively Finnish, and there are reasons for this.

The Isostatic Institute is where it is because the Finns, when they set out to do something, do it properly, and the levelling survey made of the then Tsarist Grand Duchy between 1892 and 1910 was one of the most accurate of its time. Not content with that, in 1937 the independent Finland decided to do it all over again. The differences between the two surveys were large, and very systematic. They showed that during the interval between the two surveys the whole of the country had been rising, with peak rates of more than 9 mm a year near the northern end of the Gulf of Bothnia. Even in the south, total uplifts of more than 20 cm were recorded around Turku on islands and skerries that were only just emerging from the sea. This pattern is a consequence of the Ice Ages, and their ending. The entire country, relieved of the burden of the ice that covered it during the last glaciation, is rebound-

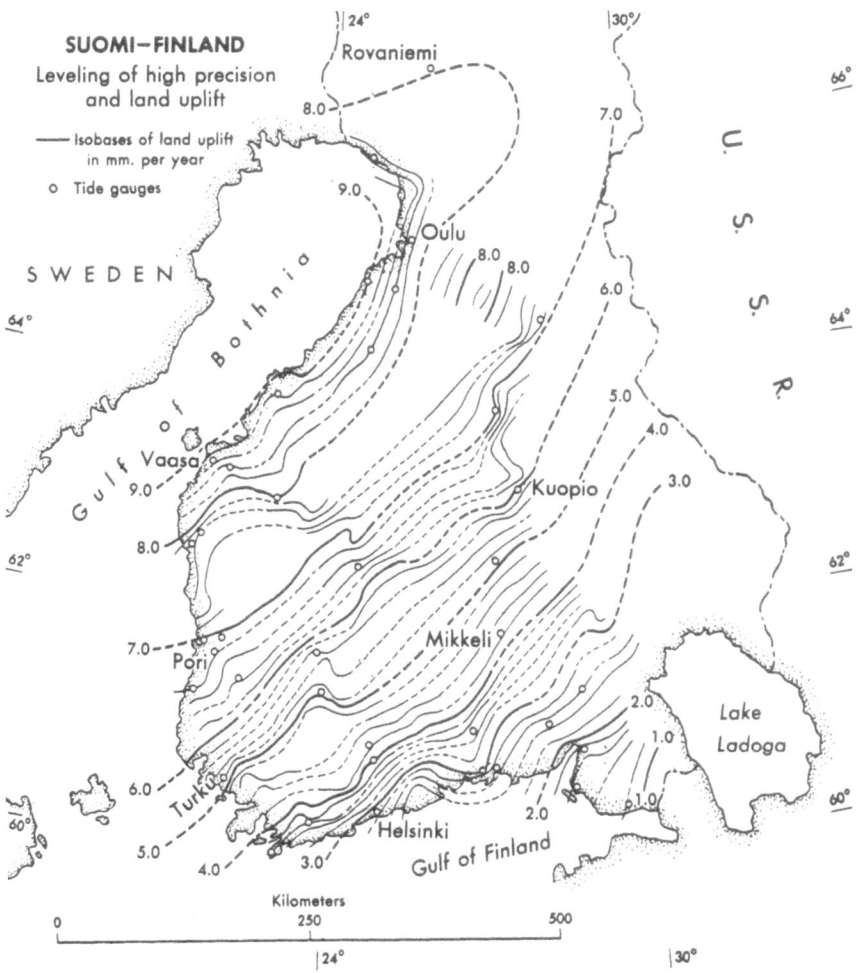

Fig. 6.6 Land uplift in Finland 1892–1910/1937–1939 (Niskänen 1939) Reproduced by permission of the Finnish Geospatial Research Institute

ing. The zero line of uplift is reached near St. Petersburg, and beyond that there is evidence from Russia for modest amounts of subsidence (Fig. 6.6).

Longer-term estimates have also been made using the locations of old shore lines, and Charles Lyell, one of the founders of British, and hence global, geological science, devoted his 1835 Bakerian lecture (Lyell 1835) to observations he made in southern and central Sweden. There he found marine fossils typical of the Baltic in beds ten or twenty metres above the present sea level, testifying to the reality of an uplift of which he had continued, in common with many others, to entertain some doubts ... partly because it appeared to me improbable that such great effects of subterranean expansion should take place in countries which, like Sweden and Norway, have been remarkably free within the times of history from violent earthquakes

but he offered no explanation. The latest studies suggest that from the time the ice began to disappear, some 12,000 years ago, the maximum uplift has been over 500 metres, in response to the removal of 2000–2500 metres of ice. The values of 'g' in this area show that the process, which should bring 'g' close to zero, is not yet complete, and the northern end of the Gulf of Bothnia may have almost 200 metres to go (Niskänen 1939). By geological standards these are very fast changes indeed, and are providing information on the mechanical properties of the crust and uppermost mantle in old, stable continental areas.

It is of no scientific significance whatsoever that the maximum uplift occurs very close to the place where, three hundred years before the start of the second levelling survey, Maupertuis was writing bad love poetry to Kristina Planström.

References

Airy GB (1855) On the computation of the effect of the attraction of mountain-masses, as disturbing the apparent astronomical latitude of stations in geodetic surveys. Phil Trans Roy Soc 145:101–104

Bennet J (1998) Derek Howse obituary. The Independent, Tuesday, 4 Aug, 1998

Bouguer P (1749) La Figure de la Terre. Jombert, Paris

Craik A (2008) Mr. Hopkin's men. Springer, London

de La Condamine C-M (1751) Journal du voyage fait par ordre du Roi a l'équateur, servant d'introduction historique a la mesure des trois premiers degres du méridien, Imprimerie Royale, Paris

Hammer S (1939) Terrain corrections for gravimeter stations. Geophysics 4:184–194

Howse D (1989) Nevil Maskelyne: the seaman's astronomer. Cambridge University Press, Cambridge

Hutton C (1778) An account of the calculations made from the survey and measures taken at Schehallien, in order to ascertain the mean density of the earth. Philos Trans R Soc 68:689–788

Lyell C (1835) On the proofs of a gradual rising of the land in certain parts of Sweden. Philos Trans R Soc 125(1):38

Maskelyne M (1775) A proposal for measuring the attraction of some hill in this Kingdom by astronomical observations. Philos Trans R Soc 65:495–499

Milsom J (1970) Woodlark Basin, a minor center of sea-floor spreading in Melanesia. J Geophys Res 75:7335–7339

Newton IS (1731) A treatise on the system of the world. English edition: Fayram F, London (1685) Newton IS. De mundi sysemate (trans: unknown)

Niskänen E (1939) On the upheaval of land in Fennoscandia. Isostatic Institute, Helsinki

Pratt JH (1855) On the attraction of the Himalaya mountains, and of the elevated regions beyond them, upon the plumb-line in India. Philos Trans R Soc 145:53–100

Pratt JH (1859) On the deflection of the plumb-line in India, caused by the attraction of the Himalaya Mountains and of the elevated regions beyond; and its modification by the compensating effect of a deficiency of matter below the mountain mass. Philos Trans R Soc 149:745–748

Smallwood JR (2007) Maskelyne's Schiehallion experiment revisited. Scott J Geol 43:15–31

Smallwood JR (2010) Bouguer redeemed: the successful 1737–1740 gravity experiments on Pichincha and Chimborazo. Earth Sci Hist 29:1–25

Sobel D, Andrews W (1998) The illustrated longitude. Fourth Estate, London

7

The Pitfalls of Pendulums

It was Christiaan Huygens who worked out the theory governing the movements of pendulums and showed that 'g' could be measured by dividing the length of a pendulum by the square of its period (Chap. 14, Coda 3). A pendulum that beats in seconds is slightly less than a metre long, and as recently as the start of the 20th Century the value of 'g' was still being quoted in terms of this length. The experimental skill needed is daunting. A change in length of one millimetre corresponds to a change in 'g' of almost exactly a thousand milligals, and the variation in gravity over the Earth's surface, from the poles to the summit of Everest or Chimborazo, is only about 7000 milligals. Milligal accuracy demands measurements of length to one part in a million (a thousandth of a millimetre) and of period to one part in two million (Chap. 14, Coda 6). It would be not be reasonable to expect that sort of accuracy from the scientists who, in the first quarter of the 19th Century, began the global mapping of gravity, but how well did they actually do?

Releasing the Pendulum

In 1672 Jean Richer discovered that gravity varied with place when his Parisian pendulum clocks ran slow near the equator. For clockmakers the effects produced by escapement mechanisms, by air resistance and buoyancy, by distortion and friction at the support, by stretching of pendulum shafts and by changes in gravity are not individually interesting. The relationships between pendulum periods and the actual mass distributions of bob and

© Springer International Publishing AG, part of Springer Nature 2018
J. Milsom, *The Hunt for Earth Gravity*,
https://doi.org/10.1007/978-3-319-74959-4_7

shaft are of even less concern, since they are dealt with in designs that owe as much to experience as mathematics. Periods were (and still are) adjusted by rotating finely machined screws to raise or lower the weights and persuade the clocks to tell the right time, but once people had become interested in accurately measuring 'g', rather than just comparing its values in different places, ordinary clocks were no longer good enough. The pendulums had to swing free.

An ideal simple pendulum would consist of a heavy but infinitely small weight or bob supported by a weightless but infinitely strong thread, but even if these things were possible the period would vary slightly with the angle of swing (the amplitude). The smaller the swing, the smaller the corrections required and this, together with the need for free swinging, drastically limits the usable observation times. For small swings, friction at the supports is more important than air resistance in gradually bringing pendulums to rest, and can be minimised by replacing the threads with rigid shafts pivoting on triangular steel knife edges resting on agate plates. Inevitably, the shafts and knife edges have mass, and the pendulums are no longer 'simple' but 'compound'. Huygens showed that his equation could still be used for such pendulums if in it the length was replaced by the distance between the pivot and a 'centre of oscillation'. In principle the position of this point could be calculated from the positions and masses of all the various components, but these would then have to be known in minute detail; it would have been this that ultimately limited the accuracy of the estimates of 'g' had not an ingenious way been found to avoid the issue, using another of Huygens' discoveries. He showed that if a pendulum were made with two opposed knife edges, each located at what would be the centre of oscillation were it to be supported on the other, then the periods of oscillation would be the same. To calculate 'g' using such a pendulum it would only be necessary to measure this period and the distance between the knife edges. The idea was suggested independently in Germany and France, but it was a British army officer who first put it into practice.

Kater

In 1802 Major William Lambton began the Great Trigonometric Survey of India when he measured twelve kilometres of primary baseline across a flat plain near Madras, and sent one of his subordinates, a Lieutenant Henry Kater, to find suitable vantage points from which to connect it to the coastal towns of Tellicherry and Cannanore. Lambton went on to become the first

Superintendent of the India Survey, a position only formalised in 1818, and in which he was succeeded in 1823 by the much better remembered George Everest. Kater, after six years working under Lambton, became ill (almost the norm for service in India) and returned to England. He retired from the 12th Regiment of Foot in 1814 as a half-pay captain and his surveying expertise was then seized upon by the Royal Society. In 1817 he was contracted by the House of Commons, via one of the Society's committees, to measure the length in London of the seconds pendulum.

In measuring 'g', Great Britain was lagging well behind France. The Paris Observatory had been founded in 1671, the year before Richer went to Cayenne, and in 1793 no fewer than seventeen pendulums were located during an audit of its instruments by the Ministry of the Interior. Kater, however, did much more than just make a few measurements in London instead of Paris. He built the world's first practical reversible pendulum. Although something similar had been suggested by the French mathematician Gaspard de Prony about twenty years earlier, his design was probably never used and, as Kater went to some lengths to point out, in a long footnote to his report to the Royal Society (Kater 1818), the Frenchman had not really understood what he was doing. What made Kater's pendulum reasonably practical was that rather than using the obvious method of altering the periods by adjusting the positions of the knife edges, he fixed the knife edges and then equalised the periods by making tiny changes in the position a small weight that could be slid along the shaft.

Kater's pendulum is shown in Fig. 7.1. The opposing knife edges are mounted near the ends of the metal shaft, with the 'Great Weight' located close to what would, in a conventional pendulum, be the lower end. During construction the small weight was moved along the bar until the two periods of oscillation were almost equal, and was then fixed firmly in place. A third weight, mounted on the slider near the centre of the bar, was then moved in very small steps by rotating a screw until the two periods were precisely equal. These increments were measured but their values were used only as aids in the equalisation process. The half-period was slightly more than one second, and to calculate 'g' it was enough to measure this and the fixed distance, of just over a metre, between the knife edges. Each 'run' began with the pendulum swinging through an arc of a little over one and a quarter degrees, reducing to about two-thirds of a degree by the end. The maximum displacements of the end-pieces were about two centimetres at the start of each run and one centimetre at the end.

The idea was a good one but the way in which Kater put it into practice was rather odd. At some stage in their school careers, aspiring physicists

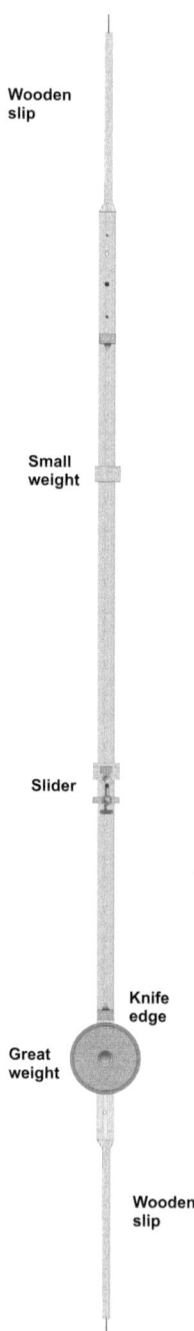

Fig. 7.1 Kater's reversible pendulum, based on an illustration in Kater (1818)

may well be asked to make and use a reversible pendulum and their designs are almost always roughly symmetrical. Kater's pendulum was very far from being so, and he never explained why. It may have been that he simply assembled it from bits that were readily to hand, and the advantage of asymmetry may only have emerged during use. This advantage, which does exist, is that the vibration time is very much less sensitive to movements of the slider in the 'normal' position, when the knife-edge further from the Great Weight is being used, than in the alternative or 'reverse', position. The reverse position can be used to make coarse adjustments and the normal position to make fine ones. This is so convenient that all subsequent designs have had asymmetrically-distributed masses, even though, as Sabine, Kater's immediate successor, pointed out, the various frictions and resistances could be very different in the two positions.

Kater was in many ways an exemplary experimentalist. He measured the distance between his knife edges against three different national standards, and before and after each experiment to ensure that no changes had taken place. He investigated the expansion with temperature and made the appropriate corrections (although, slightly oddly, he nowhere shows them actually being applied). He used the best available microscopes to make the fine measurements of length and investigated in detail the errors that might be introduced by variations in the pitch of the screw that adjusted their positions. He made experiments on the effects of air resistance and buoyancy. He used an exceptionally hard high-carbon steel he called *wootz* (which he might have come across during his service in South India, where it had been manufactured for more than a thousand years) for his knife edges. His belief that his error in the distance between the support points was less than one ten-thousandth of an inch, or a few parts per million, was possibly too optimistic but not excessively so.

For timing Kater used a pendulum clock that had been made by John Arnold (probably John Roger, who carried on the family business after his father, the original John Arnold, died in 1799), and this he calibrated against astronomical observations and another clock that had belonged to General William Roy, famous as the man who mapped Scotland. Both clocks were capable of keeping time with a gain or loss of no more than one second in a day. Kater pinned a white disc to the pendulum shaft of the Arnold clock, and shaped the narrow wooden end-pieces or 'slips' on his reversible pendulum to exactly cover the disc when both pendulums, observed through a telescope, were vertical and moving in the same direction. He called these events 'coincidences', and a half-period swing, in either direction, a 'vibration', and he began each experiment in such a way that the first coincidence occurred in the first few seconds. Why he chose to use a disc rather than a rectangle, of which much more would have been visible except at a coincidence, we shall never know. The first time the pendulums came into view after the coincidence there would have been only a very tiny sliver of the white disc appearing ahead of the obscuring 'slip', and it must have been quite difficult to see.

For the next hundred or so swings there would be progressively more of the disc showing, but beyond the quarter-period point the visible area would begin to decrease. The two pendulums would again be simultaneously vertical and coincident when the seconds pendulum had completed one vibration more than the reversible pendulum, but they would be moving in opposite directions and it would be impossible to identify the moment with any confidence. Half a period later the disc would again be totally obscured, and this time the pendulums would be moving in the same direction. This would be a measurable coincidence. The clock pendulum, beating slightly faster, would then have overtaken the reversible pendulum by a full two seconds, and it was the clock time, measured to the nearest second, that Kater recorded. He never actually bothered to count the number of vibrations of the reversible pendulum but simply subtracted two from the seconds of the clock time. The method was similar in principle to that used by Bouguer almost a hundred years earlier and by his successors in France, but failed to incorporate one item that had contributed significantly to Bouguer's success. In Kater's experiments only complete swings were counted. There was no graduated scale.

Because Kater decided that he wanted to observe at least four coincidences before the pendulum needed a fresh push, and because the swings were sufficiently visible for about 2000 seconds (slightly more than half an hour), the intervals between coincidences had to be about 500 seconds and the knife edges had to be about thirty-nine and a half inches apart. The four measurements were not, of course, independent and except for the small corrections for the finite arc of swing the analysis could equally well have been done using only the total time. Even with the four successive intervals taken together, it might seem that Kater, measuring his times only to the nearest second, could have been estimating periods to no better than one part in two thousand, but because of his choice of the distance between the knife edges, he was actually able to approach an accuracy of one part in a hundred thousand (see Chap. 14, Coda 6).

The House of Mr. Browne

Although less accurate than he thought, Kater actually did rather well. 'London', as far as his experiments were concerned, was neither the Greenwich Observatory nor the rooms of the Royal Society, as might have been expected, but the London home of a Mr. Henry Browne, another Fellow of the Society, who

> …… most obligingly allowed me the use of his house, his excellent time-pieces, and transit instrument, assisting me with indefatigable zeal …….. The house is substantially built, and is situated in a part of Portland Place not liable to much disturbance from the passing of carriages.

Traffic in Portland Place has evidently increased considerably in the last two hundred year, but even in Kater's time it might not have been as peaceful as this extract suggests. Another house in Portland Place was the official residence of the Prussian representative in London, Wilhelm von Humboldt, and he was sometimes visited there by his more famous brother, Alexander. Andrea Wulf, Alexander's biographer, notes that in the winter of 1817 *the area around Portland Place …. was one great building site because architect John Nash was implementing his grand town-planning scheme ….. there was noise everywhere as old buildings were razed to make space for new broad streets* (Wulf 2015). The work was centred on Langham Place, and Mr. Browne's house, which Kater's description identifies as the original No. 2 Portland Place, would have been close to its epicentre. It has since been replaced by the modernist northwest extension to the main BBC complex and would have been only about fifteen minutes' walk from Kater's own home in York Gate, reinforcing the impression given by the way he constructed his pendulum that short-term convenience played a significant part in his decision making. Because of this eccentric choice, the site cannot be re-occupied exactly but the British Geological Survey has assigned a value of 981, 186.6 milligal to a point a few hundred yards away and at street level. This is only thirty-seven parts in a million more than the 981,150 milligal implied by Kater's estimated length of the seconds pendulum. Put another way, his error in the length of the seconds pendulum was less than one twenty-fifth of a millimetre.

Shetland to Wight

In 1816 François Arago, a leading member of the Académie des Sciences who became, thirty years later and for just six weeks, effectively President of France, approached the British Ordnance Survey with a proposal. He wanted to make use of the recently completed Trigonometrical Survey of Great Britain to extend the 'arc of meridian' already measured in France and Spain to the most northerly points in the British Isles. As well as its primary aim of improving the existing models of the shape of the Earth, the project would also allow the 'repeating circle' favoured by the French for measuring latitude and longitude to be compared with the British zenith sector. In addition to the repeating circle the Académie undertook to provide a pendulum and an experienced observer.

Arago was no stranger to this sort of work. In 1809, while unwisely attempting to extend the meridian arc into Spain he had been arrested and imprisoned on Formentara, had escaped to Algiers, had been captured by a Spanish corsair while attempting to return to France, had been imprisoned again, this time on the Spanish mainland, had been returned to Algiers at the request of the Bey, had been shipwrecked on the North African coast while again attempting to get back to France, had yet again embarked from Algiers for Marseille and had finally arrived there only to spend months in quarantine in the lazaretto. Whilst there he began a correspondence with Alexander von Humboldt that was to trigger a lifelong friendship. In Alexander's company he visited his brother Wilhelm in Portland Place and while there may well have met Kater and Henry Browne. Despite this ultimately happy outcome, Arago must have realised that it had really not been a good idea for a Frenchman to be lighting bonfires on hilltops in the Balearic Islands at a time when Napoleon's armies were invading Spain.[1] In 1816 there was peace between England and France but he understandably decided that this time he would leave the fieldwork to someone else, and chose his colleague in the work on the Balearics, Jean-Baptiste Biot. The Ordnance Survey agreed to provide the transport to Unst, the most northerly of the Shetland Islands, and entrusted the arrangements to Captain (eventually Major-General) Thomas Colby.

The combined operation, coming so soon after the carnage of Waterloo, should have been a monument to the new, peaceful and collaborative Europe, but unfortunately the two principles decided instead to refight

---

[1] As a Catalan speaker (his family came from the Rousillon), he may have thought that he would not be noticed. His companions left the islands when the war began.

the Anglo-French wars. The fault almost certainly lay largely on the British side. Biot had something of a reputation amongst his scientific colleagues as a difficult man, but that is almost the scientific norm and, despite a poor command of English, he seems to have charmed most of the non-scientific people he met during his visit. Colby, who had been the driving force in completing the Trigonometrical Survey, was a fitness fanatic with a positive enthusiasm for rain, cold and discomfort, and for getting things done at speed. The Trigonometrical Survey had been his own particular project and he would probably have resented any intruders, and not just foreigners. The situation was made worse when the group was joined, unofficially, by the mathematician Olinthus Gregory, who definitely was a xenophobe.[2]

It was a bad start, and the conditions under which the large field party made their slow passage to Shetland in the cramped quarters of a seventy-six foot brig would have been anything but easy. So bad did relations become that the two principals made their observations in different places. Biot had originally wanted to work in the comparative luxury of Lerwick, but was persuaded by Colby to move north to Unst, where they found comfortable lodgings and an observatory site at the home of Thomas Edmondston of Buness, on the Balta Sound. This was, however, insufficiently spartan for Colby, who moved out to Balta Island at the head of the sound, where conditions were unpleasant enough to satisfy even him. The comparison between the British and French instruments was never made. When Colby had finished, he simply sailed south, leaving Biot to complete the last of his thirty-one days of pendulum observations and then to spend a fortnight finding transport back to Leith. The House of Commons, recognising that under the circumstances the results might well be unreliable, but still interested, decided to make more pendulum measurements, not only on Unst and at Leith but at four additional and widely separated stations of the Trigonometrical Survey, ending on the Isle of Wight. Naturally enough, they asked Kater to do it, and in 1819 he set out.

One obstacle to the wide use of reversible pendulums was that the periods could only be equalised by trial and error. This was just about acceptable when measurements were being made at one place only, but even then it may have been only the very coarse measurement to the nearest second that allowed Kater to get his results in a reasonable time. To repeat the whole business in many different places in impromptu observatories would have

---

[2]A more complete account of this calamitous early attempt at international scientific collaboration is given in Hewitt (2010) but the pendulum is barely mentioned. It is given much more prominence in Walker (2014). It is possible that Colby left Biot behind because he was not interested in the pendulum observations and saw no reason to wait for them to be finished.

been almost impossible, but he had a plan that drastically reduced the workload. Having already (as he believed) established a reliable value of 'g' in Portland Place, he made measurements at the same point with a similar but 'invariable' pendulum that could not be reversed or adjusted, and took that around the country. The ratios of the squares of the periods in London and at the selected sites gave the ratios of their gravity fields. So successful was this method that relative pendulums were still being used in the 1960s to calibrate gravity meter observations made over very large distances.

Accurate measurements of 'g' were of limited importance in the early 19th Century outside the narrow world of science, but nonetheless royalty, Parliament and the military all agreed with the Royal Society that they were good things to do. The Army, in particular, gave Kater its full support. He recorded that:

> A small light waggon was constructed at the Royal Arsenal at Woolwich … and a party consisting of a non-commissioned officer, two gunners, (one a carpenter), and two drivers with four horses of the Royal Artillery, was placed under my orders ….. His Royal Highness the Commander in Chief was pleased to direct that I should receive such military assistance as might be necessary for the safety of the instruments at the different stations … and an application being also made to the Admiralty for a vessel to convey me to the Shetland Islands, His Majesty's sloop of war the Cherokee, commanded by Capt. T. SMITH, was ordered to receive me at Leith, and bring me back to Scotland. (Kater 1819)

With the battle of Waterloo four years in the past, it was obviously proving difficult to find things for soldiers and sailors to do. Discipline in the Royal Navy had clearly gone to pieces since the glory days of Trafalgar and the Nile, and when Kater arrived in Leith he found *that the Cherokee had not been heard of for some time.* It would be interesting to know what a sloop captain was up to in those northern waters without the knowledge of either the harbour authorities at his home port or of the Admiralty. Happily, another sloop, the *Nimrod,* was in the harbour and, the order having gone out that Kater should be assisted in every possible way, this was commandeered for the trip.

The *Cherokee* caught up with the party after they reached Unst, and the *Nimrod* went home. Once ashore, Kater quickly located *the shell of an unfinished cottage nearby adjoining to the cow-house, in which the preceding summer M Biot had made his observations,* and proceeded to do the same, from 23 to 28 July. He then made his way south, by ship and road, to

Shanklin, occupying the four other sites on the way. By the time he arrived on the Isle of Wight it was November and the weather was so bad that he decided to wait until the spring before occupying the final station. The entire operation took almost a year, partly because, in the interests of accuracy, large numbers of measurements were made at each place. At Unst, 120 intervals, of about eight minutes each, were measured, in series of ten, and this was one of the shorter sequences. The Royal Society clearly felt that the level of royal support required some special acknowledgement, and Kater's report, running to no less than one hundred and eighty-one pages, must surely be one of the longest ever to appear in its *Transactions*. Every single observation was listed.

The measurement at Shanklin was almost Kater's last direct contribution to the study of 'g', but he never forgot the part of India where his surveying career had begun, and in 1821 he arranged for a relative pendulum "*precisely the same, in all its parts, as that used ..... at the different stations of the Trigonometrical Survey of England*" to be sent out to the East India Company's observatory in Madras (modern Chennai). The method still relied on the periodic obscuring of a white circle painted on the pendulum of a standard clock, and this caused John Goldingham, the observatory's astronomer, who made the measurements, a little difficulty.

> The clock-case was of handsome mahogany enriched with projecting mouldings, with the door in front of plate glass. The mouldings kept the pendulum at too great a distance from the part of the case where the arc could otherwise have been fastened, and it became necessary to have a support in front of the case. I therefore had a solid stand of teak wood made .... (Goldingham 1822) (Fig. 7.2)

Ensuring that the stand and the clock were rigidly attached to each other proved to be no simple matter, and in Goldingham's place I would have got rid of the moulding.

Goldingham also decided not to be content with a one second accuracy and "*therefore noticed the seconds, and parts of a second, when the disc disappeared, and also when it again appeared, both of which the Bramin put down*". He might have done this because he doubted Kater's claimed accuracy or simply because he noticed that "*there was a sensible period of time .... between the disappearance of the disc .... and its reappearance*". It was an important advance on Kater's technique, but it was not to be the last.

Fig. 7.2 Pendulum observations in Madras (Goldingham 1822). The job of the younger 'Bramin' assistant, Teroovencatachary, was "to count the clock, which he does with the greatest correctness". The head assistant, Senavassaehary, merely had to record the results

Love Among the Pendulums

England may have contributed some notable eccentrics to the history of gravity measurement but, almost inevitably, it was left to France to provide the romantic interest. This was not limited to Jean Godin and the epic journey made by his (admittedly Spanish) wife across South America. While Kater was mapping 'g' within the tight bounds of the British Isles, the French were already taking its measurement much further afield, and when, in 1817, they sent the corvette *Uranie* on a scientific voyage around the world, it carried no fewer than four invariable pendulums. These were not, however, the only unusual items taken. The captain, Louis de Freycinet, smuggled his young wife Rose on board (Fig. 7.3) and she remained with him throughout the voyage, to the wreck of the ship in the Falklands and to the couple's eventual return to France in November 1820. This flagrant disregard for naval regulations (women were not supposed to even set foot aboard French naval vessels at the time, let alone sail with them on long voyages) was not, it seems, a spur-of-the-moment decision. While the *Uranie*

Fig. 7.3 Rose and Louis de Freycinet. In these portraits, from the collection of the present Baron de Freycinet, they both appear to be in their early twenties, but Rose was actually fifteen years younger than her husband

was being fitted out, an extra storeroom was constructed on the after-deck at government expense. Once the ship was under way, it became Rose's cabin.

There are three first-hand accounts of the expedition, none of which tell the full story. De Freycinet was both captain and chief scientist, a combination that must have left him little time for any but the most essential note-taking. The first parts of his report, completed as a formidable document in eight volumes, were not published until 1826, six years after his return to France. One of these volumes was devoted entirely to the pendulum observations, which were mainly directed towards finding out whether the shape of the Earth was different in the southern and northern hemispheres (as had been suggested by some previous measurements in Capetown) and whether the lines of latitude were perfect circles.

Jacques Arago, the expedition's official artist and draftsman (and the brother of François), having more time on his hands than his captain during the long sea voyages, wrote more than a hundred and sixty letters to a friend describing his experiences. He completely ignored the boring work of gravity and magnetic measurement but provided vivid descriptions of the societies encountered at the various ports. He was particularly critical of slavery as he saw it in Brazil and elsewhere, both as an institution and because of the ways in which slaves were maltreated. Thirty years later his brother, when minister for both war and the colonies, was to secure its abolition throughout the French colonial empire. His was the first account to be published, in 1822. The Académie des Sciences, sponsors of the expedition, were obviously concerned by the absence of hard science in the diary and had it prefaced with a 27-page technical report that was nominally also due to Jacques Arago but was largely written, and signed, by seven others. Among them were the names, still familiar to any physicist, of Humboldt, Gay-Lussac and François' sometime colleague Biot.

Neither Louis de Freycinet nor Arago mentioned Rose directly in anything they wrote, but that did not stop her becoming a celebrity on her return. Although never officially acknowledged, her presence on board became widely known very soon after the *Uranie* left Toulon. Like Arago, she kept a diary in the form of letters to a friend, but these were not published until a hundred years later.[3] In them she makes slightly more mention than he did of the scientific work, but not always favourably. She was particularly incensed when, after a not very enjoyable stay in the Hawaiian Islands and at the start of the onward voyage to Sydney, where she (rightly) anticipated having a much more pleasant time ashore,[4] she discovered that …

> to my great regret, this part of our trip will be greatly lengthened by a prodigious detour to the east, which the dear commander has made, and which has as its aim research on the magnetic equator: I respect science very much, but I cannot really bring myself to like it ….

Sentiments, it can be said, that have been echoed down the years by many a wife of a field-based scientist faced with either a long separation or abandoning her own career for an expedition during which she might have too little to do and he would have too much. Coping better than most with this situation, Louis found a way to restore marital harmony by promising Rose that if ever they came upon an undiscovered island, he would name it after her. As luck would have it, they did just that. Rose Island, the southernmost fragment of land on the reef forming the uninhabited Rose Atoll in the Samoas, has the additional distinction of being the southernmost point above sea level in the territories of the United States of America.

Rose's diary is very different from Arago's. His was probably intended for publication whereas hers was not. Despite this, or perhaps because of it,

---

[3]Duplomb (1927). The book was re-issued in 2003, and the quotations that follow are translated from this edition. The original diary was purchased by the Library of New South Wales in 2014. Two English versions/commentaries have been published (Bassett 1962; Rivière 1996), but neither is easy to find.

[4]Louis had spent some time in Sydney in 1802 as a member of the Baudin expedition. His first command was the *Casuarina*, a 30 ton schooner purchased in Sydney by Baudin to chart possible harbours along the Australian coast. It was during this time he made his first visit to Shark Bay, which he mapped in great detail. His connection with Western Australia was commemorated when, in 2001, the WA Museum mounted an expedition to the Falklands to locate the remains of the *Uranie* (McCarthy 2008).

hers is in many ways the more useful. The entries are dated, and much more space is given to the nightmarish two months marooned in the Falklands that could so easily have ended badly, because Rose was clearly an exceptionally attractive young woman. A few years later, Gabriel Lafond, a man with an almost obsessive interest in shipwrecks who was making his own voyage around the world, met people who remembered the voyage of the *Uranie*. He is quoted by Bassett (1962) as saying, rather ungallantly, that while Rose herself never gave cause for a single adverse comment, she nonetheless constituted '*an apple of discord*' amongst the young men of the crew. In his opinion she would, by her mere presence, have been a threat to the discipline and good order necessary in a naval vessel.

That threat must have been even more acute during the long weeks when Rose was encamped in the midst of a hundred and twenty sailors on a desolate beach some twenty kilometres from the site of the present day (but then non-existent) township of Stanley. The wreck itself had been undramatic, with no lives lost and most of the papers and many of the scientific collections saved, but it did place her in the most vulnerable of positions. Despite this, and although her diary records miserable days of rain, terrible cold, awful food (almost entirely meat, of the pigs they had brought with them, of penguins and other seabirds, of the horses and cattle that had been introduced to the islands and allowed to run wild and, on one occasion, of a stranded 'hippopotamus'—possibly a walrus) and continually concern about Louis's health, which was clearly more fragile than her own, there is nothing in what she wrote that suggests any fears for her own personal safety. Whatever Lafond said, she must, in the preceding year and a half, have managed to gain the full acceptance and respect of her fellow castaways. They were certainly neither monks nor hermits, because Arago recorded that on Guam a disease ....

the ravages of which are most felt at Otaheite, at the Sandwich Islands, and even at Timor, has hardly been felt at the Mariannes. Our medical man met with no one instance of it; though our crew, by their imprudent confidence, often exposed themselves to its terrible effects. ... it was formerly well known here ... under the name of the French disease;

His rather censorious view of the crew's behaviour is undermined a few pages later, when he provides a vivid insight into how he himself had been spending his time on the island.

We are all on board; the anchor is to be weighed tomorrow at an early hour; it is past noon; to reach Agagna I must perform a journey of three leagues; and yet I hasten thither …. I wish to see, to hear her once more. If you had known her, you would pardon my weakness; and perhaps you would not pardon me for leaving her.

I arrived at Agagna, breathless, exhausted with fatigue; she also was still weeping. You will believe that there was sincerity in this attachment, when I shall have told you that this young woman was a savage. Oh! Yes, I was very wrong in returning to see her.[5]

Soon after the wreck, and with no immediate prospects of rescue, Louis decided to send the largest of the ship's boats to seek help. How the crew of five would have fared is unknowable, because at the time there was no settled population anywhere in the Falklands and the chosen destination was Montevideo, almost 2000 km to the north. En route they might have met with whaling or sealing vessels, of which there were many in those waters, but it would still have been a very long wait for the people they left behind. In the event, the addition of decking and other items needed for the journey took several weeks, and shortly before the boat was due to leave spirits were lifted by the arrival of an American whaler—and then dashed when its captain proved to be more interested in continuing his existing profitable business than in taking on the less well-paid role of rescuer. This depressing fact was just beginning to emerge when another American arrived, damaged and in need of repairs that the *Uranie*'s carpenters were able to carry out. The ship was flying the flag of the Spanish colonists in revolt against Spain, and was running guns from Buenos Aires to Valparaiso. After some fairly tense and often acrimonious negotiations, Louis chartered it for the journey, for an extortionate 18,000 dollars. That he had this amount available in letters of credit sheds some light on the sort of 'equipment' that had to be taken on such voyages. Once at sea he managed to buy the ship for an extra 2000 dollars, renamed it the *Physicienne*, deposited its original crew and cargo in Montevideo and sailed it back to France, pausing on the way for a second set of measurements at Rio de Janeiro.

On his return, Louis was court-martialled (and exonerated) for the loss of the *Uranie* but no action was taken over his flouting of naval regulations and misuse of naval funds where Rose was concerned. The French navy, the 'Royale', faced with the dual impossibilities of approving his (wildly popular) actions or subjecting him to a second court-martial, decided to turn a truly Nelsonian blind eye and ignore the whole affair. Nothing official was ever said, but some of the drawings made by the artists on the trip were

---

[5]Arago (1824). The extracts quoted are from letters 101 and 103. Three leagues would have been a journey of about ten kilometres. Each way.

circulated in two versions, one unofficial in which Rose appears and one approved, from which she is mysteriously omitted. A painting survives of the landing in East Timor in which she and the official greeting her are covered with light cross-hatching, presumably as a guide to the copyist as to which parts of the drawing should be omitted (Fig. 7.4).

In 1832, Louis contracted the cholera that was sweeping Paris. Thanks to Rose's devoted nursing he survived, but she then became ill herself and did not. She was just thirty-eight years old.

Sabine in the Arctic

Once Kater had abandoned pendulums and had instead become involved in the vexed questions surrounding the national standards of length and mass, it fell to another military man, Edward Sabine, to continue his gravity work. Now almost forgotten, Sabine was a major figure in 19th Century science. He spent much of the period between 1818 and 1823 travelling

Fig. 7.4 Rose arrives in Dili (Timor), as drawn by Alphonse Pellion, a talented midshipman who ended his career as a vice-admiral (McCarthy 2008). The faint cross-hatching over Rose and the Portuguese officer marks figures that were to be omitted from the official version. The drawing by Jacques Arago (see Frontespiece), the official draftsman to the expedition, shows Rose wearing very different clothes, and her description of the event suggests that it was Arago who was right

the world with invariable pendulums, and an island off the northeast coast of Greenland is still known as Little Pendulum Island. The adjacent larger island, originally Inner Pendulum Island, is now Sabine Island. For his first measurements, however, he used clocks.

Born in Dublin in 1788, Sabine joined the Army in 1803. He had a relatively uneventful early career in Gibraltar which, although under threat throughout the Peninsular War, was never actually besieged. In 1813, however, he was transferred to North America, where the British were fighting the Americans in the 'War of 1812'. It was an eventful journey, marked by capture by an American privateer and eventual rescue by a Royal Navy frigate. Once in Canada, and despite having taken an instant dislike to Quebec ("*a more wretched, narrow, filthy place I have rarely seen*"), he served with credit until the end of hostilities. On his return in 1816 to an England unprecedentedly at peace, he was elected to the Royal Society, where his elder brother Joseph was already a member. Since he had at the time no significant record of scientific work, the election may have been simply a manoeuvre to justify his appointment as astronomer to the Ross expedition that in 1818 was heading to Canada in search of a Northwest Passage to the Pacific.

There is an interesting, and only dimly glimpsed, back-story here. The elder Sabine, Joseph, was an enthusiastic amateur ornithologist and exactly the sort of Fellow that the President, Joseph Banks, liked to have in Royal Society. Edward (who was a member of the Linnean Society and after whom an Arctic gull is named) shared his brother's bird-watching interests, as did their brother-in-law, Mr. Henry Browne of Portland Place. These two, however, would have been sympathetic to the Society's rebels, who felt that the more mathematical sciences were being sidelined. Edward was an officer in the artillery, the most mathematically inclined part of the army since it required calculations of matters ranging from the trajectories of projectiles to the breaking strains of bridges, and Henry dabbled in both physics and astronomy. He had made his fortune as Chief of Affairs at the East India Company's settlement in Canton and had been able, on his return to England in 1795, to buy a large estate at North Mymms in Hertfordshire and the town house in Portland Place in which he installed his observatory. It may well have been his idea that Edward should join the Ross expedition, and he who arranged his election to the Royal Society and proposed the gravity and magnetic experiments. He forms the 'missing link' between Kater and Sabine.

The expedition (Fig. 7.5) was not a success. The ship entered the Davis Strait between Greenland and the Canadian islands in appalling condi-

Fig. 7.5 Ross in the Arctic. Given that the expedition consisted of two ships only, Ross' choice of this engraving to illustrate his official report (Ross 1819a) suggests a man well capable of seeing an entirely imaginary mountain range

tions, with visibility reduced to almost nothing by thick fog. On one of the rare occasions when the fog cleared, Ross (but no one else) saw a range of mountains blocking the way ahead (Ross 1819a) and turned back. Sabine was able to make only a small number of magnetic measurements (using an instrument borrowed from Henry Browne) and one estimate of the gravity field with a pendulum clock. It is arguable, and was argued at the time, that although ordinary pendulum clocks could not be used with any accuracy to measure 'g' itself, they were actually more reliable than free-swinging pendulums when it came to measuring the ratios of 'g' at different places. This was not the view of the French, who were strong supporters of relative pendulums because, although clocks could be observed for days rather than hours without any intervention on the part of the observer, the effects of the mechanisms that kept them going were uncertain. On a later, two-year, expedition Sabine took two invariant pendulums and two clocks and compared the results, and concluded that the French were right and that free-swinging pendulums performed better. This may, however, have been simply because, being observed over shorter time periods, there was a greater chance that they would produce results that were self-consistent when measured in whole numbers of seconds.

On his return to London Sabine wrote bitterly of his "*mortification at having come away from a place which I considered as the most interesting in the world for magnetic observations, and where my expectations had been raised to the highest pitch, without having had an opportunity of making them*". He protested long and loud and began an argument via pamphlet with Ross over which one of them had actually made the few measurements that had been recorded. In his younger days, Sabine seems to have conformed to Sheridan's earlier caricatures of choleric and impetuous Anglo-Irishmen, and his first blast (Sabine 1819) was almost as long as Ross' original report. The counter-attack (Ross 1819b) ran to 50 pages.

Whatever the rights and wrongs of the dispute, it was Sabine who was invited to join the next, two-ship, expedition to the Davies Strait. Ross was in disgrace for his failure to push forward, and had been replaced by William Parry, who had been his second-in-command. Presumably John Croker, the First Secretary of the Admiralty, after whom Ross had diplomatically named his mountains, was not best pleased when they turned out to be imaginary. This time Sabine took two pendulum clocks, but was able to make measurements only on Melville Island, where the ship was trapped in the ice and had to over-winter. He wrote that:

If any hope had been entertained of being able to do more during the winter than merely prepare for the return of more favourable weather, it was ended by the severity of the cold, far exceeding expectation, with which November set in. From this date until the close of March, the highest degree registered by a thermometer, suspended in the air, was +6° of FAHRENHEIT, and in no one of these months did the mean temperature rise above −18°

All that could usefully be done was identify a suitable site for the measurements and erect a shelter in preparation for the spring, when scientific work would at last became possible. Even that was attended by disaster.

The matting with which the outer walls were covered accidentally caught fire, and notwithstanding the endeavours of the persons who were present, the fire was communicated rapidly to the roof; it was fortunately extinguished by the exertions of the officers and men from the ships, before the clocks or any part of their apparatus had received injury …. an artilleryman … in his anxiety to place the instruments out of danger, exposed his hands incautiously, and was in consequence so severely frost bitten, as to render necessary the amputation of three fingers of the left hand, and two of the right.[6]

The unfortunate man could not have enjoyed the remainder of his stay, and the double pay that had been promised to all the expedition members must have seemed poor compensation.

The ships were so far north that the sun remained below the horizon for seventy-two days, but strenuous efforts were made throughout the winter to keep everyone amused. There were hunting parties, educational classes, and regular performances at the 'Theatre Royal'. Sabine edited a newspaper, *The North Georgia Gazette, and Winter Chronicle*, which was published in London when the ships returned and widely acclaimed. Despite the cold, the 'Dandies' (lower Frontespiece) seemed to have kept their sense of humour, and clearly had a better time of it than the crew of the *Uranie*, who were spending part of the same period stranded in the Falklands.

1960s Australia in the Antarctic

In 19th Century polar exploration, overwintering was common, even if not always intended. It is not clear from Sabine's accounts whether Parry had intended to stay in the Arctic all winter or, indeed, whether Ross had planned to return to England after a mere eight months at sea. In the 1960s, on the other hand, geophysicists who were seconded to the Australian National Antarctic Research Expedition (ANARE) knew in advance that they would be away for almost two years. They went south on ships that left Tasmania early enough in one year for them to do a useful summer of work, sat out the winter and came back the following year after another useful summer. Much of the work was in static observatories anyway, and could continue through the long night.

In many ways it was a good bargain for those who went. All living expenses were, of course, paid and all accommodation was provided. Moreover, there were large tax reliefs on the money earned whilst away so that, as with the crew of Parry's expedition, the winter boredom could be relieved by contemplating the financial package accumulating back home. For many of the more junior, it was a way of building up a nest egg with

---

[6]Both excerpts from Sabine (1821) which, despite a title suggesting greater things, dealt only with the two expeditions to Baffin's Bay.

which to begin married life. Whether the girl, or wife,[7] would still be around after two years was, of course, problematic. Sometimes she had disappeared, and sometimes the nest egg had disappeared with her.

The main problem was that during the winter it was dark almost all the time and there was actually very little to do. Typically, an observatory geophysicist might manage to spend a few hours a day changing paper charts in the recorders and taking a first look at the data, but only if he really dawdled over the job. Without digital recording and without computers portable enough to be shipped south, there was a limit to what else could be done. This was well known, and many went south with the intention of finishing their doctoral theses, or preparing the seminal paper on the Wooloomooloo Sandstone, or writing the great Australian novel, but often when they returned they had to admit that *mostly, we just slept*. Some found it difficult to adjust to life back in the real world, and some never managed it, and just went back again.

There was, during the 1960s, just one Australian attempt to measure 'g' in the Antarctic. It became the stuff of legend and, like all the best legends, is unverifiable and has lost nothing in the telling. It is said that the designated geophysicist was given the gravity task at the last minute, in addition to everything else that he had to do, and bitterly resented the fact. That, mysteriously, the instrument allocated to him was dropped from the deck of the ship during the unloading and shattered on the ice below. That when the manufacturers in the USA received it for repair, they laughed and declared the thing impossible. And that, on being told that the Australian government had no budget for its replacement but an unlimited budget for repairs, they retrieved from the tangled mess the metal plate on which the serial number had been stamped and attached that to a new instrument, and billed its full price as a repair. And that the paperwork passed through every financial scrutiny without a question being asked.

The Intrusion of Geology

In France gravity pendulums had been promoted quite specifically for use in determining the shape of the Earth, and it was with this goal in mind that measurements were made by Bouguer in South America and

---

[7]It was to be many years before ANARE felt able to take the revolutionary step of allowing women to go south, leaving husbands or boy-friends behind.

Maupertuis in Lapland. It is this idea that underpins the stories of Arago's arrest on Formentara and Biot's ill-fated trip to Shetland. It was the reason why de Freycinet was so anxious to measure gravity in both Capetown and Sydney, towns at almost exactly the same latitude where he hoped to discover whether or not the parallels of latitude were indeed perfect circles. In England, however, there was a growing suspicion that there might be local reasons for changes in 'g' that would make the idea unusable. One of the first people to put these doubts down on paper was a British naval officer named Basil Hall.[8] In 1820 he was commanding the 20-gun *HMS Conway* on attachment to Thomas Hardy's South America squadron tasked with protecting British interests, and British subjects, on a continent in revolt. Measuring 'g' was not part of his duties but he had with him one of Kater's invariable pendulums and found the time to use it on Isla Pinta in the Galapagos, at San Blas on the west coast of Mexico and at Rio de Janeiro. When discussing the Galapagos, he added the recommendation that measurements be made

> with the same pendulum at stations remote from the Galapagos, but resembling them in insular situation, in size and in geological character; such as the Azores, the Canaries, St Helena, the Isle of France, and various stations amongst the eastern islands of the Indian and Pacific oceans.

This is not the first printed suggestion that geology might have an effect on gravity, but it is one of the clearest, and is well ahead of its time in recognising the unusual environment of isolated oceanic islands. Because de Freycinet did not publish his results until 1826, Hall did not know that 'g' had already been measured at Mauritius (the 'Isle of France'), Guam and Hawaii before he himself arrived in the Galapagos, and that its value in those places had been unexpectedly high.[9] In his own personal diary he also recommended that measurements should be made in the Falklands, not knowing that de Freycinet had done this also (the geological effects meas-

---

[8]Hall's father had been a notable early geologist, and was for a time President of the Royal Society of Edinburgh. His son followed in his scientific footsteps, becoming a Fellow of the Royal Society in 1816.

[9]Hall was far-sighted in other respects. When describing his first, unsuccessful, attempt to measure 'g' in Valparaiso, where he ran out of time, he drew attention to '*the advantage which … would arise from having the whole experiment performed in England, by the person who is afterwards to repeat it abroad, not under the* hospitable *roof of Mr. BROWNE ….. but in the fields, and with no advantages save those he could carry with him. He would thus in good time discover omissions in his apparatus, which are not to be supplied abroad, and be aided in surmounting difficulties before he had sailed, as I did, beyond the reach of appeal*' (Hall 1823). His advice is as relevant today as when it was first written down, and all too often ignored.

ured by Kater, Sabine, Hall and de Freycinet are discussed more fully at the end of Chap. 14, Coda 6).

In 1822, while Hall was still in South America, Sabine went back to sea in the 18-gun sloop *Pheasant*, taking pendulums to Portuguese and British possessions on both sides of the Atlantic. His voyage was less dramatic than that of the *Uranie*, but pendulum measurements were very much the main objective and the requirements at the landfalls were correspondingly more exacting. In established ports there was generally little difficulty in finding buildings that could be turned into observatories, but where necessary Sabine was not above invoking the power of the British Crown. Faced with initial obstruction on São Tomé, then ruled by a three-man junta unsure as to whether it was or was not in revolt against Portugal, he

..... addressed a letter directly to the Junta; in which, after recapitulating the circumstances, and referring to the presence of a ship of war as sufficiently indicating the interest of the British Government, I requested, in the event of the Junta persisting in a refusal, its communication in writing; as Captain Clavering would not feel justified in quitting the island without an official document, which should enable the affair to be brought in due course under the consideration of the Court of Portugal, with which it would rest to judge between the Government of St. Thomas and the Marquess de Souza, and to decide by which of these authorities the request of the British Government, communicated with all due formality, had been frustrated.[10]

The thinly-veiled threats were effective, but for some of the crew the letter was a death sentence. Europeans venturing into the tropics in the early 19th Century were putting their lives seriously at risk, and all three of the marines assigned to the pendulum party on São Tomé fell ill and died before the work was completed. Sabine noted that, when these fatalities were added to that of a marine who had assisted him in Sierra Leone, it had been his misfortune

.... to witness the death of every individual landed for my assistance in Africa, with the exception of my servant, whose recovery from a relapse which occurred at St. Thomas's, was long very doubtful; it will readily be imagined,

---

[10]Sabine (1825). This monograph is now very hard to find. It was printed in London by John Murray at the expense not of the Royal Society but of the Board of Longitude, and the conflict between the Board and the Duke of Wellington's office over who should actually foot the bill might account for the small print run. The results were reproduced in Airy (1826) but without any details of the experiments.

that we rejoiced in departure from a climate, which has shewn itself so generally fatal to European life.

A visit to Jamaica later in the voyage passed off without fatalities aboard the *Pheasant* but the crew were appalled by the death rate amongst the garrison, at that moment very high even by Caribbean standards. Sabine commented, with true military understatement, that the loss of life *certainly appeared very considerable to persons unhabituated to the great and almost unceasing mortality of the West India Islands.* It must have been with great relief that he left Jamaica for New York,[11] and it is entirely understandable that he should have written from there to London suggesting that any expedition mounted in the following year should be to the Arctic. The navigational hazards would be great, but the risk of dying of fever would be very much less.[12] His letter bore fruit, and on the 11th of May 1823 he set sail for Norway, in a vessel that he already knew well from his early visits to northern waters. It was the *Griper*, an unlovely name for an unlovely ship whose sole merit was that its hull had been specially strengthened for work amongst the ice.

By the end of 1823 Sabine was back in London, having in two years measured 'g' in twelve places ranging in latitude from Bahia, twelve degrees south of the equator, to Spitzbergen, just ten degrees from the North Pole. He used his results to prepare a table (Fig. 7.6) in which he correlated 'excess' gravity with the local geology, and from which he felt able to conclude that

The scale afforded by the pendulum for measuring the intensities of local attraction, appears to be sufficiently extensive, to render it an instrument of possible utility in inquiries of a purely geological nature. It has been seen that the rate of a pendulum may be ascertained by proper care to a single tenth of a vibration per diem; whilst the variation of rate, occasioned by the geological character of stations, has amounted in extreme cases to nearly ten vibrations per diem; a scale of 100 determinable parts is thus afforded, by which the local attraction, dependant on the geological accidents, may be estimated. (p. 341)

---

[11]Sabine's measurements in New York were made in Columbia College, forerunner of today's Columbia University. The university is host to the Lamont Doherty Geological Observatory founded by Maurice Ewing, which was one of the trailblazers in 20th Century marine gravity.

[12]Sabine's assistants seem to have been uniformly unfortunate. Not only was there a high death rate amongst marines in his service, but Douglas Clavering, the captain of the *Pheasant*, used in 1822, and the *Griper*, used in 1823, sailed from Sierra Leone in command of the sloop *Redwing* in 1827 and was never seen again. Henry Foster, who assisted Hall as Master's Mate on the *Conway* and Sabine when a midshipman on the *Griper* and who later made pendulum observations on his own account and became a Fellow of the Royal Society, drowned in a river in Panama aged only thirty-five.

STATIONS.	$s +$ y.Sin.² Lat.	Lengths individually determined.	Individual determinations in excess or defect.	The excess or defect in mGal	GEOLOGICAL CHARACTERS.
São Tome .	39.01568	39.02074	+ .00506	+ 127	Basaltic rock.
Maranhão .	39.01607	39.01214	− .00393	− 99	Alluvial.
Ascension . .	39.01953	39.02410	+ .00457	+ 115	Compact volcanic rock.
Sierra Leone	39.02009	39.01997	− .00012	− 3	{ A soft and rapidly disintegrating granite.
Trinidad . .	33.02258	39.01884	− .00374	− 94	Alluvial.
Bahia . . .	39.02589	39.02425	− .00164	− 41	A deep soil on a sandstone basis.
Jamaica . .	39.03485	39.03510	+ .00025	+ 6	Calcareous rock.
New York .	39.10167	39.10168	+ .00001	- 2	{ A stratum of 100 feet of sand, on serpentine.
London . . .	39.13954	39.13929	− .00025	− 6	Gravel and chalk.
Trondheim	39.17738	39.17456	− .00282	− 71	Argillaceous soil on mica slate.
Hammerfest .	39.19566	39.19519	− .00047	− 12	Mica slate.
Greenland . .	39.20344	39.20335	− .00009	− 2	Sandstone.
Spitsbergen .	39.21151	39.21469	+ .00318	+ 80	Quartz.

Fig. 7.6 Pendulum results, 1822–1823 (Sabine 1825) showing the local geology and the differences from the values that would have been expected assuming an Earth ellipticity of 1/289.4. The modern accepted value is 1/298.26. A difference of one vibration per day corresponds to a difference of about 23 milligal. St. Thomas is modern São Thome, Maranham is Maranhão and Drontheim is Trondheim

This statement is remarkable for its anticipation of a technique that was not to see serious use for another hundred years, but also remarkable for being based to a significant extent on errors and misinterpretations.

In preparing his table, Sabine assumed that 'g' on a perfect ellipsoidal Earth would differ from its value at the equator by an amount proportional to the square of the latitude angle (which was almost, but not quite, true), and on that basis he calculated the straight-line relationship that best fitted his own results (see Chap. 14, Coda 6). He then looked at the individual results and calculated their differences from the values implied by the straight line. The most obvious feature was that the largest differences were for two of the tropical stations (São Tomé and Ascension) on the eastern side of the Atlantic (+127 and +115 milligal respectively) and for two of the tropical stations (Maranhão and Trinidad) on the western side (−99 and −94 milligal respectively). These differences he might well have attributed to parallels of latitude being ellipses rather than circles, and it is much to his credit that he decided that they must be due to differences in geology. He also noted that a fifth high value (+80 on Spizbergen) was on an island where hard 'basement' rocks were exposed at the surface, and suggested that the observations at the remaining eight sites could be fitted into a rough sequence based on the densities of the near surface rocks.

The real situation was rather different. 'g' is certainly high on both Ascension and São Tomé, and much higher than at Maranhão or Trinidad, but this is almost entirely due to the different thicknesses of oceanic and continental crust, which were not even suspected until Pratt and Airy began to think about isostasy, and were not fully understood until the advent of plate tectonics. They have little to do with surface rock-type, and everything to do with Ascension and São Tomé being oceanic islands. On the western side of the Atlantic the sites were very different. 'g' on Trinidad is certainly low, but this is due to the island's position at the southern end of the deep gravity low associated with the Antilles Trench, also a consequence of plate tectonics. Only at Maranhão, where low gravity is produced by a thick pile of sediments of which recent alluvium makes up a significant part, did Sabine's interpretation in any way correspond to geological reality.

In the north it was Sabine's data rather than his geological understanding that was at fault. The value on Spitzbergen, which he thought high, is actually very close to what would now be expected for sea-level gravity at that latitude, and the measurements at Trondheim and, to a lesser extent, at Hammerfest, were quite badly in error. Because these provided data points at the extreme northern end of the data set, they had a disproportionate influence on the slope of the best-fitting straight line, and it is this that was the source of the comparatively large difference between Sabine's estimate of Earth ellipticity and the modern accepted value.

In the end, it really didn't matter. Sabine had his geological revelation and, eventually, there would be people who would put it to use, but there remains the intriguing question of why things went so badly wrong in Trondheim. Enough readings were made, with enough pendulums, for errors in the actual counting of coincidences or in the subsequent calculations to be unlikely. It seems more probable that it was the timing that was at fault. At each port of call, the expedition's chronometers had to be re-set against astronomical observations, and Sabine himself made the point, in relation to the observations at Hammerfest, that the further north they were, the more difficult the measurements became. The repeating circle, an acknowledged source of error at several sites, does not seem to have been used at Trondheim, but the sextants that were used required care and dedication if times were to be accurately measured, and at the end of a long voyage these might have been a little lacking.

There may have been another factor. Trondheim was not one of the ports originally chosen for observations, and was visited only because the slow-sailing *Griper* was heading back south too late in the season to make its proposed visit to Iceland. Trondheim was a suitable alternative but Sabine arrived there without any of the usual permits and letters of introduction. Despite this, he recalled that although it had been their good fortune

> ...... to have experienced at each of the inhabited stations which we had visited the most marked hospitality and kindness, ... at none were our obligations in these respects greater than at Drontheim.

A villa belonging to the Wensels, a prominent merchant family, was placed at their disposal for the measurements, and its position was established with great accuracy by an officer of the Norwegian Engineers. All the necessities for a high-quality series of measurements were in place, and yet things went wrong. While it might have been that Sabine's heart was simply no longer fully in the work after a long and gruelling voyage, it is also possible that it was the justly famous Norwegian "*hospitality and kindness*" that took its toll.

Linking the Bases

After his two years of almost continuous travelling, Sabine settled down, marrying Elizabeth Leeves, a remarkable woman who made her own contribution to geophysics by translating Alexander von Humboldt's massive four-volume *Kosmos* into English. This caused something of a stir in Germany, where the *Augsburger Allgemeinen Zeitung* for 12 April 1849 commented, with male-chauvinist astonishment, that although few men in its country of origin understood the book, in England it had been translated by women! The English publishers were even less enlightened in this respect, entitling the work "*Cosmos: sketch of a physical description of the universe by Alexander Von Humboldt; translated under the superintendence of Lieut.-Col. Edward Sabine*". The acknowledgement of 'Mrs Sabine' as the real translator was hidden away in the Preface.

Tied down by marriage he may have been, but Sabine continued to work with gravity, and to make short trips overseas. In 1827 he used invariable pendulums to compare 'g' at Henry Browne's house, at Greenwich and at the Royal Observatory in Paris, where he collaborated with de Freycinet. The two must have had many tales to tell each other, especially as they also shared an interest in the Earth's magnetic field. Sabine would doubtless also have been keen to hear the inside story of the Frenchman's pioneering approach to dealing with the prospect of marital separation in the cause of science.

The link to Greenwich was made using an invariant pendulum that had been improved to such an extent that on one occasion sixty-two coincidences were observed before a fresh push was required. The interval between coincidences, however, was only about 400 seconds. To everyone's surprise (because the possible magnitudes of the experimental errors were still not appreciated), the pendulum appeared to swing more rapidly at Greenwich than at Portland Place, despite the latter being both lower down and further north. For the link to Paris two pendulums were used, one of which had been borrowed from the German astronomer Heinrich Schumacher just before it was due to be shipped to his observatory in Altona.[13] Sabine's report is itself a testimony to the extent of international collaboration in the scientific world of the time, since he noted that he had also

> .... obtained permission to employ a pendulum belonging to the Board [of Longitude], which had been made at the same time as M. SCHUMACHER'S, to replace the one formerly lent to Captain HALL, and since supplied at the request of the Russian Government, to Captain LÜTKE of the Russian Navy, on a scientific voyage to the Pacific. (Sabine 1828)

Two years later he revisited one of the problems created by Kater's rather eccentric way of going about things and again used the reversible pendulum to obtain a link between Mr Browne's house and the Greenwich Observatory (Sabine 1829). Before doing so, he made two modifications. He removed the wooden end-pieces, which had been found to absorb water from the atmosphere to such an extent that the period changed measurably with humidity, and instead observed the coincidences between the white disc and the ends of the brass shaft, which were blackened for the purpose.

---

[13]Altona, now a suburb of Hamburg, was then within the Duchy of Holstein, a personal fief of the kings of Denmark, and the observatory had just been constructed on the orders of Frederik VI. The duchy was ceded to Prussia after the war of 1864 between Denmark and the German Confederation. The Danish kings seem to have made a habit of funding the building of observatories for charismatic astronomers on territory that they were to lose only a few years later.

He also had the auxiliary weight removed, which meant that parts of the bar had to be filed away so that the periods could still be equalised by adjusting the slider.

Another innovation was that some of the measurements were made with the pendulum in an evacuated chamber to minimise the effects of air resistance and buoyancy, and when this was done it became possible to observe more than a hundred and twenty coincidences in a single experiment. Some effort was also made to estimate times to a quarter or a third of a second. The corrections for the finite arcs of swing were determined experimentally, for each of the knife edges, and all the other corrections that been found necessary over the years were painstakingly applied. The new measurements produced results that were consistent with gravity being less at Greenwich than at Portland Place, but Sabine made no comment on this. Instead he turned his attention to improving the accuracy of the observations, and in a series of (relatively) short papers reported on experiments on the effects of temperature and of taking measurements in air rather than in a vacuum.

Sabine's Errors

If Sabine is remembered at all today, it is for his work on magnetism, not gravity. During his lifetime he was richly honoured for both, but he was also capable, as are all scientists, of making very silly mistakes. Some years after the publication of his pendulum monograph, he had to admit to errors in calculating the mean lengths of seconds pendulums in New York and Hammerfest that amounted, when expressed in today's units, to eleven milligal in both cases. Humiliatingly, he had to confess that he had not noticed them until they were pointed out to him by Kater.

Worse, however, was to come. In five of his Atlantic locations, he made a major mistake in reading the repeating circle used for the astronomical observations that determined the clock rate. His own account, while correct as far as it goes, could best be described as an exercise in damage limitation. He said that:

> In the account of my pendulum experiments ….. the rate of the clock with which the pendulums were compared was obtained at five stations, viz. at Bahia, Maranham, Trinidad, Jamaica, and New York, by means of a small repeating circle of six inches diameter, belonging to the Board of Longitude. The correct value of the divisions of the level of this instrument having been

ascertained by Captain KATER .... the observations made with it at the stations mentioned above have been recomputed.[14]

The corrected results were then given, but there was no further explanation. What actually happened was mercilessly dissected by Charles Babbage, now remembered as the father of the computer but then better known for his conflicts with the Royal Society, in crisis thanks to its forty-two years under Joseph Banks. The turmoil usefully led to the formation of rival specialist groups such as the Royal Astronomical Society and the Geological Society, despite bitter opposition from Banks. As far as the founding of the Royal Astronomical Society was concerned, one member of the break-away group, Francis Baily wrote that an

> ..... attack was made by Sir Jos Banks on the Astronomer Royal, who, if report be true, made a very spirited reply. As a similar, and indeed a more violent, attack was made at the establishment of the Geological Society, and also of the Royal Institution, and which only tended to unite more firmly the original members, we hope that a similar result will also be produced here. (Dreyer and Turner 1920, p. 3)

In Babbage's view, however, the formation of new societies was not enough, and it was time for root and branch reform of the Royal Society itself. The unfortunate Sabine was a particular target, having no fewer than twenty-five of the two hundred and twenty-eight pages of one of Babbage's polemics devoted to him.[15] His most heinous sin had probably been to accept appointment as one of the three members of the committee which, following the abolition of the Board of Longitude, was to advise the Admiralty "*on all questions of discoveries, inventions, calculations, and other scientific subjects*". Each member received the sum of one hundred pounds per year, which particularly upset Babbage, who had earlier been prevented by Banks from obtaining a similarly well-paid position on the Board itself. For Babbage

> ..... it remains then to consider Captain Sabine's claims, which must rest on his skill in "PRACTICAL ASTRONOMY AND NAVIGATION,"—a claim which can only be allowed when the scientific world are set at rest respecting

---

[14]In a Postscript to Sabine (1828).

[15]Babbage (1830). Most of the comments on the support for science in Britain are as valid now as they were then.

the extraordinary nature of those observations contained in his work on the Pendulum.

The attack centred on Sabine's use, or misuse, of the repeating circle. What he had done, and he seems to have been very careless, was assume that each graduation on its circular scale represented one second of arc, when it actually represented 10.9 seconds. As Babbage said this *"rendered necessary a recalculation of all the observations made with that instrument"*, adding that this was *"a re-calculation which I am not aware Captain Sabine has ever thought it necessary to publish"*. In that second claim he was wrong, as we have seen, but he made a third and even more serious accusation, which was that the results must have been doctored, because there was no legitimate way that the claimed consistency could have achieved with the scale interpreted in that way.

Sabine had, by this time, become old enough and wise enough to avoid pitched battles. He simply ignored the attack, and it attracted little attention. At this remove in time, and without access to the original instrument, there is no way of knowing for certain whether or not Babbage was right about the doctoring, but in one respect at least Sabine was a very suitable target. One of the scandals Babbage had identified in the affairs of the Royal Society was the election to it of Fellows with no perceptible qualifications. The process, as he described it, was that

> A.B. gets any three Fellows to sign a certificate, stating that he (A.B.) is desirous of becoming a member, and is likely to be a useful and a valuable one. …..At the end of ten weeks, if A.B. has the good fortune to be perfectly unknown by any literary or scientific achievement, however small, he is quite sure of being elected …. If, on the other hand, he has unfortunately written on any subject connected with science …. the members begin to inquire what he has done to deserve the honour; and, unless he has powerful friends, he has a fair chance of being black-balled.

Sabine's election was a case in point. It took place just three years after he had returned to England from service in North America. There had been little time for him to make any contribution to science, and there is no evidence that he had done so. The 'certificate' supporting his application stated only that he was *"a gentleman attached to science and to natural history"* who was *"likely to become a useful and valuable member"*. There were twelve signatories in all and, predictably, they included his elder brother Joseph and his brother-in-law Henry Browne. One of the others, coming just after Basil Hall, was more surprising. It was Charles Babbage.

The Pendulum Re-designed

Errors in the measurements of 'g' in the early part of the 19th Century had little impact on the development of gravity studies, but some eighty years later a smaller error made by Prussian scientists had consequences that are still being felt today. The pendulums used in this exercise were designed by Wilhelm Bessel, who was one of the scientific giants of the early 19th Century. Although he was in no way responsible for the errors, he was, perhaps, the reason why in Prussia the study of gravity came to be considered worthwhile.

To geophysicists Bessel is the man who took Kater's barely practical reversible pendulum and turned it into a workaday instrument, but he was much more than that. To mathematicians he is the inventor of the functions that bear his name and which first terrified them and then liberated them as they learned how to use them to solve otherwise unsolvable equations. To astronomers he is the man who first used parallax to measure the distance to a star, who produced a catalogue of more than two thousand stars, and who had a hand in the discovery of Neptune and Uranus. To geodesists he is the man who defined the shape of the Earth so well that his ellipsoid is still the foundation of the mapping systems of more than a dozen countries, and to educationists he is one of the great reformers, first of the university system of East Prussia and through that of the universities of much of central Europe. Yet his formal education ended at the age of fourteen, when he was apprenticed to a firm of overseas traders, and he himself never went to university at all.

Bessel might well have remained in commerce all his life, and would have been much wealthier if he done so, because his skill with numbers won him rapid promotion. Overseas trade, however, led him to an interest in navigation and thence in astronomy, and while still in his teens he abandoned the business world for a poorly-paid post as an assistant in a private observatory near Bremen. There he made so much of a name for himself that, aged only 25 and having already turned down similar offers from Leipzig and Greifswald, he was head-hunted to found the East Prussian astronomical observatory at Königsberg. His lack of the necessary paper qualifications was solved by the immediate award of a doctorate by the University of Göttingen. All that the university had needed was a personal recommendation from Karl Gauss. For what more could any institution have possibly asked?

By inclination an astronomer, Bessel came to gravity by a circuitous route. As head of the observatory he was responsible for the East Prussia trigonometric network that linked Western Europe to the network in the (then Russian) Baltic States. One of his innovations was to use a baseline that, in defiance of the custom of the time, was less than two kilometres long, arguing that it was better to have a short baseline that had been very accurately measured than a long one that was less accurate (Viik 2006). At this time baselines were usually at least 10 km long. The East Prussian network extended along the southern side of the Frisches Haff, passing close to the tower in Frombork where, four hundred years earlier, Copernicus had fought his lonely battles with the motions of the stars and the planets. It was the necessities of this survey that led Bessel to the mathematics of converting distances measured on ellipsoids to distances on spheres, and thence to defining an Earth ellipsoid, and thence to an interest in gravity. To manufacture his instruments he turned to the Repsold family company in Hamburg, and it was they who, after his death, produced to his design the rather different, and much better, reversible pendulums that replaced Kater's for use world-wide.

A contemporary of Kater and Sabine (he was born 1784 and died in 1846), Bessel made two major improvements in reversible pendulums. The first came when he abandoned the idiosyncratic Kater design for one that was externally symmetric but still weighted asymmetrically (Fig. 7.7), and showed that by doing so the effects of air resistance could be almost eliminated. He also showed mathematically that if the periods of oscillation from the two knife edges were almost equal (which was easy enough to achieve during manufacture), and were accurately measured, then the length of the equivalent simple pendulum could be calculated. The tedious process of physical equalisation by trial and error was no longer necessary. The small size of global gravity variations became an actual advantage, because the one pendulum could be used anywhere in the world.

Disaster in Potsdam

Once he had moved to Königsberg, Bessel remained there for the rest of his life, but he also played a part in establishing the Prussian Geodetic Institute in the Berlin suburb of Potsdam, where he made some of his pendulum experiments. Drawing on that tradition, an unprecedented number

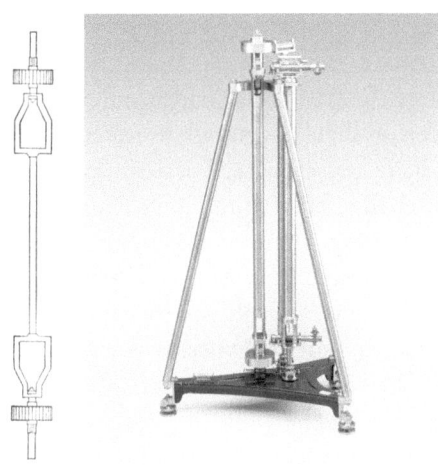

Fig. 7.7 Wilhelm Bessel, in an 1839 portrait by Christian Albrecht Jensen now in the Ny Carlsberg Glyptotek, with his basic design for a reversible pendulum and one of the Repsold instruments based on it, from the Alumnos de la Escuela Técnica Superior de Ingenieros de Telecomunicación de la Universidad Politécnica de Madrid). Sadly Bessel did not live long enough to see one built

of precision measurements of 'g' were made between 1898 and 1904 at the Institute, using Bessel-Repsold pendulums. Friedrich Kühnen and Philipp Furtwängler, who did the work, used a half-second pendulum and a seconds pendulum belonging to the Institute itself, another seconds pendulum borrowed from the astronomical observatory in Galileo's old university town of Padua and two of different weights borrowed from the Imperial and Royal Military-Geographical Institute in Vienna. Each consisted of a steel shaft with a cylindrical weight at one end and a near-identical but hollow cylinder at the other. Some interchanges of knife edges were possible between pendulums, and in some experiments the agate plates were attached to the pendulum shafts and the knife edges were fixed to the supports. In 1906 the results were published in a massive monograph (Kühnen and Furtwängler 1906). After combining the data from almost two hundred individual determinations, each taking about four days, the value of 'g' finally presented to the world was 981,274 milligal.

So impressed were the geodesists and geophysicists of the time by Prussian rigour and precision (the previous standard had been Austrian) that they agreed, rather too quickly, to adopt this as the international reference value. It was, of course, valid only for the place where it had been measured, but its scope could be extended globally using the relative pendulums that had been shown to provide a more accurate, and certainly a quicker, way of establishing a global gravity network than reversible pendulums. The Potsdam reference system was born, and remained in use for almost seventy years, even after the base value had been conclusively shown to be almost exactly fourteen milligal, or about fourteen parts in a million, too high. Better, it was felt, to use an agreed system, even one known to be a little bit wrong, than to change everything every time some laboratory, somewhere, came up with something slightly better. Now that it is possible to look back at the chaos that ensued, and still to some extent persists, after the change was finally made, one can only say that the people who resisted it for so long were right to do so, and every user of pre-1970 gravity data can be grateful to them.

The corrections needed when using reversible pendulums were well known by the time the Potsdam measurements were made. They included corrections for the amplitude of swing, for thermal expansion, for air friction and friction at the supporting knife edges, for flexure of the support, for pendulum asymmetry and for any differences in air density between the times when the pendulums were swung in their normal and reversed positions. All these were applied, yet the final result was not only wrong, but was not really a great advance on Kater's, obtained eighty years before. With the equipment at their disposal, Kühnen and Furtwängler should have done much better, and in terms of actual measurement they did. It is an astonishing fact that their final recommended value was higher than any of those that they actually measured. Had they simply taken their results at face value and applied only the most basic of statistical analyses, they might instead have published a value that could still be accepted today. Unfortunately, despite having arrived at an apparently acceptable answer, they had not been sure that they had done enough. Perhaps they were reluctant to reduce the 'Vienna' datum by the 30 milligal that their results demanded, but whatever the reason, they searched for other possible sources of error and, thinking that they had found one, they panicked.

How their supposed error first came to their notice is not known, but at some stage they may well have drawn plots similar to those in Fig. 7.8. In Fig. 7.8a the results of all the experiments with seconds pendulums are shown. Inspection suggests a likely value of about 978,261 milligal, and a simple statistical analysis gives 978,262.78 milligal. Including the half-second pendulum results, which are not shown, gives a value of a little over 978,257 milligal if they are regarded as equally valid, but their wider scatter could justify a lower weighting. Any one of these approaches would have produced a value close to the currently accepted 978,260 milligal.

The plot in Fig. 7.8b provides a little more information, by showing which pendulums were used, and demonstrating that the three values

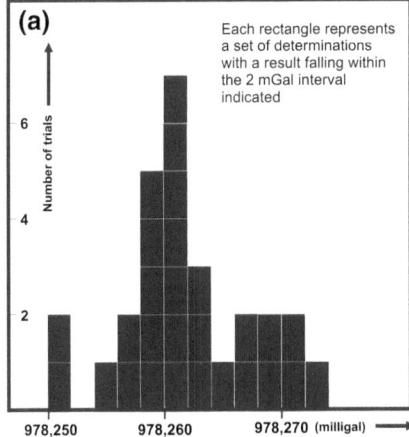
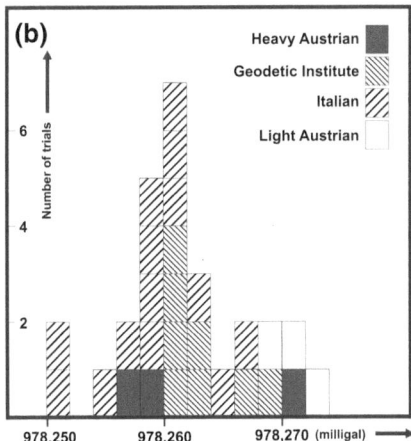

Fig. 7.8 The Potsdam results. Histograms at 2 milligal intervals of the results from the seconds-pendulums only. Each rectangle represents one set of experiments, consisting of anywhere between three and twelve individual determinations. The shading in the plot on the right identifies the instruments used

obtained with the Light Austrian pendulum were consistently higher than those obtained with the other three. By itself, this should not have been enough to make Kühnen and Furtwängler take the decisions that they did, but it might well have led to them draw another plot, this time of 'g' values against pendulum mass. In Fig. 7.9 all the pendulums are included, and the wider scatter for the half-second pendulum is obvious. The 'Austrian Light' pendulum is confirmed as producing consistently high values and, looked at with the eye of faith, the results from the other pendulums, all with slightly different masses, could also be interpreted as showing the same trend.

Neither of the pair was able to think of a reason for a mass effect of that size, but they nonetheless assumed that it existed, and that a correct result would only have been obtained with a pendulum that weighed nothing at all! To make and use such a pendulum would, of course, have been impossible, but it was possible to extend the results back to zero mass using the well-established mathematical technique of least-squares analysis and the assumption that the effect was directly proportional to mass. The 'K&F line' in Fig. 7.9 is drawn to fit all the values and reaches zero mass at the 981,274 milligal value that was adopted. Had the errors been assumed random rather than systematic, and the data certainly allow that, then a much more accurate answer would have been obtained.

It is easy to be critical and hindsight is always perfect, and the dilemma faced the Prussians was a real one, but even so it is hard to believe that

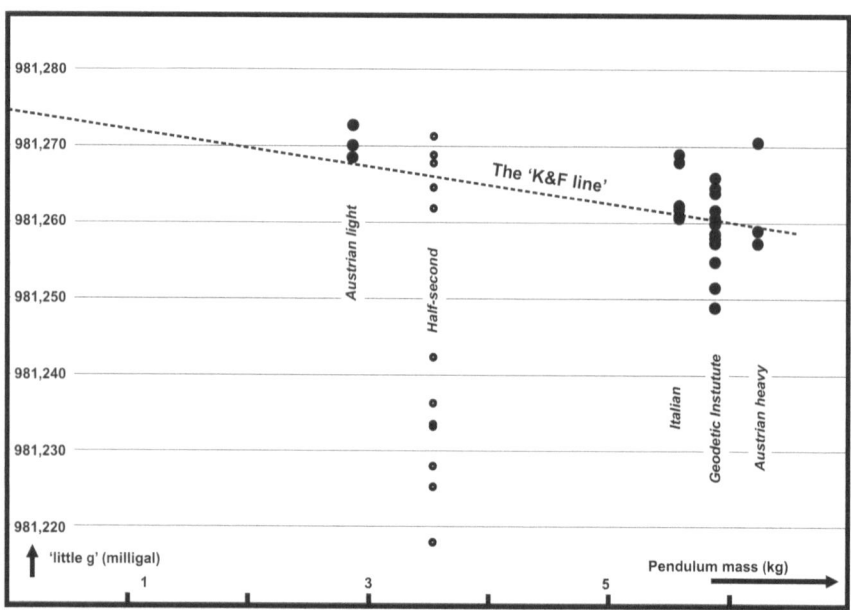

Fig. 7.9 The 'g' estimates from all the Potsdam determinations plotted against pendulum mass. The results from the half-second pendulum, which were excluded from the histograms in Fig. 7.8, show a much wider scatter than those from any of the seconds-pendulums. The 'K&F line' extends the results back to a pendulum of zero mass

many other experimenters would have done what they did. There were so many decisions to be made, any one of which could justifiably have gone the other way. First, it had to be supposed that the effect was systematic, and not merely due to random error. Then it had to be assumed that it was due to a dependence on pendulum mass, and that the correct answer would only have been obtained with a pendulum that weighed nothing. And then it had to be assumed that the effect was linear, so that the zero-mass value could be obtained using a best-fitting straight line. They assumed all these things, and one consequence was that the results obtained with the 'light Austrian' pendulum had by far the greatest influence on the final value. They were three times as important as the results from the half-second pendulum (which, having the widest scatter, might otherwise have been discarded), five times as important as the results from the Institute's own seconds pendulum and nine times as important as those from the pendulum from Padua. The 'Austrian heavy' was effectively ignored, even though it was consistent with the Geodetic Institute's own pendulum and was, as we now know, the nearest to being correct. This was eventually pointed out by scientists from the US Bureau of Standards (Dryden 1942) and also, although later, by Alfred Berroth from the Prussian Institute itself (Berroth 1949). The reactions of his colleagues to his disloyalty have not been recorded, but in a recent pictorial history of the Potsdam science complex (Bormann 2004) the achievements of Kühnen and Furtwängler (both of whom died in 1940) are recorded with pride. Berroth is not mentioned.

The Pendulum Abandoned

In the Bureau of Standards discussion of the Potsdam results, Hugh Dryden concluded that

> There is a great need of an absolute determination by some other method than that of the reversible pendulum to determine definitely the magnitude of the systematic errors present .... Under present world conditions, however, other scientific problems are much more urgent, and this one must await more normal times.

Given that this was written less than a year after the attack on Pearl Harbour, it is hard to disagree. After the war was over the problem was re-visited, and methods based on measuring times of fall began to come to the fore. These are in principle even more challenging than pendulum methods because each drop is a separate event, after which the instrument has to be re-set. That they are now preferred is due partly to the number and complexity of the corrections that have to be made to pendulum observations, but also because very accurate measurements of even very short times are now possible. In 2004, the Bureau International de Poids et Mésures decided that in future ballistic free-fall should be the standard way of measuring gravity.

Even the briefest examination of the specifications for the Micro-g FG-5, the current state-of-the-art free-fall instrument, shows just how far Galileo, Mersenne and Riccioli had to go when they tried to do the same thing. A corner-cube reflector is dropped in a near-perfect vacuum and is preceded by a falling 'elevator' that clears the few remaining air molecules out of the way. Time is measured using interference fringes referred to a rubidium-vapour atomic clock and the entire trajectory is laser-monitored. The chamber is spring-mounted to reduce vibration. Because gravity decreases by almost a third of a milligal for every metre increase in elevation in air, the drop distance is a mere 20 cm. The claimed accuracy is one thousandth of a milligal, which may be a little over-optimistic, and observations must to be extended over several days to achieve anything close to this. The scientists of the early 17th Century would have been astonished by what has been achieved, but for all its technical sophistication the basis of the method is still the same as Riccioli's: a mass falling under gravity.

References

Airy G (1826) On the figure of the earth. Philos Trans R Soc 116:548–578

Arago J (1821) Souvenirs d'un aveugle: voyage autour du Monde. Garnier Frères, Paris. English edition: Arago J (1824) Narrative of a voyage round the world. Treutell & Wurz, London

Babbage C (1830) Reflections on the decline of science in England. Fellowes, London

Bassett M (1962) Realms and islands: the world voyage of Rose de Freycinet (1817–1820). Oxford University Press, London

Berroth A (1949) Das Fundamentalsystem der Schwere im Lichte neuer Reversionspendel-messungen. Bull. Géodésique 12:183–204

Bormann P (2004) History of research institutions on the Potsdam Telegraph Hill (Telegrafenberg). In: 19th European Seismological Commission General Assembly, Potsdam, Sept 12–17, 2004

Dreyer JLE, Turner HH (eds) (1920) The history of the 1820–1920. Royal Astronomical Society, London

Dryden H (1942) A re-examination of the Potsdam absolute determination of gravity. J Res Natl Bur Stan 29:303–314

Duplomb C (ed) (1927) Journal de Madame Rose de Saulces de Freycinet, d'après le manuscrit original accompagné de notes. Société d'Editions Géographiques, Maritimes et Coloniales, Paris

Goldingham J (1822) Observations for ascertaining the length of the pendulum at Madras in the East Indies, Latitude 13°4′ 9.1″. Philos Trans R Soc 112:127–170

Hall B (1823) Letter to Captain Kater. Philos Trans R Soc 113:211–285

Hewitt R (2010) Map of a nation. Granta, London, p 2010

Kater H (1818) An account of experiments for determining the length of the pendulum vibrating seconds in the latitude of London. Philos Trans R Soc 108:33–102

Kater H (1819) An account of experiments for determining the variation in the length of the pendulum vibrating seconds, at principal stations of the Trigonometrical Survey of Great Britain. Philos Trans R Soc 109:337–508

Kühnen F, Furtwängler P (1906) Bestimmung der Absoluten Grösze der Schwerkraft zu Potsdam mit Reversionspendeln. Königlichen Preussischen Geodätischen Instituts, Berlin

McCarthy M (2008) Rose and Louis de Freycinet in the Uranie. Report 236, WA Museum Department of Marine Archaeology, Perth

Rivière MS (1996) A woman of courage: the journal of Rose de Freycinet on her voyage around the world 1817–1820. National Library of Australia, Canberra

Ross J (1819a) A voyage of discovery for the purpose of exploring Baffin's Bay. John Murray, London

Ross J (1819b) An explanation of Captain Sabine's remarks on the late voyage of discovery to Baffin's Bay. John Murray, London

Sabine E (1819) Remarks on the account of the late voyage of discovery to Baffin's Bay published by Captain J. Ross, R.N. John Booth, London

Sabine E (1821) An account of experiments to determine the acceleration of the pendulum in different latitudes. Philos Trans R Soc 111:163–190

Sabine E (1825) An account of experiments to determine the figure of the Earth, by means of the pendulum vibrating seconds in different latitudes. John Murray, London

Sabine E (1828) Experiments to determine the difference in length of the seconds pendulum in London and Paris. Philos Trans R Soc 118:35–77

Sabine E (1829) Experiments to determine the difference in the number of vibrations made by an invariable pendulum in the Royal Observatory at Greenwich, and in the house in London in which Captain Kater's measurements were made. Philos Trans R Soc 119:83–102

Viik T (2006) F. W. Bessel and Geodesy. Paper delivered at the Struve geodetic arc international conference Haparanda and Pajala, Sweden, 13–15 August 2006

Walker D (2014) Balta Sound and the figure of the Earth. Sheetlines 99:5–17

Wulf A (2015) The invention of nature: the adventures of Alexander von Humboldt, the lost hero of science. John Murray, London

8

Change of a Change

Aristotle rejected the idea of change of a change, and would not have been happy with the concept of acceleration, let alone of acceleration due to gravity. He might have been happier with the definition of 'g' as the gradient (rate of change) of gravitational potential energy, but would then have been scandalised by the idea of it itself having a gradient. Surprisingly, the first use of gravity in geology was made by measuring such gradients. The changes in 'g' due to changes in geology are very small, and the changes in its gradient are so very small that it is hard to believe that they can be measured to any useful level of accuracy. But they can.

The Cavendish Experiment

Twenty years after Maskelyne left Schiehallion, Henry Cavendish showed that it was not necessary to go to a cold, wet, foggy mountain in Scotland to estimate the total mass of the Earth. It could be measured much more conveniently in the comfort of a nice warm laboratory situated, in his case, in part of a large house at the edge of Clapham Common in London.

© Springer International Publishing AG, part of Springer Nature 2018
J. Milsom, *The Hunt for Earth Gravity*,
https://doi.org/10.1007/978-3-319-74959-4_8

Cavendish has been described as not merely a great scientist but a fanatical one, and a neurotic of the first order.[1] Since he was also a member of one of the richest and most powerful families in the country (it included the Dukes of Devonshire), he could easily afford the luxury of a well-equipped personal laboratory. The costs would have been trivial compared to those involved in catering to some of his other eccentricities, the most notable of which was a pathological shyness, particularly where women were concerned. He reputedly had an extra staircase added to the back of his house to avoid any chance meeting his housekeeper, with whom he communicated by written notes. His social life was limited to the meetings of the Royal Society (at which there would be no risk of encountering women) and even there he was famously introverted, sometimes retreating into a far corner of the room if anyone attempted to speak to him. He was, however, an active correspondent, and in 1771 he was exchanging letters with Maskelyne that led eventually to the Schiehallion expedition. His contributions to the planning of the expedition and to the subsequent calculations are poorly documented but were almost certainly significant. There was, however, no question of his joining in the actual fieldwork. The months of close contact with other human beings would have been torture, and the final boisterous party might well have been the end of him.

Cavendish was also in frequent written contact with a Yorkshire clergyman and noted scientist, John Michell, and it was together with him that he devised the laboratory-based counterpart of Maskelyne's experiment. Today Michell is remembered chiefly as the first person to suggest the possibility of gravitational black holes, but he may deserve less credit for that[2] and far more credit for other things he did. His paper written in the aftermath of the Lisbon earthquake (Michell 1760) was one of the first serious works on seismology. It introduced the idea of earthquake waves and their association with movements along geological faults, and even contained a very respectable estimate of the location and depth of the primary shock. He was, however, somewhat isolated from the scientific mainstream, spending the last

---

[1]Much of the information given in this section is taken from Falconer (1999).

[2]Michell's idea was based on the well-known fact that the gravity field at the surface of a sphere of constant density is proportional to the radius of the sphere. From this starting point, and from his belief that light was made up of particles and not waves, he went on to argue that a sufficiently large sphere might have a gravity field strong enough to prevent light from ever leaving it. This has some elements of 'black-hole' theory, but the bodies involved were envisaged as being very very large rather than 'singularites'.

twenty-five years of his life as rector of Thornhill, near Leeds.[3] The most a contemporary diarist felt able to say about him (in manuscript MSSXXXIII in the British Library's Cole Collection) was that

> John Michell, BD is a little short Man, of a black Complexion, and fat; but having no Acquaintance with him, can say little of him. I think he had the care of St. Botolph's Church [Cambridge], while he continued a Fellow of Queens' College, where he was esteemed a very ingenious Man, and an excellent Philosopher. He has published some things in that way, on the Magnet and Electricity.

Cavendish had certainly known Michell at Cambridge and according to some accounts also visited him in Yorkshire. This seems a little unlikely. It is hard to imagine anyone as averse as he was to human contact undertaking such a formidable journey, even though his enormous wealth would have made it as comfortable as the times allowed. He certainly did have a considerable input into the design of Michell's most famous experiment, but, disarmingly, he began his account by disclaiming any credit for it, or for the construction of the apparatus. Instead, he wrote that:

> Many years ago, the late REV. JOHN MICHELL, of this Society, contrived a method of determining the density of the earth, by rendering sensible the attraction of small quantities of matter; but, as he was engaged in other pursuits, he did not complete the apparatus till a short time before his death, and did not live to make any experiments with it. After his death, the apparatus came to the REV. FRANCIS JOHN HYDE WOLLASTON, Jacksonian Professor at Cambridge, who, not having conveniences for making experiments with it, in the manner he could wish, was so good as to give it to me.

Cavendish might also have added that a few years earlier the French physicist Charles-Augustin de Coulomb had used a rather similar instrument, which today we would call a torsion balance, to measure electrostatic forces. It was Coulomb who established that for wires made of uniform material the twisting force (the torque) is proportional to the angle of twist and to the fourth power of the radius and inversely proportional to the length of the wire. Cavendish may have known of this work or have discovered the proportionality independently, but his description of his own instrument (Fig. 8.1) was succinct and to the point.

---

[3]Appropriately, he was born in Eakring in Nottinghamshire, which thus has the distinction of not only being the site of one of the UK's first onshore oilfields but also the birthplace of one of its earliest real geologists.

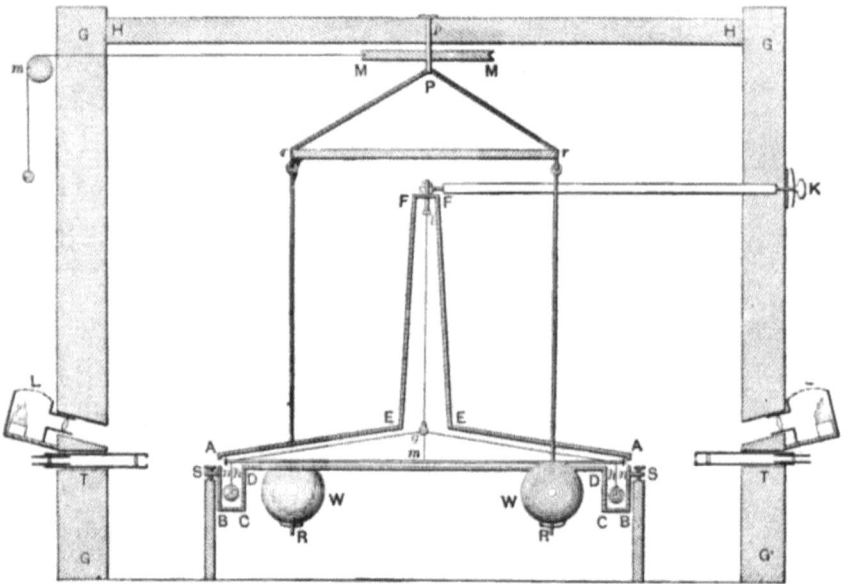

Fig. 8.1 The Cavendish apparatus. A simplified version of the drawing of the instrument in Cavendish (1798)

The apparatus is very simple; it consists of a wooden arm, 6 feet long, made so as to unite great strength with little weight. This arm is suspended in an horizontal position, by a slender wire 40 inches long, and to each extremity is hung a leaden ball, about two inches in diameter; and the whole is inclosed in a narrow wooden case, to defend it from the wind.

As no more force is required to make this arm turn round on its centre, than what is necessary to twist the suspending wire, it is plain, that if the wire is sufficiently slender, the most minute force, such as the attraction of a leaden weight a few inches in diameter, will be sufficient to draw the arm sensibly aside. (Cavendish 1798)

What he was doing was what Newton had thought impossible. He was measuring the gravitational forces between small masses (by his own estimate, masses with no more than one fifty-millionth of the mass of the Earth) and comparing them with the forces exerted by the Earth's own gravity field. In order to do so he almost completely rebuilt Michell's original apparatus, or, as he himself put it, he had to *make the greatest part of it afresh*. Most importantly, he placed it entirely within *a room which should remain constantly shut*, with the weights being moved and the motion of the suspended arm being observed (by telescope) from outside.

The first step was to determine the force needed to rotate the arm through a known angle, which was done by measuring its natural period of oscillation when the weights were as far as possible from it. Although what he referred to as the 'stiffness' of the wire could be calculated directly from this measurement, Cavendish chose not to do this. Instead, he calculated the half-period of a simple pendulum of the same length as the 'arm' of his apparatus from the (then only approximately known) length of a seconds pendulum in London and used his result to express the stiffness as a fraction of 'g'. It is at this point (and only at this point) that the attraction of the Earth, which was needed for estimating its mean density, entered into his calculations, but the fact was well concealed in his written description.

With these preliminaries completed, the large weights were moved into positions close to the suspended balls and the deflection of the beam was measured. Intuitively it might be expected that the deflection would be very small and difficult to measure, but this was not the case. On the contrary, in the first experiments it was so large that the original wire had to be replaced by a stiffer one to prevent the arm colliding with the walls of the case that shielded it from air movements.

With the deflection measured, a modern scientist might have given a small cheer and made a fairly straightforward calculation. Newton had said that the force between two point masses or uniform spheres, M and m, separated by a distance r would be proportional to Mm/r^2. Since the experiment measured this force, via the angular deflection A, and since r, M and m could all be measured, the constant of proportionality, 'Big G', could be calculated. From there it would have been a simple matter to calculate the mean density of the Earth using the known value of 'g' in London and the known radius of the Earth.

Cavendish, however, went off in a completely different direction. He first calculated the volume of water that would have the same mass as one of the weights, and then obtained the ratio of the attraction of a sphere of water with that volume to the attraction of a sphere of water one foot in diameter (the 'spherical foot'). He then introduced into the calculation the mean diameter of the Earth, and after deriving some more ratios eventually arrived at a value for the mean density of the Earth without ever having calculated 'Big G' at all!

Most of the remainder of his paper was devoted to discussions of the possible errors, and the possible reasons for the difference between his own value of the mean Earth density, averaged over the results of twenty-nine distinct experiments, as 5.48 times that of water and Hutton's Schiehallion estimate of "*4½ times that of water*". This difference was, he stated, "*rather more than I should have expected*", and he was very clear about the need for further investigations into its causes. Thanks to John Smallwood, and as discussed in Chap. 6, we now know that these lay almost entirely in the calculations done by Hutton, which omitted all gravity effects not directly associated with the mountain, and only to a much smaller extent with the work done by Cavendish. The standard deviation for his results was only 0.221, or about 4%.

Armed with a value for the mass of the Earth and a value for 'g', 'Big G' can be calculated from the Cavendish results, although there are some uncertainties because he did not record the weights of the balls. The result is a numerical value[4] of 6.74×10^{-11}, which is within 1% of the present-day estimate of 6.67428×10^{-11}. This is especially impressive since 'Big G' remains to this day the least precisely known of all the physical constants, but its value was not what interested Cavendish. His was the approach of a geophysicist trying to learn more about the Earth, not of a pure physicist trying to learn more about gravity, but his call for further experiments went largely unanswered. The gravity measurements made during the next hundred years by Kater and Sabine, and their counterparts in other countries, were directed almost entirely towards determining the shape, not the mass, of the Earth. The new science of geodesy was emerging, and geodesists have their own very specific reasons for worrying about 'g'.

Baron Eötvös

In 1881 the members of the Hungarian Academy of Sciences decided that they ought to know how the gravity field changed throughout their country, which was then very much larger than it is today, including almost all of modern Croatia and Slovakia and large parts of Poland, Romania, Serbia and the Ukraine. To do this they turned to the extravagantly-named Vásárosnaményi Báron Eötvös Lóránd (Baron Roland Eötvös of Vásárosnaményi). Aged thirty-three, the baron had already had a varied academic career, having disappointed his family's initial hopes that he would become a lawyer (then, as now, a much surer path to power and prosperity) by becoming interested in science and studying at Heidelberg under some of the greatest physicists of the time—people such as Helmholtz, Kirchoff and Bunsen, names familiar to any physics student. At the time the Academy made its proposal, Eötvös had already been credited with breakthroughs in the understanding of surface tension in liquids, but he seems to have abandoned that subject completely for his new interests, and from 1886 onwards published only on gravity and, to a lesser extent, on

---

[4]The units in which 'Big G' is measured are metres cubed per kilogram per second squared, which is enough to put sensible people off gravity studies for life.

magnetics. He wrote his scientific papers mainly in German but he was in all other ways a true Hungarian. He is almost certainly the only scientist mentioned in these pages who routinely travelled to work on horseback—a journey of eleven kilometres each way.

Being financially independent, the baron did pretty much as he pleased scientifically, and instead of following the crowd and making the measurements using pendulums, he took a completely new approach based on his interest in the equivalence principle. Newton had stated that the acceleration given to a body by an applied force is proportional to its mass and also that the gravitational attraction between two masses is proportional to those masses. What he assumed, and what no-one since has been able to prove, was that the 'inertial' mass in the first of those equations was the same thing as the 'gravitational' mass in the second. This is still a live issue in physics, where the very concept of mass is becoming ever more nebulous and requires something as incomprehensible as the Higgs boson to give it existence.

The baron's idea was that if gravitational and inertial masses were actually different things, then there would be differences in the directions of the accelerations produced by gravity and by other forces, and that these would be detectable by their effects on the rotation of a suspended horizontal rod. By 1891 he was making his first measurements to test this idea, with a torsion balance that he had designed himself as a modification of the Cavendish original. What he brought to the project was a quite extraordinary experimental

Fig. 8.2 *Left* Lóránd Eötvös, photographed in about 1890, when he would have been about 40 years old and fully engaged in developing the torsion balance. *Right* Field party on Sag Hill in western Hungary, 1891. The instrument is in the rigid shelter on the left, where it is shielded from air movements. The motion of the beam is being observed telescopically through a window in the shelter's side. Photos reproduced courtesy of the Lóránd Eötvös Museum, Budapest

skill, and a quarter of a century later his former colleagues were still referring to the almost unbelievable (*'fast unglaublich'*; Pekär 1928) sensitivity of his instruments. It was as a further development of this work that he made the first practical measurements of gravity gradients outside a laboratory when, in 1890, he made estimates of the mass of the Ság Hill in western Hungary (Fig. 8.2). One consequence of the high sensitivity was that it was crucially important for the instrument to be shielded from the slightest air movements. In part this was achieved by enclosing all the components in a rigid case but it was also necessary to take rigid shelters into the field for the benefit of the instruments. The observers had to remain outside, no matter what the weather.

From studying the gravity effects of objects that were visible, Eötvös moved on to the more subtle effects of objects that were not, making him, so it has been claimed, the first ever exploration geophysicist. While this honour (if it is an honour) must surely belong to the anonymous Swedish miner who at some time in the Middle Ages pioneered the use of a lodestone needle to detect iron ore, Eötvös was almost certainly the first person to use gravity directly for geology when he took his instrument out on to the winter ice of Hungary's largest lake (Fig. 8.3) and reported that

> My unknown country spread out far below the frozen surface of Lake Balaton. I have never seen it and shall never see it, only my instrument sensed it, still how hard it was to part with it when the ice started to melt (Király 1993)

Fig. 8.3 The Lake Balaton torsion balance survey. A re-drawn version of the original map showing the results obtained on the frozen lake in 1901 and 1903. The arrows indicate gradient magnitudes and directions

The language is poetic, but the Baron was also a poet. He was, indeed, a man of very wide interests and talents. He studied fine art under Gusztav Keleti and was a competent artist (although he clearly recognised his limitations and concentrated on painting landscapes rather than people). He was twice minister for education in the Hungarian government, and in the 1880s he designed the new building that was to house the Institute of Physics in Budapest. Although he was not a great self-publicist, and reluctant to publish anything about his instruments and experiments until he had got them exactly right, his name seems to be everywhere. Inevitably, the Geophysical Institute in Budapest is named after him, and one of the universities, and the unit in which gravity gradients are measured, and the correction that has to be applied to gravity measurements made in moving vehicles. He also gave his name to one of the fundamental laws of surface tension and, less predictably, to a peak in the Dolomites near Cortina. This latter honour commemorates the fact that he was, throughout his life, an enthusiastic mountaineer. He was still climbing in the High Tatra of southern Poland in his late sixties, and initiated his two daughters into this particular hobby. After what must have been strenuous childhoods, they both became notable athletes. He died in 1919, as Hungary was being dismembered and was losing almost all of its mountains in the aftermath of the First World War.

Torsion Balance Surveys

The instrument that Eötvös used on Lake Balaton was simple in concept, but very difficult to manufacture to the tolerances required. It consisted of a beam with a weight at one end and a balancing weight suspended from the other (Fig. 8.4a). The whole assembly was supported in a horizontal position by a fibre that was attached to the centre of the bar at its lower end and to a calibrated screw at its upper end. There are obvious similarities to the Cavendish balance, but also very clear differences. The most obvious is the absence of the large spherical masses (no longer required because it was the Earth's gravity field that was being investigated), but the placing of the two small masses at different levels could be considered the most important. As discussed in Chap. 14, Coda 1, it was this that allowed all aspects of the gradient (in mathematical jargon, the full tensor) to be measured. Instruments that lacked this refinement and had the weights at the same level came to be distinguished from torsion balances by being known simply as gradiometers.

It seems that Eötvös himself was not very interested in commercialising his instruments, leaving that to others, but he was interested in using them to solve problems in 'pure' geology. He investigated the subsurface distribu-

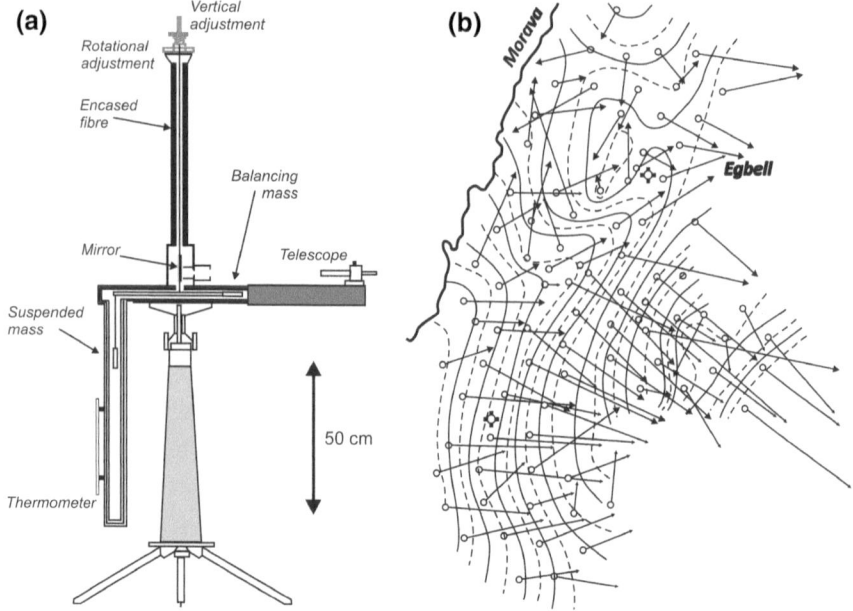

Fig. 8.4 **a** The Eötvös torsion balance (redrawn from Király 1993). **b** Results of the 1916 survey over the Egbell (Gbely) oilfield in modern Slovakia. The main domal structure is clearly defined by the combination of arrows indicating the magnitude and direction of the gradient and contours of its magnitude (redrawn from Howarth 2007)

tion of mass beneath the Hungarian plain and studied isostasy in many of the mountainous areas that then lay within Hungary. He also predicted a commercial use, saying that

> Geologists seem to agree that the most substantial discharges of gas occur in the immediate vicinity of gas-bearing anticlines, and overlying sediments. Experience gained in America (Ohio) and observations in Transylvania where the subsurface geological structures could be determined from superficial indications further endorse these assumptions. Such geological indications, however, are absent in the sand and humus-covered surface of the Great Hungarian Plain. He who searches for gas-bearing anticlines in this or similar areas should not fail to take note of conclusions drawn from torsion balance observations.

An anticline is an upwardly-convex dome in the rock strata, and Eötvös was right. Such features can create traps for the oil and natural gas known generically in the industry as 'hydrocarbons'. In rocks these fluids are contained in the pore spaces between the mineral grains and, being lighter than the water that would otherwise be there, they migrate upwards until they encounter a barrier. They can thus be trapped in a subsurface dome that is capped by an impermeable layer, which is usually shale, and remain there until liberated through a drill hole. It was over a trap of this sort at Egbell (Gbely) in what is now Slovakia that the first test survey was made. Eötvös presented the results (obtained by two of his assistants) on a map by using arrows to represent the magnitudes and directions of the horizontal gradients and contour lines to provide a smoothed and interpolated version of the magnitude information (Fig. 8.4b). This method was so effective that it became a global standard for torsion-balance work. At Egbell the gradient arrows clearly identified the location of the main dome, and led to further drilling and the discovery of oil as well as gas. Some accounts suggest that the work was supported by the D'Arcy Exploration Company, which eventually became BP, but this would have been remarkable since the United Kingdom and Austria-Hungary were at war at the time. It is more likely that D'Arcy only became involved after the war.

There were snags. A single reading with the early instruments might take a day (or a night, often favoured because conditions were quieter), so a complete survey was a long-drawn-out affair, and in the early days it really needed Eötvös himself to be on hand to make the measurements. Few people could match him in both the skill needed to handle the instruments and in understanding the theory, and it was this combination that made him very good at recognising when something was going wrong. An added obstacle to rapid survey work was that the early models were very heavy and hard to move. I have one of the few survivors in my garage. Imperial College no longer wanted it, and I volunteered to take it over, on behalf of University College. It took three people to lift it into the store-room at the college, and it remained there until I retired, when it was gently suggested that I should take it with me.

That, of course, was not one of the Eötvös originals, but a later version built in London by the Oertling Company in 1925 (Fig. 8.5), and I have a theory as to why it might have been in Imperial College. Another item that I inherited from them was a book, declared redundant by the college library, describing the work of the 1929 'Imperial Geophysical Experimental Survey' in Australia. The library may have had more copies than it needed because one of the participants had been John Bruckshaw, who eventually became Professor of Geophysics in the college's Royal School of Mines and who in 1928 was in the process of abandoning pure physics (he had worked at Manchester University on X-ray diffraction with W. F. Bragg) for exploration geophysics.

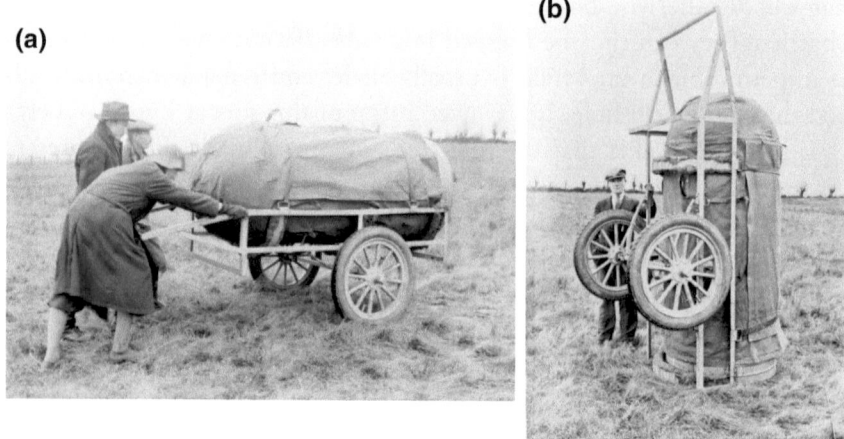

Fig. 8.5 Torsion balance survey in the UK, c. 1927 **a** on the move **b** erected for a reading. The instrument used in Australia by the Imperial Geophysical Experimental Survey was very similar. Photographs supplied by British Geological Survey under permission letter EA17/104

The Experimental Survey evidently provided a very thorough introduction to geophysics, since many of the people involved played significant parts in its development in Australia in the decades that followed. They included Jack Rayner, who became the BMR's Director, and Bob Thyer, who eventually became its Chief Geophysicist (but only after the post had been boringly renamed Assistant Director—Geophysics). Less predictably, one of the 'Temporary Assistants' was Norman Fisher, later the BMR's Chief Geologist and eventually Jack Rayner's successor as Director. He would have been a student at Queensland University at the time of the survey. The 'experimental surveyors' tried many different methods, and one of their greatest successes came when a torsion balance and a gradiometer were used in Victoria to map deposits of brown coal, a substance only one step up from peat and with a very low density. They were thus able to make a significant contribution to global warming.

Salt

If the Straits of Gibraltar were to close, the Mediterranean would dry up, and it might do so in as little as a thousand years. All that would be left would be a few hypersaline lakes similar to today's Dead Sea, and even these might in the end disappear, leaving thick layers of salt behind. This may sound unlikely, but it happened about six million years ago, in what is known to geologists as the Messinian salinity crisis. Nor was this a unique event in geological history. Most major sedimentary basins have their salt layers that can, like those in the Mediterranean, be several kilometres thick. Because there would not be enough salt in one Mediterranean to produce such a layer, these thicknesses can only be explained by repeated cycles of basin filling and drying. The sequence would in the end be terminated by a more permanent influx of water, producing a sea or an ocean within which sands, silts and shales would be deposited.

As the overlying sediments thicken, strange things can happen. Salt is slightly more than twice as dense as water, and normally also slightly denser than the overlying sediments when these are first deposited. However, as more sediments are laid down the density of the older layers increases as the spaces between the mineral grains are gradually closed down by the increasing pressure. Salt deposits have no pore spaces to close and retain their original density, so that the layers immediately above them become denser than the salt layer. Salt is also weak, and can flow under pressure. Bumps, called pillows, may form on the originally flat upper surface of a salt layer, and by their very presence create pressure differences that drive the process even further. The pillows may then cease to bend the sediment layers above them and instead force their way up through them, becoming diapirs or salt domes. If the process goes still further, the domes may merge, forming continuous walls (Fig. 8.6).

Even then, the story is not over. Common salt, sodium chloride, which geologists call halite, is not the only mineral found in the diapirs. They also contain much less soluble materials such as calcium sulphate. During the drying-out process this is the first chemical to be deposited, as gypsum and

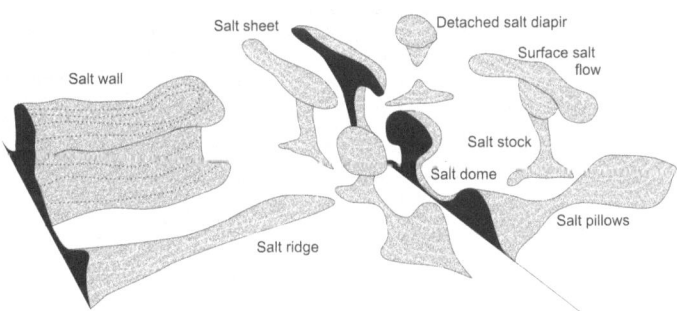

Fig. 8.6 The wonderful world of salt. It is easy to understand why structures such as these, formed of low density rock, have very definite effects on 'g'

anhydrite, but some will remain in solution and will eventually be incorporated in the halite mass. A rising salt dome may cut through sediment layers that contain abundant water, and if they do so, the halite may dissolve. The gypsum and anhydrite will not, and may remain to form a dense cap-rock over the crest of the dome.

Oil and gas can neither flow through salt nor be stored within it, but traps can be created in the uplifts over domes or where porous rock layers have been tilted and then cut by the rising diapirs. Many of the oil fields found the early days of oil exploration were associated with such features, and these were the targets of much of the exploration work done in the 1920s and 1930s, and especially of the work done in the Gulf States of Texas and Louisiana. Most of the early domes, including the country's first major oilfield at Spindletop, were discovered because the rising salt had pushed up the ground surface to form an easily recognised, even if quite subtle, topography. At Spindletop the ground had been raised by not much more than two metres, but that had been enough.

It didn't take the oil companies long to realise that there might be other domes that were even less obvious, and they began to interest themselves in the reports coming out of Central Europe of a new, geophysical, way of finding them. In 1921 Everett DeGolyer, at the time vice-president of the Amerada Petroleum Company and the country's most prominent petroleum geologist, bought two torsion balances from the Ferdinand Suss company in Budapest and, as part of the package, arranged for instruction in their use to be provided in Budapest by Desider Pekär, once assistant to Lóránd Eötvös himself. The man sent to Budapest by Amerada was Donald Barton, who became one of the most important and innovative of the early petroleum geophysicists and Member No 1 of the Society of Exploration Geophysicists, but who died tragically early, aged only 50, in 1939.

In June 1922 Barton reported back to DeGolyer from Hungary that

> As I see it now, the taking of observations at a station is really the simplest part of the whole procedure of Eötvös Gravitational measurements. It is the incidental work in preparation of the station and calculation of the results that takes the time and labor. This afternoon I roughly calculated up the number of operations entered into the calculation of a single station's observations. They were: Number of operations of addition and subtraction: 232. Number of operations of multiplication or additions: 99. Number of entries to be made in forms: 320
>
> These calculations must be made immediately the next morning in order to get a rough idea of how the gravitational structure is running so that the succeeding stations may be intelligently located. (Robertson 2000)

A tedious business indeed. Making the observations may have been simpler than doing the calculations but still made considerable demands on the field crews, who had to work day and night. The actual measurements were made at night, when there was the least background noise, but the site had to be prepared and made ready for use in daylight. A typical day began with breakfast soon after 5 a.m., after which all the apparatus was loaded on to wagons and the party, consisting of ten horses, five drivers, four workmen and three observers, moved off. Once at the chosen site, the surrounding area had to be surveyed out to a distance of a hundred metres so that terrain corrections could be made, and the area within three metres of the instrument had to be physically prepared to be '*as level as a tennis court*' with a slope of less than one degree. While this was being done the observers took magnetic measurements. One hopes that the man who was going to be up all night observing the torsion balance was allowed some time off, because his observational work began at about 9 p.m. and continued until breakfast the next day. After breakfast, he had the calculations to do.

The effects of salt domes on 'g' are not simple. Pillows will only form if the salt is lighter than the rocks immediately above it, but can continue to rise diapirically even after reaching levels where it is denser than the surrounding sediments as long as the pressure at the base of the salt mass is less than the pressure due to an equivalent column of sediment. A dome may thus be marked by a gravity low due to the deep salt of relatively low density, with a central high produced by the uppermost parts, where the salt is denser than its surroundings. There may also be a central high due to a dense anhydrite cap. Once back in the US, Barton selected the Spindletop Dome for the first tests, and Spindletop is one of the domes where a caprock is present. The tests proved that the dome could indeed be detected gravitationally.

That was promising, but several other known domes and prospects were then surveyed with disappointing results, and Amerada were on the verge of abandoning the method, partly on the grounds of expense (and if their field procedures were anything like those used in Hungary, they must have been very expensive) when a survey of a topographically unremarkable area on the Nash Ranch in Texas defined a gravity maximum almost identical to that at Spindletop. Drilling in November 1924 showed the cause to be a salt dome but it was not until more than a year later that oil was discovered. Unlike Spindletop, there was no oil over the crest of the structure and the discovery was made on its flank.

The Nash Dome is general accepted as being the first geophysical 'wildcat' discovery of a commercial oilfield (the Egbell anticline having been already

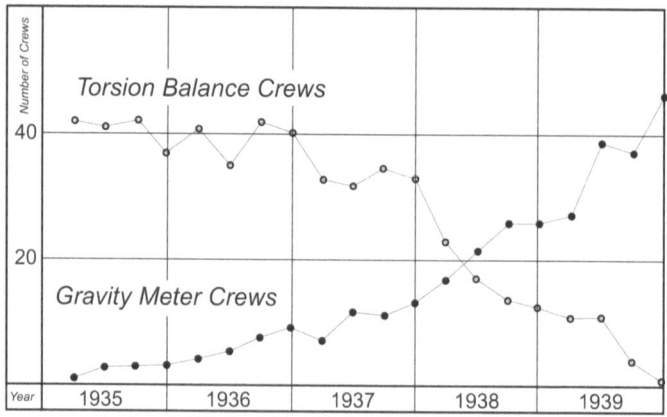

Fig. 8.7 Quarterly reported gravity crew counts for the US Gulf states in the years 1935–1939. Redrawn from Eckhardt (1941)

a gas producer when the surveys were made), and was sufficient to earn the torsion balance a place in exploration for the next fifteen years. The instruments were, however, considerably modified and improved during that time. In October 1936 the American Askania Corporation placed an advertisement in the journal *Geophysics* showing an oil company client greeting an 'operator' with the memorable line '*Every time I see you, your equipment gets smaller and smaller*', to which the operator replies '*Yes, this new torsion balance is the smallest, lightest and fastest instrument on the market*'.[5]

Sadly for him, by the time the advertisement appeared the use of torsion balances was declining rapidly, because the new gravity meters that were just becoming available were even smaller, lighter and faster and could be used to make measurements at dozens of different points in a survey day, while the fastest torsion balance crews could only manage six. By January 1939, American Askania was placing full-page advertisements in *Geophysics* for their 'mechanical-electrical' gravimeters. It would be nice to think that the operator in the torsion balance advertisement found fresh employment using the new instruments, and the fact that the number of 'gravity' crews operating along the Gulf coast of the United States remained almost constant during the changeover (Fig. 8.7) suggests that he probably did.

More than fifty years later, measurements of gravity gradients again became fashionable—but that is another story (Chap. 11).

---

[5]The advertisement is reproduced on page 2309 of the 50th Anniversary edition of the journal *Geophysics*, published in December 1985.

References

Cavendish H (1798) Experiments to determine the density of the earth. Phil Trans Roy Soc 88:429–526

Eckhardt EA (1941) History of gravity method of prospecting for oil. Geophysics 5:231–242

Falconer I (1999) Henry Cavendish: the man and the measurement. Meas Sci Technol 19:470–477

Howarth RJ (2007) Gravity surveying in early geophysics. II From mountains to salt domes. Earth Sci Hist 26:229–261

Király P (1993) Eötvös and STEP. Poster presented at the satellite test of the equivalence principle (STEP) symposium, Pisa, April 1993

Michell J (1760) Conjectures concerning the cause, and observations upon the phaenomena of earthquakes. Phil Trans Roy Soc 51:566–634

Pekär D (1928) Die Entwicklung des Eötvösschen Originaldrehwagen. Die Naturwissenschaften 51:78

Robertson H (2000) A historic correspondence regarding the introduction of the torsion balance to the United States. Lead Edge 19:652–654

9

The Rise and Fall of Springs

Making measurements with even the simplest of torsion balances or gradiometers takes a long time and the results are often difficult to interpret, but neither is it quick or easy to measure 'g' accurately with a pendulum. Temperature has to be carefully controlled, and even in vacuum chambers corrections have to be made for the buoyancy and drag effects of the residual gases. Knife edges are, at microscopic levels, actually rounded, so that the points of contact shift very slightly during each swing, and even double pendulums, swinging in opposition, cannot be balanced so perfectly that there is absolutely no flexing of the support. Pendulums have now been largely replaced for measuring total or 'absolute' gravity by instruments that measure the rate of fall of masses in vacuum, but these also require complicated corrections, and readings still take several hours. Neither pendulums nor falling weights can measure 'g' to much better than one milligal unless observations are made for days on end, and it is many years since either was used for field surveys. For those, we have gravity meters.

An Impossible Ambition

The distinction between weight and mass is fundamental, but most people get through their lives without ever worrying about it. Even scientists, once such people appeared, took a century or so to notice that the difference, which can be illustrated by weighing a pound of something with a spring balance and then with scales and a set of standard weights. Take everything

© Springer International Publishing AG, part of Springer Nature 2018
J. Milsom, *The Hunt for Earth Gravity*,
https://doi.org/10.1007/978-3-319-74959-4_9

to a different latitude and the scales will still show a pound to be a pound, because weight is balanced against weight. Modern travellers, worried about their baggage allowances, are more likely to use spring balances, and these, if sufficiently sensitive, will reveal the truth, which is that the change in 'g' alters the weight of the mass but does not change the springiness of the spring. A modern gravity meter is basically a spring balance in which the weight of a fixed mass is changed by taking it to places where 'g' is different.

The first person to suggest a spring-balance gravity meter may well have been John Herschel. The sole product of the late marriage of William Herschel who, along with his sister Caroline, was a notable astronomer, he followed in their footsteps. As a student at Cambridge (where he eventually graduated as Senior Wrangler), he became friendly with Charles Babbage and together they founded 'The Analytical Society'. After leaving Cambridge, they were amongst the breakaway group of Royal Society Fellows who founded the Royal Astronomical Society. It was a controversial move, particularly in the opinion of Joseph Banks, then still President of the Royal Society. On 29 February 1820, at the Astronomical Society's first meeting, the Duke of Somerset was elected President, but by the next meeting Banks had persuaded him to decline the offer of this '*hostile position*'. William Herschel then became President, but died soon after. By 1830 a Royal Charter was being sought, and an application was duly made in the name of Sir James South, the then President.

It was presumably hoped that this new society would allow astronomical science to be pursued in a more relaxed and collaborative atmosphere, but that was not to be. South had wanted desperately to go down in history as the first president of the chartered society, and was correspondingly furious when, because of delays in the grant and the Society's rule that no President could serve for more than two years consecutively (presumably to avoid the problems that had afflicted the Royal Society under the long incumbency of Joseph Banks), the Presidency passed elsewhere. He was in any case a famously acerbic individual who was

in the habit of strolling up and down his garden in the evening, shouting his grievances at the top of his voice to some friend, while people from the neighbourhood were regularly enjoying themselves on the other side of the wall by listening to his ravings.[1]

---

[1]Much of the information here has been taken from Dreyer and Turner (1923). The quotation celebrating South's irascibility appears on p. 55.

He cannot have been an easy colleague. I have in front of me, but sadly cannot afford to keep, his personal copy of '*The Philosophical Experiments and Observations of Robert Hooke*', a collection of notes and correspondence published some twenty years after Hooke's death. One feels some trepidation at even touching a book that collects the writings of one such combative individual and was owned by another. In Terry Pratchett's *Discworld* it would undoubtedly have attacked anyone who dared to open it. It may have been the endless bickering in the new society that in 1833 prompted John Herschel to pack up everything and decamp with his wife and family to Cape Town for the five years that he was later to describe as the happiest in his life. Before he left, however, he would have seen the great textbook that he simply entitled '*Astronomy*' published not only in England but in America. In it he set out his ideas for a new way of measuring 'g'.

Figure 9.1 shows how he thought it could be done. D, he said, is *a smooth plate of agate ... which can be adjusted to perfect horizontality by a level*. To use the instrument the length and the strength of the spring would have been adjusted *so that the weight F shall be sustained by just swinging clear of the agate plate in the highest latitude*. To make a measurement, small additional weights would have been added until contact was just made. Once this had been done in one place, the spring and the weights would have had to be carefully unhooked and put in separate boxes before being carried to the next site, where the whole contrivance would be equally carefully reassembled and rebalanced. Herschel seems to have thought it would be possible to measure gravity differences to about one part in ten thousand (about a hundred milligal) with such an instrument but seems to have been not entirely convinced that it would work at all. In a footnote he admitted that

> Whether the process above described could ever be so far perfected and refined as to become a substitute for the use of the pendulum must depend on the degree of permanence and uniformity of action of springs, in the constancy or variability of the effect of temperature, on their elastic force, on the possibility of transporting them, absolutely unaltered, from place to place &c.

In writing this he showed himself aware of most of the problems of using such an instrument, of which the difficulty in deciding when the mass was 'just touching' the base plate is merely the most obvious, and it is unlikely that he ever attempted to produce one or test it. The mere thought of assembling and disassembling it each time a measurement was made is the stuff of

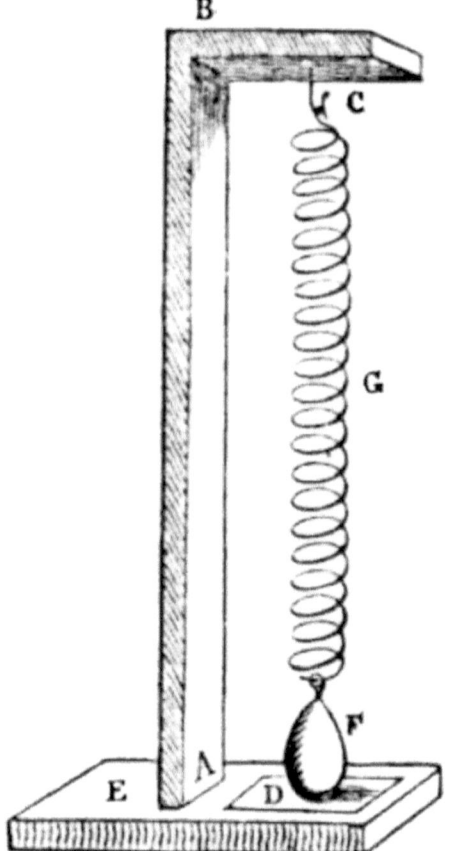

Fig. 9.1 The first spring balance gravity meter (Herschel 1834)?

nightmares for any modern user of a gravity meter. However, Herschel also recognised that

> …. the great advantages which such an apparatus and mode of observation would possess, in point of convenience, over the present laborious, tedious and expensive process, render the attempt to perfect such an instrument well worth making.[2]

The basic idea was certainly sound, and in the 1880s the Royal Society offered a prize to anyone who could put it into practice. Even Lord Kelvin tried to do so, but no-one succeeded. Then, a few years later, Lorand Eötvös developed his practical and portable torsion balance and the measurement

---

[2]The two quotations are from a footnote on p. 125 of Herschel (1834).

of 'g' in gravity mapping was abandoned for almost fifty years in favour of measuring gravity gradients.

Spring-Balance Basics

Unworkable as it was at the time, Herschel's design embodied most of the main features (and many of the shortcomings) of modern gravity meters and its very simplicity makes it easy for these to be understood, and for the necessary jargon terms to be defined. For a start, the instrument would have been relative, not absolute. When pendulums or falling weights are used, the calculations are based on simple underlying formulas that do not involve the properties of the materials from which the apparatus is made. These properties can affect the results, and corrections may have to be made for imperfections, but they are not fundamental to the method. The properties of Herschel's spring would, however, have been fundamental to his measurement, since the extension caused by a given change in 'g' would have depended upon them. The instrument would have needed calibration, and that would itself have been a difficult business. Nonetheless, and although the idea of making measurements by adding or subtracting weights may seem a little bizarre, this was one of the methods by which the gravity meters used in the second half of the 20th Century were calibrated.

The main problem with building any gravity meter is that very small effects have to be measured against a vastly stronger background. Because changes of as little as a tenth of a milligal can be important in exploration work, gravity meters need to be sensitive to changes of only a few parts per billion. A spring-balance instrument would seemingly have to simultaneously have a strong spring to support the mass but a very weak spring so that the response to a very small gravity change would be measurable. One early author suggested that the spring of a 'working' gravity meter would have to be about forty feet long. Herschel himself never specified a length for his spring, but the practical limit would probably have been about a metre and would have changed by only one-thousandth of a millimetre for a change in g' of one milligal. To achieve anything approaching this precision the base-plate would have to be very accurately horizontal and the support would have to be precisely vertical. All real gravity meters share this need for very accurately levelling and although in the most modern examples the final adjustments are made automatically, the operators still have important roles to play. With experience a typical instrument can be levelled in well under a minute but novices have been known to be still trying (and failing) after a full half-hour.

Herschel's meter would also have had a limited range. Once the mass was in contact with the base-plate it could go no further, so the greatest gravity field that could be measured would be the one that brought the mass into contact with the plate when no extra weights were attached. The smallest value would be defined by the sum of all the weights available that could be attached simultaneously. A gravity meter built to this very simple design would also have suffered from a defect that has been a feature of all later variants. It would not have given a constant reading if read repeatedly in the same place, because the inevitable temperature changes would have affected the elastic properties of the spring, and these would also have changed, over longer time intervals, as the spring weakened with age. To again use the modern jargon, the instrument would have suffered from both short-term and long-term 'drift'. Drift corrections are routinely estimated in gravity meter surveys by making repeat measurements at fixed points at the beginnings and ends of each survey day (and sometimes at regular intervals during the day) and then assuming that any changes observed have taken place uniformly with time. This assumption may be wrong, because many types of meter also suffer from 'tares', in which a re-ordering at molecular level of the spring materials leads to very high drift rates over intervals of just a few minutes.

Herschel's gravity meter would have been a 'null' instrument, in which the measurements would have been of the force needed to bring the mass (often called, for reasons lost in antiquity, a 'proof' mass) to its reference position in contact with the plate. This use of a fixed position is a feature of most, although not quite all, of the gravity meters that have been used since.

The Earth Tides

Drift causes the reading of a gravity meter to change, even if it remains in the same place, but there is another reason why such changes occur. The meter may be working and not drifting at all, but be accurately recording changes in the Earth tides.

Galileo came to grief over tides, because he could not accept the possibility of events on Earth being influenced by events in the 'heavens'. He was wrong, and the gravity fields of the Sun and Moon not only pull the seas backwards and forwards and also distort the solid Earth, but they act directly on any device used for measuring 'g'. The effects are small, never amounting to more than a few tenths of a milligal, but they are measurable and sometimes have to be corrected for, particularly when the value of 'g' is being established at a reference base. Fortunately, tidal forces are predictable

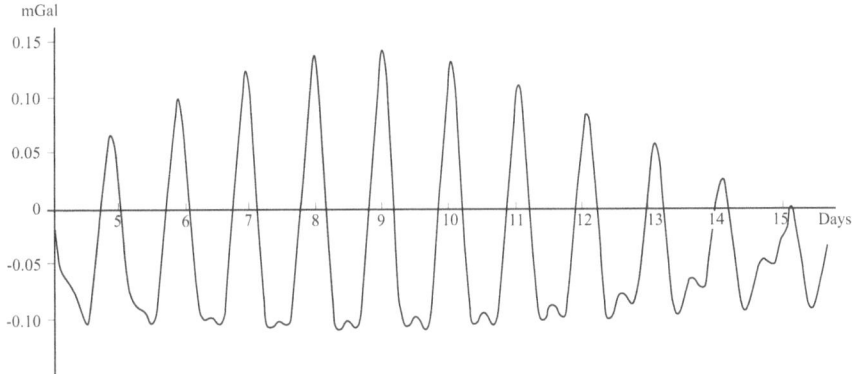

Fig. 9.2 Earth tides, for half a month. The total effect is a combination of the effect of the moon and the much smaller effect of the sun, and is at its greatest when the Sun, the Moon and the Earth are in line, at full Moon and new Moon (spring tides). When the lines to the Earth from the Sun and the Moon form a right angle it is at its weakest (neap tides)

via an equation of considerable size and ferocity, and the residual uncertainties in the overall effects as recorded by a gravity meter, which are due to the way the Earth distorts in response to those forces, are very small indeed.

It is common practice in routine surveys for gravitational tidal effects to be corrected as part of the instrumental drift but Fig. 9.2 shows that a constant rate of change with time cannot be assumed and that for the most accurate work, and especially for surveys at certain times of day, drift and tide should be dealt with separately.

1985: Southern Europe

It is probable that rather more of the results of gravity surveys are massaged, ever so gently, by field crews than the people who commission the surveys would like to think. Where a single value seems incompatible with measurements at nearby points, a transcription error may be assumed and a more plausible value may be substituted. Where the measured drift is slightly outside the contract specifications, a slightly adjusted reading may be written down, to avoid the rejection of an entire day's work. Modern meters that record everything automatically have reduced both the opportunities for taking such actions and the need to do so, but all sorts of short cuts were being taken before they arrived on the scene. A survey could easily become an, often quite good natured, battle of wits been the clients' representative, also known as 'the QC' (for Quality Control) or 'bird-dog', and the contractor's field crew.

Visualise: a flat, hot and dusty plain, stretching to the horizon, beneath which there may be masses of metallic ore that could change 'g' by several milligal. Imagine also that the only place where the field crew can rest at night with access to comfortable beds and cooling drinks lies at one edge of the plain, and that the nearest gravity base station, which contractually has to be visited at the beginning and end of every survey day to measure the meter drift, is near to the opposite edge. Suppose that the stations to be occupied are on a two-hundred metre grid and that the work is being done with a gravity meter that has a very low normal drift rate (of the order of a couple of hundredths of a milligal per day) but which is vulnerable to occasional tares that may sometimes exceed a milligal.

In the field notes, on which the plot in Fig. 9.3 is based, Day 6 begins with the base reading B1, followed by readings taken at stations spaced two hundred metres apart along one of the survey lines and at almost the same elevation. At the end of the day, a final base reading B2 is noted down and the field crew heads back to the comfort of their hotel.

Day 7 begins with another reading, B3, at the base, which differs from the last reading recorded on Day 6 by almost two milligal, implying a massive overnight tare. However, the instrument appears to be behaving itself and a day's worth of readings are taken. The initial and final base readings (B3 and B4 respectively) are similar, apparently confirming the overnight tare. However, the differences between successive stations along the grid are

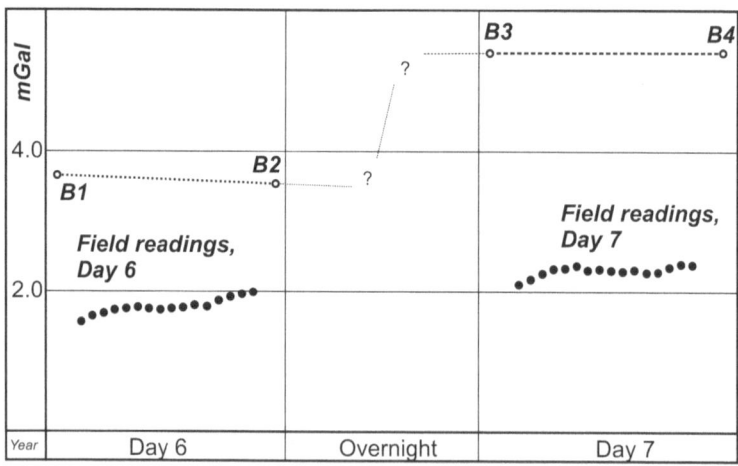

Fig. 9.3 Field readings, corrected for meter calibration but no other effects, plotted for two consecutive days of the survey described in the text. Open circles are readings at the remote base station, closed circles are field readings along one continuous traverse

not only all small, but they are similar to those recorded on Day 6, implying that the overnight tare indicated by the base readings could not have happened. This problem is only noticed when the calculations are being made back at head office, by which time the field crew has left the area. The results are eventually given to the client, with a note to the effect that the readings made on Day 6 are incorrect and should be ignored. No charge is made for them.

Because the data as plotted are inconsistent with everything known about the meter being used, the results are reassessed by the client. An assumption is made that allows the Day 6 stations to be used after all. The extra information is obtained without a penny having to be paid for it.

This much is fact. What follows is speculation.

At the end of Day 6 the observers were an hour's drive from the base, and ten minutes' drive from their hotel. It was still hot, and they were dry. The temptation to rely on the low drift of the meter was overwhelming. A final base reading B2 was invented and the hotel bar, rather than the base station, was occupied.

The observers might have done this more than once, but only on this one occasion were they unlucky. A tare did happen, but on Day 6 during the journey between the base station and the first station on the traverse line and not overnight between Day 6 and Day 7. Because the field crew had failed to notice that anything was wrong, the field sheets were sent for processing, which was done by a computer program rather than by a real person. Only when the map came to be drawn was it realised that there was a problem, and by then it was too late. The contractor could only make his excuses and waive payment, but the field sheets still had to be sent to the client.

A true story, with one useful result. During the 1990s it was becoming increasingly difficult to find test questions at university level that could not be answered by a quick internet search. Working out a probable sequence of events that could have produced this particular set of field notes was certainly such a problem, and very few of the undergraduates who were confronted with it managed to find a solution in the allotted fifteen minutes. However, the same exercise was also given to students on a parallel Master's course, most of whom had spent two or three years after graduation in jobs that involved working for, or contact with, geophysical contractors. Almost without exception, these people arrived at the probable answer in less than five minutes.

It is sad to contemplate the levels of cynicism that can be engendered by very brief experience of the real world.

The First Gravity Meters

During the 1920s and 1930s a number of ingenious instruments were proposed in which very small changes in gravity were converted by spring systems into much larger and more measurable changes in something else, and some of these were used quite extensively. The most historically important was the Gulf gravity meter, so called because it was developed by the Gulf Oil Company's Research and Development Center, but also very suitably named because it was mainly used along the Gulf of Mexico coastline in Texas and Louisiana.

In the Gulf gravity meter the spring was made of steel ribbon rather than wire and untwisted as it extended. A very small extension produced a relatively large rotation at the lower end of the spring, which was attached to the mass and to a small mirror. The rotation was measured optically, by reflecting a narrow beam of light from the mirror. In the production models the rotation amounted to about ten arc-seconds for every milligal of gravity change, and measurements could be made to a tenth of a milligal. The principal disadvantage was a very limited range, of about 30 milligal, because the reflected light could only be observed over a five-degree arc, but it was a simple matter to rotate the support to bring the reflection back into view when moving to a new area. The first prototypes were built in 1932 and the first production model entered service in 1935. Two years later there were eight instruments in use, and by the beginning of 1940 more than 200,000 stations had been occupied.

Other meters were designed and built using more conventional wire springs, and all the successful examples used Herschel's original idea, of measuring the adjustments required to return the proof mass to its original position following a change in gravity. There were two types. In the 'stable' or 'static' versions, the seeming inevitability of very small extensions was accepted, and the designers concentrated on making these measurable. Figure 9.4a shows the Hartley gravimeter, which was built with a relatively strong main spring and a much weaker restoring spring. As the main spring extended or contracted in response to changes in 'g', it dragged the hinged beam with it, and this was returned to its original position by rotating a calibrated screw that raised or lowered the upper end of the weaker spring. Very small changes in 'g' could be measured, because the restoring spring was not only weak but was much closer to the hinge and had much less leverage. As with the Gulf gravity meter, the return of the beam to its original position was observed using an optical system that provided further amplification of very small changes.

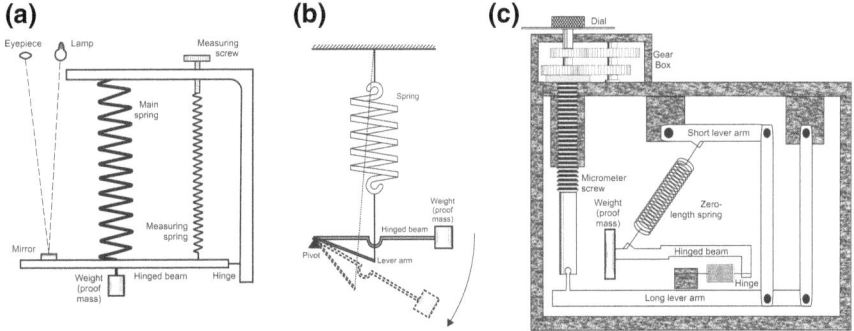

Fig. 9.4 Gravity-meter systems: **a** The Hartley static gravity meter **b** An astatic gravimeter system **c** The LaCoste meter with zero-length spring

Figure 9.4b illustrates the principle behind the 'unstable' or 'astatic' metres. The hinged beam is rigidly attached to a lever arm, and it is to this that the spring is connected. A change in the gravity field will cause the beam/lever arm assembly to tilt but, because it is attached to the lever arm, the spring will rotate through a greater angle than if it were directly attached to the beam. The system is highly non-linear (it depends on the sines and cosines of all the angles involved, and on the lengths of the spring, the beam and the lever arm), and can be made very sensitive to small gravity changes. Some very early gravity meters used this principle but it took one very ingenious innovation to make such instruments truly practical.

The Zero-Length Spring[3]

Gravity meters are expensive, costing tens of thousands of dollars, and only a few thousand of any model were ever made. With such small numbers, a close relationship can develop between maker and user, especially when the user owns more than one instrument, and several of my colleagues in Australia in the 1960s visited Lucien LaCoste in Austin, Texas (Fig. 9.5), where he made gravity meters. They all agreed on one thing. The old man (as he seemed to them; he was not yet sixty) was fanatically keen on tennis. Without that seemingly irrelevant fact, the LaCoste-Romberg gravity meter, which dominated its field for some forty years, might never have been created.

---

[3]Much of the information in this section comes from Clark (1984)

Fig. 9.5 Lucien LaCoste at the University of Texas at Austin, at ease on the wing of his Packard. It has been suggested that he may have loved the car even more than he loved playing tennis, but this is probably untrue Photo from Oakes (2017)

It is a good story and, unlike those told about Galileo, is amply documented.

When LaCoste was choosing a university for his undergraduate studies, he selected the University of Texas at Austin because of the presence there of Daniel Penick, who taught Greek. LaCoste wanted to study engineering, but teaching the classics was not the only thing Penick did. He was also a remarkably effective tennis coach. Under his tuition, the LaCoste backhand improved enough for him to become a member of the university's highly successful tennis team. Somehow, he also managed to graduate, in Electrical Engineering.

After his first degree, LaCoste chose to stay in Austin for postgraduate studies (and more tennis), and, as his only geophysics course, he enrolled in a class run by a seismologist called Arnold Romberg. Students at Austin were routinely given small practical assignments and Romberg decided to ask this engineering graduate to design a better seismograph. He was possibly surprised when LaCoste succeeded in doing just that, and probably even more surprised when he came to see him a few weeks later complaining that his tennis partner had not turned up, leaving him with nothing to do but think of an even better solution, involving something he called a 'zero-length spring'. This was a spring in which, over a range of lengths, the tension was proportional not just to the extension from some rest length, but to the actual length. It turned out that this unlikely-sounding object could be made and that, incorporated into the system sketched in Fig. 9.4c, it could indeed make a better seismograph. Just how it worked is discussed in Chap. 14, Coda 7. LaCoste wrote a paper about it and sent it to Romberg, who made only one change, which was to remove his own name as co-author. Almost certainly he was undervaluing his own contribution by doing so, because LaCoste was later to say of the first prototype that ....

> Before the morning was over, we had a contraption which looked terrible but behaved beautifully - a combination that Romberg always sought.

The paper (LaCoste 1934) was duly published and caught the eye of Reginald Copeland, one of the directors of the American Seismograph Co, who saw that the design could be used for a gravity meter. For a time it seemed that his company might become the manufacturer, but something went wrong with the relationship and LaCoste and Romberg formed their own company and began building and marketing gravity meters under their own names.

Unfortunately for the two innovators, they had not thought to patent their idea, and by the time they tried to do so it was too late. Patents are only granted on inventions that are not already in the public domain, and their idea was in the public domain, because they had put it there themselves. They had had two years (reduced still further nowadays) to file, and it took them longer than that to realise that what they had produced had commercial possibilities. They did eventually manage to get a patent, but only by introducing some very minor and inessential modifications. It was quite easy to produce gravity meters that used the same principles but left out

Fig. 9.6 The LaCoste gravity meter in action. Anti-clockwise from the bottom right. *1* An early model LaCoste meter, as used in APC-IEC surveys in Papua in the late 1930s. *2* The meter being transported through the Papuan swamps. *3* The meter set up ready for a reading. *4* The modern equivalent. Photos, except for (4), courtesy of Oilsearch Ltd

the inessential elements, and LaCoste later estimated that, while almost 98% of working gravity meters used his ideas, only 25% were actually his meters. However, the company was profitable enough to support his tennis habit well into his seventies, and in 1984 he described himself in the following terms:

> Lucien LaCoste likes to think of himself as self-employed but he could equally well be considered unemployed. He claims to be working on various problems which he has accumulated over the years. However, most – if not all - are probably beyond his abilities.[4]

Rarely for a pair of scientific collaborators, LaCoste and Romberg never fell out, despite having long lives (Romberg died in 1974, aged 92 and LaCoste in 1995, aged 87). During the Second World War they left the production of the gravity meters, which originally weighed more than 40 kg (Fig. 9.6, all illustrations except [4]) but by 1942 had become one-man

---

[4]The two quotations in this section are from a short introduction written by Lucien LaCoste himself to the reprint of his original 1934 paper (LaCoste 1988). The self-description is very much in the style of the humorist James Thurber, who was something of a cult in the 1950s and 1960s.

portable, to Romberg's son and worked together on designing flight simulators for fighter aircraft that were so advanced that they continued in use for a decade after the war had finished. When peace broke out they went back to making gravity meters for surveys on land, and designing and building variants for use on ships, in aircraft and even down boreholes.

2007: Of Patents and Promoters

Patent law is a strange thing. Patent lawyers are interested in two things only. Can the thing actually be built, and has anyone suggested it before? It was at the second of these hurdles that LaCoste and Romberg fell, but there is a story that Arthur Clarke, who founded the British Interplanetary Society, was even less fortunate, falling at the first, remounting and then falling beyond hope of recovery at the second. He may have been the first person to realise that if a satellite was placed in the appropriate orbit around the Earth, it would circle the Earth in exactly one day and it would therefore be 'geostationary', remaining fixed above one point on the Earth's surface, and he was almost certainly the first person to try to patent the idea. His application was, however, rejected because artificial satellites did not then exist, and his device could therefore not be built. Being unaware of the sad tale of LaCoste and Romberg, he then fatally used a geostationary satellite in a science-fiction story. When satellites began to be placed in orbit, he applied again for a patent but was refused because the idea was already in the public domain. That it was he who had placed it there interested the lawyers not at all.

The story may not be true, but what is true is that patent lawyers worry not at all about whether an invention will actually work, and the world of geophysical exploration has always been awash with devices that do not. The first ever article in the monthly journal *Geophysics*, which is still the world's most important periodical for exploration geophysicists, was devoted not to science but to '*Doodlebugs*' (Blau 1936). The editor's introduction noted that:

> The term 'doodlebug' is coming more and more to mean proposed methods of geophysical prospecting that are neither based upon scientific fact nor upon known or proven properties of oil, minerals and geologic formations. The geophysicist is often consulted concerning the reliability of such a proposed method, and his task is then to explain scientifically just why the proposed method fails and is unsuited for the intended purpose.

Things have not changed very much over the years, and some of the doodlebugs even get to be patented. Seventy years after the article was written, a company applying for oil exploration leases in a country that had, up to that time, seen very little by way of exploration, came up against an obstacle. The minister responsible for issuing the leases had somehow been convinced that the only way ahead was to use a fantastic new instrument which, it turned out, had been patented. A rival company was planning to use it, and were therefore on the inside track.

The promotional literature for the device claimed that it was a new and unique way of finding oil, but at least the patent was clear about what it really was. It was a gravity meter. It would, at the best, do only what existing gravity meters were already doing, but it did have one unique feature that the others lacked. It couldn't possibly work. The design was based around a small float in a fluid reservoir. The idea was that, as the force of gravity changed, so the float would rise slightly higher, or sink slightly lower in the reservoir. This change could be measured, and converted into an estimate of gravity field.

As Archimedes could have told the minister, no matter how much the gravity field changed, nothing useful would happen. If the weight of the float changed, so would the weight of the fluid that it was displacing. The flotation line would remain exactly the same.

Of such stuff are (some) patents made.

Meters Made of Glass

LaCoste-Romberg gravity meters used steel springs. Steel is an excellent conductor of heat, and its strength is also quite severely affected by temperature, so that the instruments had to be maintained at constant temperature. The heavy batteries needed to supply enough heat to keep them warm for full survey days represented more than half the weight of the entire system, and keeping the batteries fully charged was a continual problem in remote areas. Sam Worden, a scientist working for Texas Instruments, decided to solve the temperature problem without recourse to batteries, and in late 1947 he applied for a patent on an instrument that, while still using a zero-length spring, was made entirely of fused quartz, or glass.[5] In the words of his patent, the

---

[5]These are generally known as 'quartz meters', even though quartz, once fused or melted, is a glass and no longer quartz at all. The 'quartz' idea may have been popularised by manufacturers who felt that their customers might not take 'glass' meters seriously.

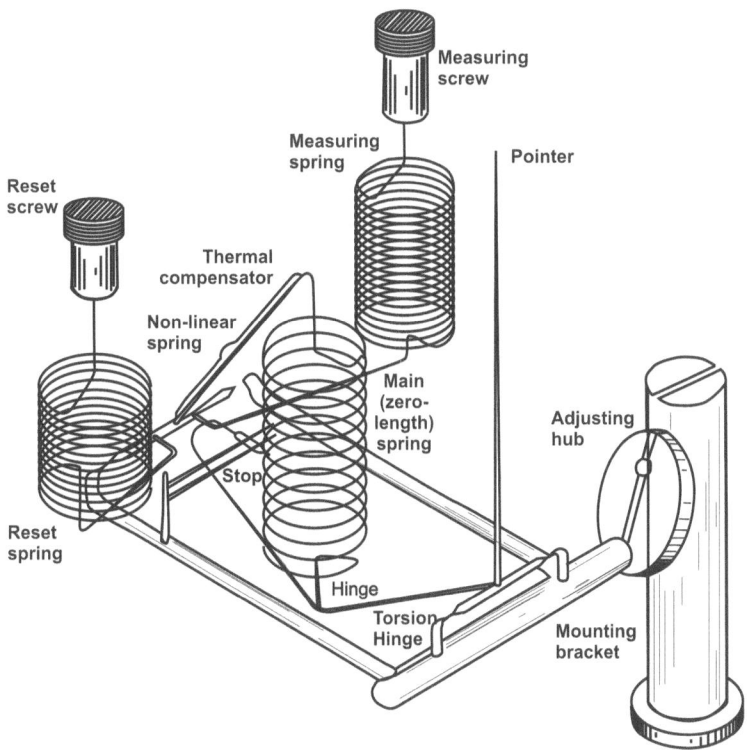

Fig. 9.7 The essential elements of the Worden gravity meter. The entire spring assembly is made out of a single piece of glass. The measuring spring, which can be raised or lowered by rotating the measuring screw, eventually gave these meters a range of about 200 milligal, although in early versions it was only 60 milligal. Drawing based on Worden company literature

conventional gravity meters now commonly used weigh from fifty to seventy-five pounds, whereas the meter embodying the present invention will weigh approximately five pounds. (Worden 1954)

Popular gravimetry had arrived

Spinning springs from glass is more art than science, especially when there are three of them, each with its own exacting specifications, to be attached to a glass framework and a glass proof mass to make a single instrument (Fig. 9.7). Only a few people have ever managed it, and it is commonly assumed that the reason that science museums throughout the world can proudly display examples of the insides of Worden gravity meters is that these are the ones that didn't quite work and are therefore essentially worthless. However, quartz meters have real advantages. The physical properties of glasses change relatively slowly with temperature, they are unaffected by magnetic fields and they are very poor conductors of heat. The temperature variations that are the main cause of meter 'drift' can be kept at levels acceptable for most purposes by enclosing the entire assemblies in vacuum chambers. Field parties began to go out on survey with instruments that looked like oversized thermos flasks and were often assumed by passers-by to be containing their lunch. Drift was reduced still further by the inclusion of an ingenious compensator made of two different materials with different rates of thermal expansion. As the patent rather charmingly put it:

> It will be apparent to those skilled in the art that change in temperature will cause the relative lengths of the arms to change …… in such a way as to effectively compensate for the effect of temperature variations …

If constant temperature was considered really essential, it could be achieved using heaters powered by ordinary torch batteries.

Sam Worden got his patent in 1954, but by then he had already begun manufacturing. The Wordens and their near-clones (the Sharpes, Scintrexes and Sodins) were much lighter than the LaCoste meters of the time, and for a few years they dominated the market. However, in 1960 LaCoste and Romberg struck back, with the 'G' (for geodetic) meter, which became the state-of-the-art instrument of choice for gravity surveys on land for forty years. The last of the more than 1200 built was delivered in 2004. Many are still in use, but the people who can operate them quickly and accurately are becoming rarer each year

Controlling temperature by enclosure in a vacuum chamber does have one major disadvantage. The mechanism has to be so completely isolated from the outside world that it cannot be clamped when not in use, making it vulnerable to tilting and shock when in transit, whereas the LaCoste company claimed that one of their meters, when clamped, could survive any accident up to and including a helicopter crash. True to the spirit of scientific enquiry, the Australian government put this to the test on one of their aid programmes in Indonesia and found it to be correct. The unclampable quartz meters were much more delicate, and a nightmare to transport. Almost all their users have on at least one occasion been involved in massive arguments with an airline that insisted that the meter should travel in the hold, at the mercy of the baggage handlers, and not in the cabin. Even the purchase of a seat for the instrument as well as one for the operator was not always enough.

Practical Gravity

During the second half of the 20th Century 'g' was being measured almost entirely with gravity meters built around the zero-length spring. During that time, millions of individual readings were taken. States, provinces and countries established their own databases, some containing hundreds of thousands of data points. Major oil companies had archives of similar size, although these were sometimes lost or degraded during mergers and take-overs.

At every reading point or 'station', the process would have been much the same. A three-legged concave aluminium dish would have been placed on the ground and made firm. The gravity meter, also on three legs (for stability, because three points define a plane, and a three-legged stool does not rock), would have been placed on it, and would have been shuffled around until both of the level bubbles were approximately centred. Fine adjustments would then have been made by rotating the screw-threaded instrument 'legs' until the bubbles (or, later, their electronic equivalents) were precisely centred. At this point the spring system would have been precisely vertical and the measurement could be made. The lever arm, or some proxy for it, would be observed through a telescopic eye-piece, and would be centred using a calibrated screw. The reading would be shown on a dial or digital counter, and would be recorded, along with the time and some form of station number. The survey as a whole would have been divided into loops, each beginning and ending at the same station. The difference between the initial and final reading would then be used to estimate the rate of instrumental drift.

There were, of course, differences. Many of the readings during those decades would have been made with LaCoste meters, with their steel springs, but probably rather more would have been made with meters with quartz springs (Fig. 9.8). Most of the LaCostes used for this work were the 'geodetic' G-meters that they could be taken to any point on the Earth's surface and be used immediately, without reset. The quartz meters, in which the lever arms were moved by weak auxiliary springs, had limited ranges, usually of about two hundred milligal but sometimes very much less. When they were taken to places where the gravity field was very different, a second auxiliary spring was used to make adjustments that were either uncalibrated or calibrated to much lower accuracies, to bring them within range. With limited-range instruments the dial readings were converted to milligal differences using a single conversion factor, but the adjustments to the G-meters were made using a very long screw acting on the single spring. Its pitch inev-

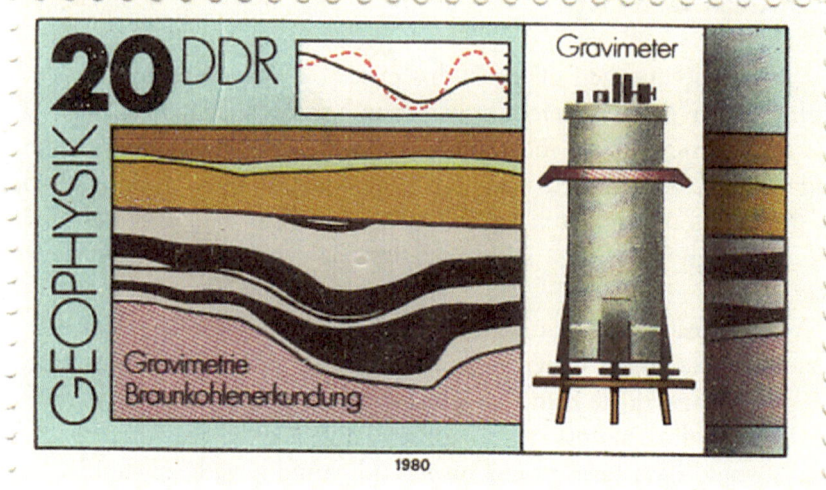

Fig. 9.8 Gravity and philately. The stamp was issued by the German Democratic Republic shortly before it ceased to exist when Germany was unified. The make of meter is not specified but is clearly one of the types using a vacuum-flask for thermal insulation. Note the three-legged support with levelling screws. The continuous curve shows the sort of change in gravity that could be produced by the 'brown coal' geology, but the origin of the dashed curve is a mystery

itably varied slightly over the range, and different conversion factors were listed for each hundred milligal interval.

Each instrument had its advantages and drawbacks. Set against the advantage of being able to clamp a LaCoste meter during transit, making it almost immune to shock, was the disadvantage that it had to be maintained at constant temperature by heaters powered by heavy batteries. Keeping the batteries charged in remote locations was a challenge, but the reliance of quartz meters on the very low thermal conductivity of quartz and on insulation within vacuum chambers to keep temperatures reasonably constant almost always led to much higher drift rates. In the end, the choice often came down to availability and personal preference. Availability, in turn, was often dictated by the fact that the LaCoste meters were about twice the price of most of their competitors.

So—the accident in Eastern Papua that propelled me into a career in gravity was not really the fault of the man in the field. The people who should have been blamed were the ones who decided to send him out, not with a LaCoste G-meter valued at about $30,000 but with a Worden which, while costing only half as much, was much less field-worthy. And the meter was not actually broken (it might have been better if it had been). It had merely been tilted enough for the various springs to become entangled and, as Fig. 9.7 suggests, this could be very easily done. That it had happened was not immediately obvious because, in an area where the results were unpredictable (as they certainly were in the Bowutu Mountains), the instrument had seemed to be working, and a considerable amount of money was spent, pointlessly, on helicopter hire. Wisely, after that experience, and as well as sending out geophysicists and not geologists to do gravity surveys in New Guinea, the field parties were always equipped with LaCoste meters.[6]

1975: An Earthquake Detector

True to their origins in Arnold Romberg's classroom project, LaCoste meters (in common with all other gravity meters) are also seismographs, and detect seismic waves. During an earthquake, the reading needle travels slowly from one side of the field of view to the other and refuses to settle. The movement is distinctive and easily recognised, and there is absolutely nothing that can be done except wait until the last of the seismic waves have passed. This can be a nuisance in places such as Indonesia and Papua New Guinea, where events large enough to cause problems can be expected to occur at least once every few weeks, but in those cases the sources are usually local ones (say, within a thousand kilometres) and the wave train soon passes. Really large shocks can be detected all around the world and, because the waves travel by many different routes and at many different speeds, it can be hours before all of them have passed a distant location.

Working in Oxfordshire on a fixed-price contract (for the Gas Board) in the 1970s, we usually took full advantage of the long summer days to make as many measurements as possible, but on one day, to our landlord's disgust, we were seen to be lazing around in his hotel lounge at two in the afternoon. It was not until we had persuaded the British Geological Survey to send him the earthquake summary for that particular day that we were able to convince him that we had been prevented from working by a very large quake (a magnitude 7.9 shock near Mindanao, in the Philippines). Today, we would simply have referred him to the internet.

---

[6]A few textbook authors, some of whom should have known better, have claimed the lack of a clamp as an advantage, but an unclamped Worden is actually at all times as vulnerable as a LaCoste that has been accidentally left unclamped and, as this story shows, has to be treated with extreme care. If the springs become tangled, that is the end of the survey. Only the manufacturers can sort out the mess.

1976: A Phantom Survey

The basic gravity survey techniques of laying out traverse lines (usually straight, although in regional work they may follow roads), making measurements at intervals along them and then contouring the processed results have changed very little over the years, even though, thanks to Global Positioning Satellites, the methods of ensuring that the measurements are made in the right places have improved out of all recognition. A whole industry has grown up to do this sort of work. Gentle massaging of data may sometimes occur, as already described, but the fabrication of an entire survey would be much more difficult. I have only once come across anything that came even close. On that occasion, as with the early use of gravity meters, the target was salt, and the affair had its funny side.

Throughout much of the 1970s I made my living as an independent consultant, and at one point was asked by a contact who'd been asked by another contact, and so on through a long chain, whether I would like to spend a few months teaching at a large overseas university. The story was that the geophysicist in their geology department had died rather suddenly and, with the final exams looming, someone was needed to fill the teaching gap. The request had been passed through many many hands before reaching me because the job itself was not that financially attractive, but there were some large contracts coming up in the country concerned, I had never been to it, and it seemed a good opportunity to discover if I really wanted to spend a couple of years of my life there.

The students were delightful and very keen, but it soon became obvious that the need to have them taught had not been uppermost in the university's collective mind when it began its hurried search for a replacement. The deceased staff member had been half-way through some potentially very profitable gravity surveys for a local mining company, and the university needed to see the contract completed and the money paid. Otherwise the considerable advance payment would have to be returned—and it had already been spent!

It looked an interesting project and even though the fieldwork had supposedly been completed, I assumed that it would not be too hard to justify at least one site visit, and see a bit more of the country. It was only when the various maps and notebooks were examined in detail that a problem emerged. There had been three survey areas, separated from each other by a few kilometres, and for two of these there were very respectable maps, and computer listings of the data for the individual stations, but no field notebooks or surveyors' reports. For the third survey there were notebooks and reports, but no maps. When maps were drawn, using the information contained, they made no sense. The obvious next step was to take the departmental gravity meter and visit the field area—which turned out to be a wide and rather depressing plain entirely given over to the production of root vegetables.

The previous field work had been completed only a year earlier, the small concrete blocks that marked the gravity stations were mostly still in place and it was easy to find villagers who had worked with the geophysicists and surveyors. The new gravity readings agreed almost exactly with those in the field notes from the original survey. There had been nothing wrong with that part of the work, but another problem did emerge. Just half-way along the first line, our helpers cheerfully pointed to one of the concrete markers from the line next to it. These two lines, supposedly five hundred metres apart and only two kilometres long, actually crossed, so it was not surprising that the maps looked bizarre. The remainder of the trip was spent working out where the lines actually went, which was very, very different from where they were supposed to go.

Once all the stations had been plotted in their proper places, a reasonable map should have emerged, but the new version looked only slightly better than the old one. At this point, errors in the heights above sea level of the reading points were the only things left to investigate, and since errors of only five metres would have produced errors in 'g' equal to the one milligal contour interval, that investigation had to be done. It had not been done earlier because the distortions on the map were so large that height errors seemed unlikely to be the cause, but since the measurements had been made by the same contractors who had so disastrously mis-surveyed the line locations, the possibility seemed to be there. On the second trip into the field, the gravity meter was abandoned in favour of a surveyor's staff and an optical level.

The marker blocks were only fifty meters apart, and the first height difference measured was in error by more than six meters. A few more measurements were made, just to be sure, but there wasn't much point in staying any longer, as an entire re-survey would have to be done, and that would take weeks. The rains began that night and we left in a hurry, but even so we almost had to walk home as the dirt roads turned to mud. There would be no more visits to the area for months, and since I had less than a week left in the country there was not much to be done but write the report, blame the surveyors and leave, with a sigh of relief.

There were, however, still the other two surveys, with impeccable final maps but no supporting data, to be explained, and it was hard to believe that they could be entirely fictitious. In the first place, it is not easy to fabricate an entire gravity survey and make the results look convincing, and in the second, the truth would be almost bound to come out eventually, with unimaginable consequences. It wasn't until two days before I was due to leave that the breakthrough came. Stuck in the folder containing the previous year's examination results for that year's finalists (they did get taught, despite the trips into the bush), there were papers that showed what had really happened.

Some years before there had been civil war. Many buildings had been destroyed, and many records had been lost, and these had included the field notes from two gravity surveys. Someone in the university had found the final results and someone, perhaps the same person, had decided that the company that had commissioned the work (and which had been on the winning side) might still want the work done, and might be willing to pay for it all over again. The only problem with this plan was that the company might want to send people out to visit the field operation to see their money being spent. Since there had to be something for them to see, the third area was added, and was surveyed. But, except for the gravity measurements, very, very badly.

My last few hours in the country were spent writing a report for the company, which was posted from the airport departure lounge.

I have never been back.

Gravity on the Moon

One of LaCoste's instruments even went to the Moon, with Apollo 17, but it is something that he might have preferred to forget.

It took little more than three years from the successful Apollo 11 Moon landing in 1969 for the mood in much of the US to change from enthusiasm to apathy. It did seem, as some people had predicted, that the surface of the Moon was actually a very boring place, and it was widely suspected that the dramatic near-catastrophe of Apollo 13 had been orchestrated by NASA to revive the flagging public interest. If so, it worked only briefly and, although not planned that way, Apollo 17, launched in late 1972, was to be the last visit by the Earthmen of the 20th Century to their satellite. The rocket took with it a payload of scientific experiments, including two gravity meters.

Fig. 9.9 Detail from NASA photo AS17-142-21730 showing the Traverse Gravimeter deployed off the rover by Gene Cernan at Station 8 at the base of the Sculptured Hills, Taurus-Littrow Valley

By the time that Apollo 17 was being planned, the LaCoste-Romberg G meter had become the 'must have' piece of equipment for every organisation that was serious about measuring 'g', and it is rather surprising that the company was not asked to supply the Traverse Gravimeter that was to do work on the Moon similar to the surveys then taking place all over Earth. That job was assigned to a Bosch Arma vibrating string accelerometer, which required less in the way of calibration for the lunar surface gravity. A total of 26 measurements were made across the Taurus-Littrow valley close to the landing site (Fig. 9.9) and the results were interpreted as showing that the lavas in the valley, which were denser than the older rocks that formed the hills on either side and which had been broken up by millions of years of meteor impacts, were at most about a kilometre thick (Talwani et al. 1973). This was a little bit less than the 1.4 km estimated from seismic surveys, but there is little evidence that anybody really cared. More interesting, perhaps, was the tie made between the Earth and the Moon that showed that the 'g' at the landing site was 162,695 milligal, with an estimated uncertainty of five milligal (Giganti et al. 1973).

The other gravity meter included in the scientific package was intended to remain in one place and send back records of lunar 'g' even after the astronauts had left. It was hoped that the changes recorded would provide information on the deep structure of the Moon and might even, when compared with simultaneous measurements made on Earth, lead to the detection of the mysterious, and at the time hypothetical, gravity waves. For these purposes something better than the tenth of a milligal sensitivity of the Bosch Arma was required, and the design and manufacture was given to LaCoste-Romberg.

By this time the company was making some meters, known as D-meters, that had ten times the sensitivity of the G-meters and also, for underwater use, meters that could be operated by remote control. All that was needed was to combine these two properties in a meter that could be used on the moon and add telemetry. The main difficulty was one that would have been recognised by Basil Hall, swinging his pendulums a hundred and fifty years before. It was not possible to follow his far-sighted recommendation that, before serious work was attempted, any instrument should be tested by an observer *'in the fields, and with no advantages save those he could carry with him'*. Unfortunately, while centrifuges can be used to produce gravity fields on the surface of the Earth that are larger than 'g', there is no way to produce total fields smaller than 'g'. The necessary settings on the lunar meter had to be calculated in advance, the instrument had to be calibrated using those settings, and it then had to be despatched to the Moon, on a rocket and a prayer.

In the end, four instruments were made. The first two were built for proof-of-concept only and, the concept having been proved, were discarded. It was then decided to build not one but two lunar-ready instruments, although only one of these would actually leave the Earth. The other would be for back-up, in case a problem emerged with the meter chosen to go in the days immediately before the launch. At the time this would have seemed a sensible precaution, but it was ultimately, and in conjunction with another decision made at the same time, to prove fatal. The second decision was to give the lunar meters ranges of only 100 milligal, as compared to the 200 milligal of the usual D-meter and the 7000 milligal of the G-meter.

Meter Lunar-3 was delivered in July 1971 and Lunar-4 in December 1971. It was Lunar-4 that went to moon and was set in place on its surface by the mission commander, Eugene Cernan. Control was then transferred to the team back in Houston. The telemetry and the remote levelling worked perfectly, but the gravity field that the instrument was trying to measure proved to be outside its limited range. Lucien LaCoste was called in for advice and the source of the problem was identified. It lay in his own calculations, which had been made, with complete accuracy, for Lunar-3 but had then been used in setting up Lunar-4. The error was small, and had the full 200-milligal range of the standard D-meter been used, all would have been well. The 100 milligal range was simply not enough. Eventually an operator in Houston was able to get the instrument working after a fashion, but not with anything approaching the sensitivity required. The Lunar Surface Gravimeter was the only scientific instrument on Apollo 17 that did not work as intended, and one of only two failures in Lucien LaCoste's otherwise remarkably successful career.[7]

The Making of a Monopoly

In the end, and as far as land gravity meters are concerned, it was the quartz meters that triumphed. Ironically, this was partly because they were so difficult to make using designs based on zero-length springs.

Soon after the Worden meters came into general use, a challenger emerged in the Sharpe Instruments Company of Toronto, where the quartz sensors were created by a glass-working genius called Wolf Sodin. The meters were popular, and in 1967 the company merged with Siegel Associates, which specialised in instruments for mineral exploration, to form the Scientific Instruments Research and Exploration Company, or Scintrex. Marketing was initially of a model known as the CG-2, which was essentially the same as its Sharpe predecessors.

What happened next is not well documented, but at some stage after the merger Wolf Sodin began selling meters under his own name (they are still being produced—in China) and the Scintrex company had a major rethink. They decided that manufacturing gravity meters had become altogether too dependent on extraordinarily skilled individuals, and that Herschel had been right after all. Beginning in 1984 they designed and built a completely new single-spring quartz meter, the CG-3, under the direction of Andrew Hurgill. That he could succeed where Herschel and Kelvin had failed was thanks to a technology that had advanced to the point where changes in spring lengths of less than one part per billion could be measured. Unsurprisingly, Herschel's idea of having the proof mass just touch an agate base-plate was rejected in favour of electronically measuring its distance from a metal base, and electrostatic forces were used to restore it to its null position. The whole process was made automatic through a feedback loop, and the fact that this was electrical rather than mechanical allowed

---

[7]The full story, with reproductions of some of LaCoste's original notes, is told in Chapin (2000). The second of Lucien's 'failures', involving some problems with his early marine gravity meters, is discussed in Chap. 10.

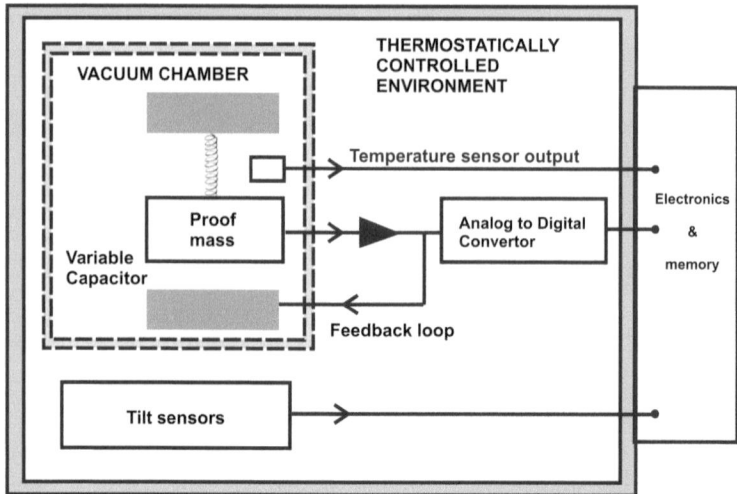

Fig. 9.10 Simple schematic of the CG-5. The single spring supports most of the weight of the proof mass, but a small amount of additional support is supplied by the electrostatic force acting between the plates of the capacitor formed by the lower surface of the proof mass and the upper surface of the base plate. This force is regulated by the feedback loop that restores the proof mass to its null position. Drawing based on Scintrex company literature

the sensor assembly to be entirely isolated from the outside world in a high-grade vacuum. The spring could not get tangled with any others, because there were no others. The movement of the mass could be closely confined, making the instrument insensitive to shock, and temperature could also be monitored electronically and continuously, for corrections to be applied.

The LaCoste-Romberg company, now without either of its founders, responded with their own automatic meter, but it was too late. The CG-3 had got there first, and had become established. Nevertheless, it is something of a miracle that Scintrex survived, because in 1997 it was the target of a successful hostile takeover on the Toronto stock exchange. The new owner, Mariusz Rybak, immediately fired the company's existing CEO, took on the job himself and began an acquisition-fuelled re-organisation of the company. Accounting irregularities appeared, auditors resigned and in May 2000 Rybak was forced out. Somehow, the recently appointed head of what had become merely the Scintrex division of Rybak's Intelligent Detection Systems managed to steer the company through to a merger with LaCoste-Romberg in 2001. Of the two competing designs for an automatic gravity meter, now within the one company, it was Hurgill's that survived.

The CG-3 was almost fully automated and its successor, the CG-5 (Fig. 9.10), which was first marketed just after the merger in 2002, has taken this still further and makes few demands on its operators. They do not even have to be near the meter when the reading is made, which is a great advantage in areas where the ground is soft and the operator's weight can cause continual tilting. Some 90% of all new gravity meters now being sold are CG-5 s, and the company also dominates the measurement of absolute gravity with its FG-5 weight-drop instrument, acquired when Rybak took over Micro-g Solutions of Denver in 1999.

The End of an Era?

Is this the end of the story?

For most purposes there would be little point in producing gravity meters more accurate than the CG-5, which is already sensitive to changes in height of less than a centimetre, the practical limit for field surveys. To replace it, a rival machine does not have to be more accurate, but it must be cheaper, smaller, lighter and, ideally, even less breakable. Several groups are trying to achieve this, and at the time of writing the most promising contender is a half-inch rectangle of silicon suspended by silicon fibres no thicker than a human hair. This has one great advantage over its rivals, which is that a low-sensitivity version is currently being field tested, in millions of examples, in the MEMS (micro-electromechanical system) accelerometers that ensure that the images on mobile phones and tablet computers are always the right way up. The high-sensitivity version being developed at Glasgow University has, almost inevitably, become known as the 'Wee-g'.[8]

This poses a question. If the equivalence principle is true, and gravity forces and accelerations are indistinguishable, then the terms 'gravity meter' and 'accelerometer' seem themselves to be equivalent. A newcomer to the technology might well echo Shakespeare and ask 'What's in a name?' As with the Capulets and the Montagues, the answer is largely a matter of history, and attitude.

As described by Walter (2007), in their early days accelerometers were designed mainly for measuring the accelerations of such things as passenger lifts and aircraft shock absorbers and the vibrations of steam turbines and underground pipes, but by the beginning of the 21st Century they had

---

[8]Although this name was not used in the formal announcement by Middlemiss et al. (2016)

spread across the world in the triggers for motor-vehicle airbags. All these uses involve large accelerations on very short time-scales. Whatever it is that tells an airbag to inflate, it has to do so very fast, and such instruments are conceptually a very long way from the gravity meters, pendulums and weight drops used for measuring 'g'.

Another difference is that while gravity meters are deliberately designed to be, to some extent, self-levelling and measure only 'g', accelerometers have been equally deliberately designed to measure accelerations in whichever direction they are pointed. And, while the emphasis in measuring 'g' has always been on accuracy at the expense of speed, in many accelerometer applications accuracy was dispensed with if it slowed the response. Stability was often questionable and drift rates were often high, but these problems are being increasingly successfully addressed as the technology improves. The Wee-g seems to have largely solved the stability problem, since it has already proved that it can measure Earth tides to microgal accuracy over periods of days. The challenge now is to make it portable and durable and, given the extraordinary range of applications that could be found for gravity meters costing only a few thousand dollars and weighing only a few ounces, it is hard to believe that these things will not be achieved. Prototypes are already being developed that are only a few inches across and draw less than a watt of power. Vacuum can be enhanced and maintained by chemical scavengers and both the thermal control and the optical read-out can be on-chip; automatic levelling can be included, and all of the read-out and control software will, inevitably, be computer controlled.

There is another strand. MEMS devices may become the norm for field surveys but greater sensitivity is needed in aircraft and satellites to measure changes in the much weaker gravity fields high above the surface of the Earth. Down on this surface this is being achieved by superconducting gravity meters that are never moved and which are recording changes in gravity due to such things as the rise and fall of groundwater and even the evaporation of dew (Van Camp et al. 2017). The technology involved is daunting, and the instruments are never going to be either cheap or portable. They require niobium spheres the size of ping-pong balls to be kept at temperatures so low that their electrical resistance vanishes and they can be supported by magnetic fields produced by currents circulating in superconducting wires. The tiny changes in current required to keep the spheres in place as 'g' changes are recorded and converted into milligals. The instruments are relative, not absolute, but the method of calibration shown in Fig. 9.11, where a 13-year old boy is used as the test mass, is unlikely to ever become standard.

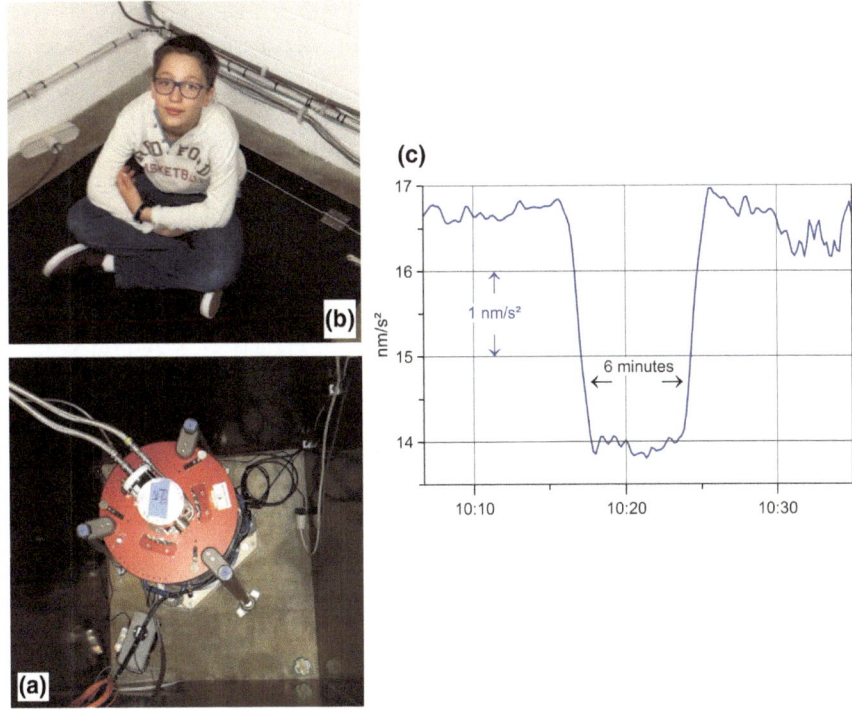

Fig. 9.11 **a** A superconducting gravity meter belonging to the Royal Observatory of Belgium, installed in a shaft at the Rochefort site. **b** The human test mass, weighing 45 kg, centred one meter above the instrument. **c** His effect on 'g', averaging approximately 0.28 μgal, measured over a period of six minutes. Photos and graph provided by Michel van Camp

References

Blau LW (1936) Black magic in geophysical prospecting. Geophysics 1:1–8

Chapin D (2000) Gravity measurements on the Moon. Lead Edge 19:88–91

Clark LD (1984) Lucien LaCoste. Lead Edge 3(24):29

Dreyer JLE, Turner HH (eds) (1923) History of the Royal Astronomical Society 1820–1920. Royal Astronomical Society, London

Giganti JJ, Larson JV, Richard JP, Weber J (1973) Lunar surface gravity meter experiment. In: Parker RA (ed) Apollo 17 preliminary science report. National Aeronautics and Space Administration, Washington

Herschel J (1834) Astronomy, Longman, Rees, Orme, Brown. Green & Longman, London

LaCoste LJB (1934) A new type of long-period vertical seismograph. Physics 5(178):180

LaCoste LJB (1988) The zero-length spring gravity meter. Lead Edge 7(178):180

Middlemiss RP, Samarelli A, Paul DJ et al (2016) Measurement of Earth tides with a MEMS gravimeter. Nature 531:614–619

Oakes M (2017) The History of the University of Texas at Austin Department of Physics. https://web2.ph.utexas.edu/utphysicshistory/

Talwani M, Thompson G, Dent B, Kahle H-G, Buck S (1973) Traverse gravity meter experiment. In: Apollo 17 Preliminary Science Report. National Aeronautics and Space Administration, Washington

Walter PL (2007) The History of the accelerometer 1920s–1996—prologue and epilogue, 2006. Sound and Vibration Magazine 40:84–92

Van Camp M, de Viron O, Watlet A, Meurers B, Francis O, Caudron C (2017) Geophysics from terrestrial time-variable gravity measurements. Rev Geophys 55:938–992

Worden SP (1954) Gravity meter. US Patent 2,674,887

10

The Challenges of Motion

Measuring 'g' with a gravity meter is a slow business, because every measurement has to be made individually. It would be much more convenient if the information could be gathered continuously. The ideal, of course, would be to work from an aircraft, with a speed of coverage increased by several orders of magnitude, but the first successful attempts were made from ships.

The Search for Alternatives

The pendulum remained the tool of choice for measuring 'g' until well into the 20th Century, but its use at sea presented problems. The experiences of Huygens and his successors with pendulum clocks on ships had not been encouraging, and non-pendulum methods were being sought almost as soon as a use for marine gravity measurements had been identified. The innovator in this area was William Siemens, head of the English branch of the Siemens Company. The company was one of the main contractors laying cables across the Atlantic and William hoped that measurements of gravity field could be used to estimate water depth. As he said:

> .... If an instrument could be devised which would be capable of indicating extremely slight variations in the total gravitation of the earth .... It would be found, I contend, that these indications would vary with the varying depth of water below the instrument, in such a definite ratio as would render it possible

© Springer International Publishing AG, part of Springer Nature 2018
J. Milsom, *The Hunt for Earth Gravity*,
https://doi.org/10.1007/978-3-319-74959-4_10

to construct a working scale, the divisions of which would represent depth of water (Siemens 1876)

Because of isostasy, which ensures that the average of sea-surface gravity over a sufficiently large area will be equal to the theoretical value dictated by latitude alone, the conversion of gravity to depth is less simple than this extract implies. Moreover, in calculating that *for a depth of 1000 fathoms, gravitation diminishes by 1/3691 of itself* (about 0.15 milligal per metre in today's units) William used the spherical rather than the flat-plate approximation for the effect of the water column and in doing so doubled the effect. This does not alter the fact that in making this suggestion he was already anticipating the method by which maps of global bathymetry are being routinely produced today (see Chap. 12). It was a remarkable insight, but the Siemens brothers, the six sons of an unremarkable tenant farmer of Crown Estates near Hanover, were a remarkable group. Werner, the eldest, co-founded the firm of Siemens and Halske which became today's Siemens AG, and the youngest, Carl Heinrich, went to Russia and built the Russian telegraph system. Carl Wilhelm, the fourth son, became so thoroughly anglicised during his time in London that he changed his name to Charles William, was known everywhere as William, was knighted, was elected to the Royal Society, and died an Englishman. Endlessly inventive, he designed machines and apparatus in so many fields that he often lacked the time to see his ideas through. This was true of his attempts to measure 'g', and they are usually ignored in the brief summaries of his life to be found in reference books and on the internet.

The first instrument he designed and took to sea was based on the standard mercury-in-glass barometer, and was duly named a *bathometer*. A conventional barometer has three main components. There is a reservoir, open to the atmosphere, a vertical glass tube from which air has been evacuated and which is sealed at its upper end, and a mass of mercury which partially fills the reservoir and which is driven up the tube to a height of about 76 cm by the pressure of the atmosphere on the reservoir. To make his bathometer, William modified all three. The reservoir was sealed so that it was no longer affected by atmospheric changes, and the vertical tube was widened in two places to form the upper and lower bulbs shown in Fig. 10.1b. And, most importantly, two additional layers, of dilute alcohol and juniper oil, were added to the liquid column, in amounts chosen so that the interface between the mercury and the alcohol lay in the middle of the lower bulb and the upper surface of the juniper oil lay in the middle of the upper bulb. The interface between oil and alcohol was in the narrow tube linking the two bulbs.

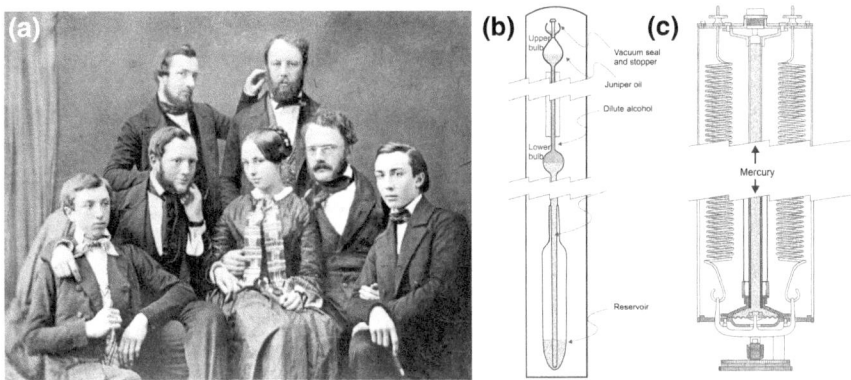

Fig. 10.1 **a** The Siemens brothers (with Werner's wife). William Siemens is seated, second from left. **b** Siemen's first bathometer. **c** The second bathometer. Photo reproduced with permission of the copyright holders, the Siemens Historical Institute. Drawings based on figures in Siemen (1876)

William's assumption was that when a cable-laying vessel headed out from port into deeper water, 'g' would decrease, and a small amount of mercury would flow from the reservoir to maintain the weight of the liquid column. The changes in the level of the mercury/alcohol interface would be tiny, amounting to only about a millimetre for a change in 'g' of 1300 milligal, but the change in level of the alcohol/oil interface would be proportional to the ratio of the surface areas of the two interfaces, which could easily be made a hundred or more. To an observer the effect of changes in 'g' would be expressed almost entirely as changes in the level of this interface, since the level of the upper surface of the oil column would also change very little because of its position in the upper bulb. This layer was actually not essential but it stabilised the system and minimised some of the corrections required.

In 1859 William took the instrument on sea trials in the Bay of Biscay that were moderately successful, but it was difficult to use and temperature control was a major problem (he ran out of ice). This was, however, a very busy period in his life. He had a business to run and 1859 was the year in which he got married and became a British citizen (swearing allegiance, as he reportedly said, to two different women on the same day), and for sixteen years he made no further attempts. During those years the problems involved in laying cables in waters of unknown depth increased, and in 1875 he tried again.

For his second attempt, William kept the mercury column and the name *bathometer* but changed almost everything else. There was no longer any distinction between column and reservoir. The mercury column was enclosed in a steel tube and acted as a constant mass, and the gravity 'sensor' was a flexible steel diaphragm across the lower end of an inverted cup-like extension at the base of the tube. The widening from tube to cup converted the very small changes in the pressure at the base of the mercury column into larger changes in force on the diaphragm, since the pressure acted over the whole of its area. William's description was written in perfect English (it is hard to believe it was not his native tongue) but left tantalising gaps in his reasoning. It seems likely that the diaphragm was not strong enough to support the mercury column by itself and that the pair of formidable-looking springs shown in his diagram (Fig. 10.1c) were added to do this, acting via a crosspiece pushing up against a boss at the diaphragm centre. The crosspiece would rise or fall slightly with changes in gravity, and its position could be measured by moving an electrode up or down to make contact with its lower surface.

William made some calculations of the magnitudes of the likely effects of changes in 'g' on the height of the yoke, but it is clear that the instrument would have required calibration against known gravity fields. His sea trials were made on the cable ship SS *Faraday*, and the initial results appeared promising, but he was running out of time. He died a few years after his report to the Royal Society, which could be read as a plea for someone else to carry on the work.

No-one took up the challenge immediately, but at the end of the 19th Century another 'barometric' instrument was designed and built at the Geodetic Institute in Potsdam by Oskar Hecker, a seismologist. His idea was that atmospheric pressure could be measured both with a standard barometer and by finding the temperature at which water boiled. The first measurement involved gravity (acting on the mercury column), the second did not. With some manipulation the two together could provide an estimate of 'g'. Considering the accuracies required in making the measurements and the detailed corrections needed to make sense of them, success seems improbable but, amazingly, results were obtained that were accurate to about 30 milligal This is not very good by modern standards, or in terms of the sizes of the changes that were there to be measured, but Hecker used his instrument on expeditions into the Atlantic Ocean in 1901 and to the Indian and Pacific Oceans between 1904 and 1905 and considered the results good enough to publish.

In retrospect, of the three bathometers actually tested, it is William's first design which seems to have the greatest potential as the basis for a successful instrument, but there was no follow-through on any of them. Hecker's surveys did, however, have one very important consequence.

The Eötvös Effect

Hecker's bathometer may have been an instrumental dead end, but Lorand Eötvös, by this stage well into retirement, read his report (Hecker 1910) and realised that a correction should have been made for the movement of the vessel over the Earth's surface, because a ship's east or west velocity alters its effective rate of rotation about the Earth's spin axis, and so alters the part of 'g' that is due to centrifugal force.[1] Documents preserved at the Eötvös Institute in Budapest show that Hecker was less than delighted when this was pointed out to him, but his colleagues persuaded him that the idea should be put to the test. In 1908 new measurements were made in the Black Sea using two ships, one moving east, the other west. At that latitude, and if the ships were moving at 10 km/hr, the difference between the two sets of results would have been about 100 milligal. Measuring this should have been within the capabilities of even the Hecker gravity meter, and evidently it was.

Hecker's initial resistance seems to have been as near as Eötvös ever came to experiencing the brutal trivialities of academic infighting, and even in this case, and most unusually, his opponent conceded without much of a fight. Just to be sure, in 1915 Eötvös built a device that provided a visible proof. It consisted of a balance with equal fixed weights attached to opposite ends of a horizontal arm. When this was rotated about a vertical axis, the weight moving towards the west became very slightly heavier, because its speed of rotation about the Earth's spin axis decreased, and the weight moving towards the east became lighter. Intuitively, it might be thought that the difference would be too small to be measured, but Eötvös was very good at producing instruments that were sensitive to very small changes, and in this case his secret lay in the choice of rotation velocity. When the rotation time was made equal to the natural period of oscillation of the system, the effect was enhanced by resonance and became visible.

Until GPS satellites made it possible to measure true velocities of ships and aircraft, rather than merely their speeds through water or air, the Eötvös effect was the major limitation on the accuracy of marine gravity measurements and made airborne gravity (with its inevitably greater velocities) almost unthinkable.

---

[1]There is also a much smaller effect associated with any north-south velocity.

The Submarine Pendulum

Recognition of the Eötvös effect may have been an essential first step towards the proper processing of any measurements of 'g' made at sea but it did nothing towards making such measurements possible. In the end, it was only by abandoning new designs and returning to the pendulum that anything was achieved. The first successes were recorded by Felix Vening Meinesz, who began his career in the Gravity Survey of Netherlands in 1915 with a dissertation on the defects in the existing methods of using pendulums on land. As he himself later said, it was the fact that so much of his country was almost under water anyway that sparked his interest in marine gravity. Even in the 'land' areas it had been difficult to find sites stable enough for pendulums to be used, because the typically soggy Dutch ground vibrated noticeably with each swing, taking the whole apparatus with it. Vening Meinesz solved this problem by mounting two pendulums on one suspension and setting them to swing simultaneously but in opposite directions. This greatly reduced the (mainly horizontal) ground movements and led him to speculate on whether the same method could be used on board a ship. His more important advance, however, was in recognising that this might be done with better chances of success if the ship were submerged below the level of the wave disturbances (he generously credited this idea to the then Director of the Netherlands Government Mines Department, Frederik van Iterson). A further advantage of submarines was that the electric motors that powered them when submerged produced much less vibration. The Dutch Navy agreed to help and, after some preliminary tests around the naval base at Den Helder, the first serious sea trial was made on a voyage from the Netherlands to Gibraltar in 1923.

It was a failure. Even when the vessel was submerged to well below the stipulated 30 metres, the rough weather in the Bay of Biscay meant that the allowable half-degree limit on angular deviation during a reading was always exceeded. It was the British navy that came to the rescue, when the dock-yard in Gibraltar designed and fitted a suspension platform. With this modification, measurements could be made with a mean error of only 4–5 milligal which, although large by today's standards, was not noticeably worse than pendulums were then achieving on land. Useful marine gravity surveys became possible, as long as the variations were large enough. Further modifications reduced the errors to about a milligal, and by 1926 the stage was set for something more ambitious. The Dutch Navy was asked whether it would mind sending the next submarine destined for Indonesia (then the Netherlands East Indies) out the long way via Panama rather than the much shorter route via Suez, with the seemingly esoteric aim of determining whether the Earth was a tri-axial or bi-axial ellipsoid, i.e. whether the shape produced by slicing it along one of its parallels of latitude would be a true circle or an ellipse.

Amazingly, the Navy agreed, although whether they consulted the submariners before doing so is not known. It is hard to imagine that the geophysicists were welcome shipmates, since their apparatus, placed between the two periscopes and standing more than two metres high, took up a significant amount of space in what was already a very crowded environment. The eighteen ordinary seamen lived in in a room about 25 ft. long and only 6 ft. broad, and the twelve petty officers were scarcely any better off, being allocated a room 20 ft. by 7 ft. Behind this was what was known (possibly ironically) as the 'long room', 14 ft. by 7 ft., for the officers. The hatches could only rarely be opened and life on board therefore involved living *at rather close quarters and in a more or less unpleasant atmosphere*' (Vening Meinesz 1931).

To make things worse, suspending the pendulums on an axis parallel to the ship eliminated only the effect of roll. Pitch still had to be kept to within a half-degree limit, which could be done only by submerging to a sufficient depth but even then asked a great deal of the two men steering the vertical rudders. It was also impossible to keep the platform sufficiently steady unless all the crew kept very still for the whole of the half hour or more that it took to make a single measurement. In 'detailed' surveys where the stations were only 50 km apart, the ship would have to dive and re-emerge several times a day, and that also was a strain on all concerned. The sight of 'Dr. Felix' making his stately way towards them across the quay must have struck terror into the hearts of even the most resolute submariners. As his biographers have testified, he was, in every way, a big man, and such people are seldom entirely welcome in the cramped quarters below the waves.

For all their disadvantages, throughout the period between the First and Second World Wars pendulums based on those developed by Vening Meinesz and deployed in submarines were almost the only instruments used for measuring gravity at sea. Writing shortly before the outbreak of the Second World War, Maurice Ewing, who later founded the Lamont Geological Observatory at Columbia University and became its first director and therefore arguably the most important marine geophysicist of his day, listed twenty-one submarine surveys that were known to him. He devoted little more than a page to non-pendulum methods, which was about as much space as he gave to measurements made with pendulums on ice floes (Ewing 1938).

One of the reasons that marine gravity came to be seen as worthwhile was due to pure chance. Once Vening Meinesz arrived in Java, he used his pendulums to investigate the surrounding seas, and the map he produced (Vening Meinesz 1932), and particularly the deep gravity lows by which it was dominated, played an important part in changing ideas about the Earth's crust. However, the Indies had not been chosen because of geological problems that were crying out for solutions (although there were plenty of those) but because they had been colonised by the Dutch. Had the technique been developed in Britain rather than the Netherlands, the first surveys might have been made in the seas around India or Australia. The maps would have been far less spectacular, and would almost certainly have attracted much less by way of global interest.

To say that luck played a part in Vening Meinesz's success is not, of course, to undervalue his contribution or his scientific status. Once he had the map, he enlisted the greats of Indonesian geology, people such as Umbgrove, Verbeek and van Bemmelen, to help him explain it. Working within the 'fixist' ideas that then held sway throughout most of US and European geology, they evolved a theory of 'mega-undations' that attributed the deep isostatic gravity lows associated with the marine trenches and troughs off the southern coasts of Sumatra and Java and around the Banda Arc into the Molucca Sea to deep downwarping of the crust into the mantle. This was a long way from plate tectonics, but once plate tectonics became accepted, mega-undation theory was quickly adapted to it. The identification of the 'downwarps' with the subduction zones of plate tectonics required little more than a recognition of the asymmetry of features that earlier theories had assumed to be more or less symmetrical.

Marine gravity measurement has come a long way since the 1930s, and modern maps, such as the lower map in Fig. 10.2, show vastly more detail. The main gravity low off Java and Sumatra is divided into two parts (associated, respectively, with the trench and the forearc basin) separated by a high along the forearc ridge, and there are other complexities in the extreme east. Vening Meinesz could never have hoped to define features at this sort of scale with a mere two hundred and sixty measurements, but what he did achieve was enough to produce a step-change in the way geologists and geophysicists thought about the Earth.

Plate Tectonics

Almost as soon as there were any even vaguely realistic maps of the Atlantic, there were also people commenting on the apparent match between the western coasts of Europe and Africa and the eastern coasts of North and

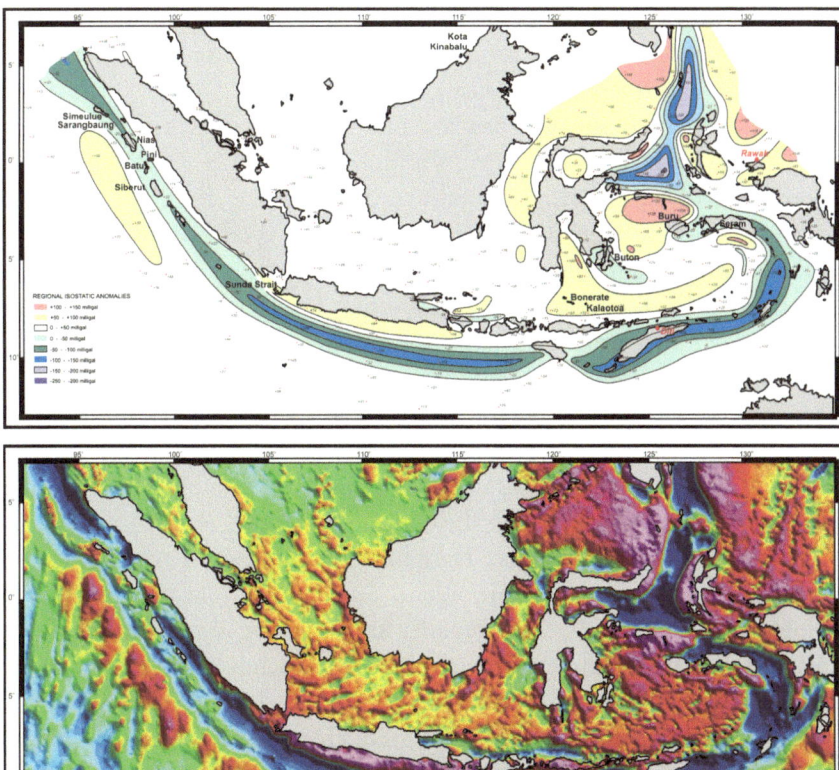

Fig. 10.2 Isostatic gravity anomalies in Indonesia. Above: Contours at 50 milli-gal intervals of isostatic gravity based on submarine pendulum measurements by Felix Vening Meinesz and corrected for the effect of the Airy isostatic model as modified by Weikko Heiskänen to allow for crustal rigidity. The points at which measurements were made are marked by dots accompanied by the estimated value in milligal. Redrawing based on the map published in Heiskänen and Vening Meinesz (1958). Rawak and Dili, two of the points where De Freycinet made pendulum measurements, are marked in red. Other named islands are those mentioned either in this chapter or Chap. 13. Below: The modern equiva-lent map, based on the 2012 World Gravity Model described in Chap. 12

South America. It was not then a very great leap to supposing that they had once been together. Distribution patterns of fossil plants and animals that could not sensibly be explained by vanished land bridges provided more convincing arguments but it took a meteorologist, Albert Wegener, to com-bine this evidence with his own studies of past climates and produce, in 1918, a coherent theory of continental drift. It then took the massed ranks of the theoretical physicists to proclaim that drift could not happen since, as they rightly said, it was mechanically impossible for the continents to plough their way through the crust of the oceans.

Nowadays it is hard to appreciate just how little was known about the ocean basins before the Second World War. When a single bathymetric sounding involved lowering a weight to the sea floor, deciding when it had arrived (no easy matter) and then hauling it up again, soundings in water depths of more than a few hundred metres were rare. It was the needs of submarine, and anti-submarine, warfare that led to the development of the echo-sounders that were used, once the war was over, to map the global sea floor. The discovery of great continuous rises in the centres of some oceans came as a considerable surprise. An explanation was needed, and the (mainly ex-naval) US oceanographers produced one. New sea-floor, they decided, was being formed at the rift valleys that split the crests of many of the ridges, and it was this that was pushing the continents apart.

The problems were obvious. The globe is finite, and if the oceans were expanding there was a space problem. Sam Carey, maverick professor of geology at the University of Tasmania and a strong supporter of sea-floor spreading, saw a way through by invoking an expanding Earth, but the theoretical physicists objected to this too, since Sam's theory required 'Big-G' to change with time. When, in 1962, I graduated in the UK and headed out to Australia, geology was a science in an advanced state of schizophrenia. By and large, the physicist's arguments against moving continents were accepted in Europe and North America, but southern hemisphere geologists looked at the distributions of fossils on their much more widely separated continents and refused to believe that drift had not happened.

The Cold War then supplied another piece of the jigsaw. The study of earthquake locations was a product of the military need to locate nuclear tests, and it led to the setting up of a world-wide network of standardised seismograph stations. Almost simultaneously, but quite independently, Hugo Benioff in the United States and Kiyoo Wadati in Japan used the data from this network to produce maps and cross-sections showing planar seismic zones angling down from the deep oceanic trenches to depths of hundreds of kilometres. These, they suggested, could be the places where oceanic crust was being absorbed back into the mantle.

All the elements of plate tectonics were thus in place by the start of the 1960s, but to many people the processes still seemed physically improbable and the evidence unconvincing. What was needed was a smoking gun, and that came when Fred Vine and Drummond Matthews at Cambridge University explained the linear magnetic anomalies or 'stripes' that were already known to be features of much of the oceanic crust by a tape recorder analogy (Vine and Matthews 1965). Their idea was that as new crust solidified at the mid-ocean ridges, it preserved in it the direction of magnetisation at that time. If the Earth's field reversed (and by then it was fairly well established that it did), so too would the magnetisation of the next bits of crust to be formed.

With hindsight, it can be seen that the time was ripe and Plate Tectonics was inevitable, because someone was bound to recognise the significance of the magnetic stripes before very many years had passed. It should have been Ron Mason, or perhaps Arthur Raff, although it was Ron who was the sole author of the first publication (Mason 1958).[2] He had argued strongly against the almost random zig-zag patterns of most research cruises and in favour of systematic grids, and when Scripps learned that the US Coast and Geodetic Survey was planning systematic bathymetric surveys off the Pacific coast of the US, he was brought over from the UK to include magnetometry and 'do' the science. The surveys led to the first discovery of magnetic 'stripes' in an ocean (Fig. 10.3) and for many years, and even after the general acceptance of the Vine-Matthews hypothesis, these were commonly referred to as the Raff-Mason lineaments, but the name has gradually been forgotten.

In 1960 marine gravity measurements were too unreliable, and marine gravity meters were too rare and too expensive, to be routinely used. If Ron had been able to make a gravity map as good as his magnetic map, it is just possible that he would have realised what was going on, because gravity in this area emphasises the spreading centre and its offsets. It is, however, unlikely that he would have told anyone. He was a brilliant and innovative thinker, but as a scientist working in the real world he had one major failing. He had an almost paranoid dislike of committing his ideas to print. Almost his only comment when, in 1968, he took me on as a student to write about Eastern Papua was that New Guinea was a risky place to study because there was a lot of new work being done there, and I might be proved wrong in a few years' time. Rather than take that risk with his magnetic observations, Ron and his co-author published their maps along with three 'geological possibilities' or explanations, none of which gave any thought to the processes that might be at work.

---

[2]This first published report clearly showed linear anomalies but had little impact, possibly because it appeared in the first volume of a UK journal that only later became internationally important. In a doomed attempt to share the credit evenly, the complete survey results were reported in two consecutive 1961 papers (Mason and Raff 1961 for the south and Raff and Mason 1961 for the north) but it was as the Raff-Mason lineaments that the anomalies entered history.

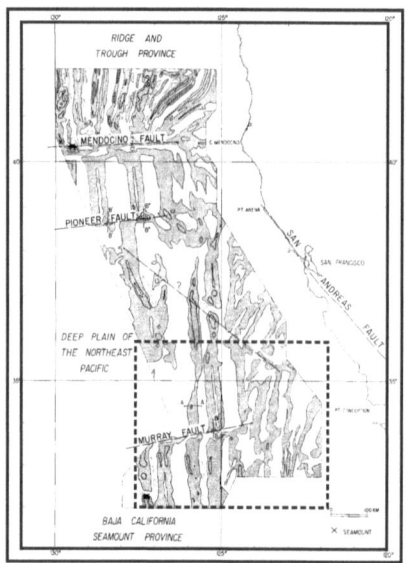 

Fig. 10.3 The Mason and Raff (1961) (left) and Raff and Mason (1961) (right) 'index' magnetic maps of the eastern North Pacific. The maps are to the same scale and overlap between 40°N and 42°N. The dashed rectangle on the left shows the area covered by the map in Mason (1958). Grey or black shading shows places where the Earth's magnetic field is above the level predicted by the global reference model. Reproduced with permission, Geological Society of America

By the time I knew him, Ron had moved onshore. His new idea was to measure the rate of sea-floor spreading directly, using a radio-location instrument called a Mekometer to determine, extremely accurately and repeatedly, the distances between points on either side of the central rift in Iceland. Today GPS systems are being routinely used to do the same thing, over much wider areas, but in the late 1960s Ron was once again very much ahead of his time. It was the sort of project he enjoyed. There were plenty of calculations to be made, and no pressure to publish because, of course, the first set of measurements meant absolutely nothing. It was only comparison with those made two or three years later that gave them meaning. When I worked in Iceland many years later, I was told that he had casually come up with an explanation for a relationship between its topography and geology that had been a puzzle for a number of years. He never published that, either.

Whether he ever regretted not inventing Plate Tectonics, I never knew, because he was not a man who readily shared his deepest thoughts. But there was at least one other person who had abundant cause for regret. A Canadian geologist called Lawrence Morley, who had been a student in Tuzo Wilson's department in Toronto in the early 1950s, studying rock magnetism, had seen the Raff-Mason map when it was first published in 1961, had come up with the solution, and written an explanatory paper, which he submitted first to *Nature* and then to the *Journal of Geophysical Research*.

It was rejected by both. He was not, he was told, presenting any new data. Apocryphally, his ideas were unkindly likened by one reviewer to speculations appropriate only to a cocktail party. He would, it was implied, be better occupied as a minerals geologist, which is what he became. The field was left open for Vine and Matthews, who did have new data to present, and were allowed to accompany their presentation with a little bit of speculation.[3] From there on, it remained only to systematise the theory, provide it with a jargon of spreading centres, fracture zones and subduction zones, ignore the Russians (who, mainly working in the heart of the Asian continent, were deeply suspicious of any theory that could only be demonstrated far out to sea), and re-write the text books.

The Russians had more problems than simple lack of familiarity with marine geology. They had a monolithic structure in their science, and at its geological pinnacle sat Vladimir Beloussov, who was not going to drift anywhere. In 1971 he came to the Pacific Science conference in Canberra and found himself in a very unfamiliar environment. Many of the up-and-coming generation of American oceanographers had turned up, people like Dan Karig and Tanya Atwater who were just beginning to make names for themselves. There were also the Australian 'active margin' specialists down from New Guinea and, most vociferous of all, the New Zealanders, proud to be for once at the forefront of things, with their spectacular plate margin and their equally significant fossil assemblages. It was not a good place for Beloussov to launch an attack on Plate Tectonics, but that is what he had come to do. The Young Turks could barely contain themselves, and the questions poured in as soon as his talk was finished. How did he explain this, without motion? What was the reason for that, if not sea-floor spreading? Beloussov, clearly a man unused to being contradicted, looked grimly out at the hostile audience and gave it his reply.

---

[3]Probably the best of all the histories of the Plate Tectonic revolution is Menard (1986). Bill Menard was an innovative marine geologist who briefly abandoned the sea to become head of the US Geological Survey, hearteningly proving that very nice people can sometimes make it to the top.

I do not accept your data

There is not a lot that you can do with that, but it seems that attitudes of mind can be even more difficult to change than beliefs. Twenty years later, at another conference, this time on Sakhalin, the one part of Russia that is at an active margin, the young Russian geologists were loud in their complaints that they were not being allowed to believe in anything other than Plate Tectonics in its purest, most uncompromising form.

1983 to 2001: Indonesia

In Indonesia neither land gravity measurements nor marine gravity measurements can tell the whole story. Both are needed, and when a geologist at London University's Chelsea College called Tony Barber created a consortium for geological research in Southeast Asia, land gravity measurements were an important part of its programmes. The group was supported by a fluctuating assembly of oil companies and survived for more than two decades. Eighteen years after it was founded one of the sponsors asked in wonderment 'What genius kept this thing together for so long?' That genius was, of course, Tony. His reward for running, unfashionably, a research group that cut across the boundaries between the London colleges was to never to reach the rank of professor, even on retirement. His mistake was to believe in cooperation, at a time when the various provosts, rectors and principals of the larger London colleges were busy dismantling the collegiate university and competing to swallow its smaller institutions (Chelsea itself disappeared into the maw of King's College soon after the consortium was founded). After Tony retired the group degenerated into a single-college project.

All that was in the future, but from the very start much of the work was done using small local trading boats. For twenty years they were our homes as well as our transports for geology and geophysics throughout much of the archipelago. Small-boat work in Indonesia, like warfare, could be described as consisting of 'long periods of boredom punctuated by shorter periods of extreme terror and occasional moments of exhilaration'. True, no-one was deliberately trying to kill us, but sometimes it seemed as if they were (but we never lost a student, or even a member of academic staff). Almost inevitably, any boat available for hire when the stupid foreigners arrived was the one that no-one who knew the port and its people would dream of using. Nobody, perhaps, has ever summarised more succinctly the situation in which we so often found ourselves than La Condamine who, rather than describe in detail the problems met by the French expedition when trying to travel by sea from Panama to Guayaquil in 1736, simply said:

We were dealing with a business man, and we were foreigners[4]

A flavour of those times.

Bonerate Islands (Flores Sea). We are boarding a boat of particularly unseaworthy appearance when my Indonesian counterpart screams with rage and turns on the owner with every sign of wishing to do him actual physical harm. This is unusual, since my friend is a man of mature years, of a generally peaceable disposition and has a senior position in the Indonesian Geological Survey. The fuss is over some dirty bottles of oily liquid that I had assumed to be spare fuel. It seems that the owner has decided to combine our charter with a little fishing, and that what they actually hold is home-made nitroglycerine. It is, of course, entirely normal to have a boat captain who sees a charter as an opportunity for a little commerce on the side. That is one of the rules of the game, but a cargo that includes bottles of unstable explosive is definitely a foul.

Bonerate Islands (again). The same unusually incompetent captain/owner, and the same unusually unseaworthy boat. Thanks to this combination, we are heading back to the Bonerate lagoon in the dark, with an engine that is more stop than go. Each time it stops, the waves carry us a little closer to the long, barely exposed, reef that surrounds the lagoon. We are far too close when the engine gives a heart-broken sigh, and stops with an air of near-finality. As we are washed sideways towards the reef, I find myself thinking that I am *probably* not going to die, but that I am going to get very badly cut about as I scramble on to and over the coral in my efforts to reach the calm, and shallow, waters of the lagoon on the other side.

The engine catches again, roars for a few seconds, and explodes. This time it really is dead, but the final burst carries the boat just far enough for the waves to wash us through the gap in the reef and into the calm and peace of the lagoon.

It is so calm and peaceful that, once away from the entrance, we are stranded, in barely enough water to float. The crew unship a long bamboo pole and begin pushing, but manage only a demented zig-zag around a generally circular course.

---

[4]'Nous traitons avec un Marchand, & nous étions Étrangers' de La Condamine (1751) Journal du voyage fait par ordre du Roi a l'équateur, servant d'introduction historique a la mesure des trois premiers degres du méridien, Paris, Imprimerie Royale, p. 10

Mostly, I don't interfere in the way a boat is being run, even if it is plainly not being run very well. My philosophy is that at least the crew are still alive, so they must (unlikely as it might seem) have been doing something right. But this time they are clearly trying to do something that I know more about than they do. I take the bamboo, adopt the approved stance, and punt us towards the shore. Three years at a riverside university have not been entirely wasted.

We still have to wade a few hundred metres when the boat finally runs aground, but it is better than having to swim all the way across the lagoon.

Kalaotoa (Flores Sea). A surprise when we arrive. We are told that there is already another European on this really remote island. It seems churlish not to visit. He is French, and he is waiting for a boat to be built. On an earlier visit he had fallen in love with the local trading boats (they are beautiful, as can be seen in Fig. 10.4, and very seaworthy when properly looked after) and had decided to order one for himself. He then left, but after several years of waiting for something to happen, had realised that the only way he would ever get his boat would be to go back to the island and watch it being built. We could have told him that. He has already been here several months.

When he hears what we have come to the island to measure gravity, he becomes quite animated. 'You mean, it is not already known? That there are things like that still, that nobody knows?' Put like that, it does seem rather exciting. It is good to be reminded of the fact.

Buru. It is not even necessary to be at sea for there to be trouble with boats. In the centre of Buru Island there is sacred Lake Rana. Much to our

Fig. 10.4 The Indonesian work-horse. A typical inter-island trading vessel

surprise, we manage to reach it. There is no hope of getting through the surrounding forest, on any reasonable time scale, and a canoe is hired so that measurements can be made around the shoreline. Only when we are afloat are we told there is a problem. The gods of the lake have to be paid if they are to look kindly upon us, and not in paper money either. Of course, none of us have any coins. As a last resort, I look in the bottom of the gravity meter case and there find gold. Or at least, two Cyprus 50p pieces, left over from an earlier trip. They are all that we have, so we throw them in.

We get back to the truck that brought us up the logging road from the coast to find that it has two flat tyres. The coinage of Cyprus is not, it seems, legal tender with the gods of Lake Rana.

Buru (again). Violating yet again the rule about securing lines of retreat, we are on the south side of the island when the Southeast Trades begin to blow in earnest. The ferry we plan to use to take us back to the port of Namlea, at the island's northeast corner and from where we can leave for Ambon in a comfortingly large vessel (a second-hand Brittany ferry), is cancelled. Our only consolation is that while we are waiting, hopefully but in vain, for the much smaller local ferry to arrive, the entire top class from the local school comes down to the jetty and sings to us. It is nice to believe that this is a demonstration of sympathy, but perhaps they always have their singing classes down there, so as not to disturb the rest of the school.

None of the local boat owners will even consider a charter. Most just point out at the now impressively high waves and shrug. Our last chance, an elderly Chinese, starts by being as pessimistic as the rest, but then he has an idea.

"I'll take you the other way".

In just five hours we are deposited in a harbour on the north coast, having run along the south coast with the swell behind us and then turned north and eventually east in calmer water sheltered by the bulk of the island. As an added bonus, we have traversed more than half of the coastline, visiting places for gravity measurements and geological sampling that we had never, in our wildest dreams, imagined that we would manage to reach. But of course, for such service, we have to pay. In the restaurant at the end of the jetty, acutely conscious of handling what, for most of the customers, represents at least six months' wages, I count the banknotes out by feel under the table. Nonchalantly, our Chinese friend dumps the impressive pile (only small denomination notes are acceptable in Indonesia's wild east) on to the table in front of him and re-counts them in full view of everyone. This must, I think, be a very safe and law-abiding place.

Two months later, a minor dispute in Ambon between a Christian and a Muslim over a taxi fare escalates and the Moluccas burst into flames. Five years later on Buton, at the western edge of the Banda Sea, I walk through new settlements where refugees from Buru are living, unable or unwilling to return home.

Simeulue (Sumatra forearc). A deep bay on the north side of the island, conveniently placed for the measurements we need to make in order to track down yet another of those ophiolites. We go into the usual routine for places without a jetty. Into the canoe, get ashore, take the reading, paddle back. But here the 'shore' does not exist, only big mangrove trees rooted in mud. There is just one patch of mud high enough and dry enough for the gravity meter, but it is far too small for me. I try, but meter and I sink into the mud together. In the end I put the meter on the mud patch, climb one of the mangroves and hang upside-down from a branch to make the measurement.

Simeulue (again). Perhaps the best boat that we have ever hired, and the most competently run. Nothing much to look at, of course, but it takes us all the way round the island in five days, with no problems about getting ashore to collect data. As we near the home port of Sinabang, there is just one tiny island left to be visited, a few miles offshore. Converging on it, and us, from the opposite direction is a glossy floating gin-palace of a yacht. We drop anchor only a few hundred yards away from each other, and suddenly the sea seems full of beautiful girls (there are probably only two, but we have been away from home a long time and imagination runs riot). The sea also rapidly fills with London University research students.

It turns out that the yacht belongs to the Italian ambassador to, I think, Myanmar. In the water, I look back at the two boats, the gin palace and the island trader, and feel happy with my choices.

Siberut (Sumatra forearc). A total defeat. A tiny offshore island, with big trees. But, we discover, no land at all, not even mud. Just trees and water. The island must have sunk very recently, because the trees are still alive with their roots deep in salt water, but this is, after all, one of the most geologically active areas on the planet. The epicentre of what is to be the great Sumatra earthquake of 2006, which generated a tsunami that killed a quarter of a million people, is not far away.

Siberut (again). A change of tactics (or a return to the methods of those first surveys in Eastern Papua). A wide, winding river offering speedboat access to the inland in the south of the island. Things go well up to the point at which the boat is swung hard over and charges full-tilt at the bank. And through, into the river again. Someone has thoughtfully cut a 10-yard slot through the neck of a meander. Completely overgrown, but it is only a matter of hitting the right spot at speed.

Exactly the right spot.

There is a village at the end of the navigable section, with a little eatery. Only the women are at home at this time of day, and just one of their menfolk, too old to go into the forest, watching us from his verandah while we eat. He wears only a few leaves round his waist and an interesting head-dress, but I recognise him at once. His spiritual cousin is a fixture in my village pub back in Wales. In five minutes, I know, he will be over with us, bumming a cigarette.

Actually, it is less than five minutes.

Back to the coast, and the shock of finding the bar full of Australian tourists. Siberut is not supposed to be on any tourist routes. It is better described as being on no routes whatsoever. This, however, is a very special company, specialising in taking people to places where no-one with any concern for their comfort, health or digestion could possibly want to go. Interestingly, all have been provided with T-shirts decorated with a map of the island that is, in those days before GPS, far better than anything that we have been able to find. We buy a shirt, and use it to plot the gravity stations.

Nias/Pini (Sumatra forearc). Even by our standards, a very small boat. Another bad sign is that, as well as the inboard diesel engine, it carries an outboard motor as a spare. This is proudly pointed out as a Unique Selling Point, but in Indonesia a spare motor means two things. Firstly, because it exists, maintenance on the main engine will have been a little—casual. Secondly, no-one will have bothered to see if the spare actually works.

Still, it is a beautiful day, the sea is calm, and we go. Progress is slower than expected, and we are clearly not going to reach Pini before dark. By nightfall the wind is blowing, the rain is raining and the canopy that is supposed to shelter us has collapsed. We are also having to bail out the water that is oozing into the boat, using just a couple of old tin cans. It is a relief to come up in the dark to a middle-sized fishing boat that is hauling in its nets. At least we are not going to drown.

After half an hour we decide that the combined smell of rotting fish and diesel fumes is so bad that we would rather risk drowning, and leave. It is also very cold. We are almost exactly on the equator, but it is one of those nights when even a soaking-wet sleeping bag is better than no sleeping bag at all.

Dawn comes at last, sullen and grey, with the thicker grey of a long coast line not very far away. Our captain immediately turns the boat and heads in the opposite direction. Why? Because it is a large island (true), and that means it must be Sumatra (false), and we hadn't asked him to take us to Sumatra. We point out that he is now taking us due north, which means that the coastline runs from east to west, and that there is only one island in the area with an east-west coastline, and that is Pini, which is where we want to go. Yet again, we have failed to recruit from the top drawer where seamanship is concerned.

We return to Nias eventually. Almost within sight of the port city of Gunung Sitoli and home, the engine does break down. Proudly, the spare motor is unearthed. There is just one problem. There is no place to mount it. Indonesian ingenuity comes into play, exploiting the fact that every boat carries a tangled mess of odd lengths of rope and strange cuts of timber. A sort of gantry is rigged to which the motor is lashed, but the propeller is still well out of the water. Or it is unless the gantry is pushed out to one side of the boat, and everybody hangs out over that side. Then the propeller is just in the water and, amazingly, the engine starts. In slow and stately fashion, we crab our way into port.

Sarangbaung (Sumatra forearc). A speedboat for the trip from Nias to the Banyak Islands. This is good in principle, but the drawback is that the 'crew' consists of the owner only, and he is not going to risk damage, of any sort, to his beloved boat by going anywhere near the coral. At one island I have to wade ashore, carrying the gravity meter, through water that is, at the start, over the top of my head.

We get half-way to Banyak but there is a great black cloud covering the islands ahead. The owner declines to go further and there is nothing for it but to stop and wait. Fortunately there is somewhere to do this, a pocket handkerchief of an island called Sarangbaung. I know nothing about it except that it is roughly square, a few hundred metres to a side, and that my geological colleagues stopped there for lunch the week before. It seems to be just another boring coral island.

We don't even have lunch with us, although we saw signs of theirs. To pass the time, I decide to make one of the world's smallest gravity surveys. A few hundred yards along the beach to the first corner. Take a reading. Head for the next corner. And there, displayed along the beach, is the equivalent of several hundred metres of vertical section of the rocks of the forearc sedimentary sequence, tilted at about 60° and displayed ready for easy mapping. It is sort of exposure that geologists dream of, and that our geologists have been fighting their way through the forests on the larger islands hoping to find and never have.

They have to go back there a year later, for rather more than just lunch.

Buton Strait (SE Sulawesi). Another less than competent captain, but one who is at least willing to make an effort to get us ashore. Unfortunately, at the entrance to the strait and on a falling tide, he chooses to do this by nosing into the soft sand of the beach, and then keeps position by running the engine. By the time the reading is taken and the shore party is back on board, the boat is well and truly stuck. We prop it up with planks and logs scavenged from the beach to stop it keeling over as the tide leaves us.

Four hours later I walk all the way round it, without getting my feet wet.

Sunda Strait. A very different boat, but a very familiar problem. Just visible on the horizon is the cone of Krakatau, a shadow its former self after the eruption of 1883. It hasn't moved for the last four hours, and neither have we. Every so often the engines roar into life, there is a hint of motion, and then everything shuts down yet again. But this is no suspect local trading boat, this is the UK's finest and newest research vessel, the R/V Darwin, on a round-the-world series of cruises to mark the centenary of the cruise of the HMS Challenger, which ushered in the era of modern oceanography. The problem is clear. The engine-room computer is programmed to look at the engine revs, and if they seem too high, shut everything down. This particular computer is deciding that they are too high if the speed goes above 5 knots, and there is no manual override.

It hasn't been easy to get even this far. Indonesia at this time is effectively still ruled by the army, and the military mind is notoriously intolerant of curiosity, including curiosity-based research. Right up to the time that the Darwin docked in Tanjung Priok, the port for Jakarta, the necessary permits had still not been issued, and we were only really sure that we would be going when two Indonesian Navy officers turned up with all their kit for a 30-day stay aboard. Perhaps Bouguer and La Condamine felt the same when Juan and Ulloa arrived. We are later told that two days previously the outgoing British ambassador had been asked by a very senior general what he would like as a leaving present and had, with extraordinary selflessness, replied that he would like permission for the Darwin to carry out the Indonesian leg of its cruise. This conversation had, inevitably, taken place on a golf course, the main job of the ambassador being to play golf with generals. So, we have been able to leave Jakarta, but we haven't got very far, and perhaps are going to go no further.

Five hours later the trouble has been narrowed down to a faulty thyristor in the control system. It is about the last thing to be suspected but, fortunately, there are spares. Thirty days of intensive data gathering (mainly of long-range sonar images of the sea floor that Vening Meinesz criss-crossed in his submarine) awaits us.

Batu Islands (Sumatra forearc). Mostly, unless there is a jetty in a sheltered anchorage, the boat that is also our home stays a little way out to sea, and the shore party is ferried ashore in a canoe. If I am going, I usually swam ashore, in view of my well-known ability to capsize any canoe unless it has a full set of outriggers (and sometimes, even then). Outriggers are not used in most parts of Indonesia. This particular beach is more of a problem because a strong surf is running. In one place there is a small reef a few yards out to sea that is reducing the surf but making it very difficult to reach the sand beyond.

We devise a cunning plan. I will swim ashore, at a place where there is no reef, walk along the beach to where there is a reef, wade out on to it and stand there. Agus, our guide from Nias, will come past me in the canoe, hand me the gravity meter. I will wade back to the beach and take the reading. Simple.

It would have worked, but at the last minute Agus gets that gleam in his eye that shows he is about to take control, shouts *Saya datang* (I'm coming), and heads straight for me. What I can see and he cannot, is a really large wave building up to a quite impressive peak behind him. Canoe and wave arrive at the reef together. I grab the meter as they go past, just before the canoe flips over, Agus is dumped on the beach and I am dragged across the reef into deeper water by the backwash, still hanging on to the handle of the gravity meter case. It is on this occasion that I make the useful discovery that the case of a LaCoste gravity meter, even with the meter and its nickel-cadmium battery inside and an average-sized European male hanging on to the handle, will float.

We sort ourselves out eventually, and I still have the blood-stained field notebook as a memento.

Guadalcanal, Solomon Islands. Not my story, but told by John Grover, sometime Chief Geologist of the British Solomon Islands, and worth re-telling. An American expedition is trying to do much the same sort of work that we were to do in Indonesia, and the approach to the island in heavy surf ends with the American geophysicist and his Solomon Island helper clinging to the upturned canoe, some way out to sea. The helper has been very thoroughly briefed about the importance and value of the gravity meter (a LaCoste 'G'). When the first rescue canoe arrives, its crew is told very firmly to take the meter ashore and leave the geophysicist behind. Magnus Gudmundsson who, given the choice, would rather have seen his supervisor disappear down a crevasse than Iceland's only LaCoste gravity meter, would have approved.

The geophysicist can't swim. It is amazing just how tightly you can cling to an upturned canoe, when you have to.

And finally—Malaysia (Kota Kinabalu, Sabah). The great thing about being an invited speaker at a conference is that you may get your accommodation free, and that can include accommodation for your wife as well. For the conference in KK, held in the normally-beyond-our-budget luxury of the Tanjung Aru Hotel, the deal is even better. We are told that if we feel like staying after the event, we will be charged only the special concessionary rate available to the Geological Survey and other government servants. However, we think it will be nice to visit some of the islands just off the coast, so we check out for a couple of days, promising to return.

On the way back from the islands, and within sight of the harbour, the heavens open. The inadequate tarpaulin that is supposed to protect the passengers does little to protect us but collects enough water to saturate any bits of us that remained dry before it collapsed. Once ashore, there is just one taxi that will even consider accepting us as a fare. It is old, it is battered, one window is broken, one door is wired in place and cannot be opened, and when it moves it trails a black cloud of toxic vapour. But, it gets us back to the Tanjung Aru.

We climb the marble steps, stumble past the impassive Sikh at the door and trail pools of oily water across the marble floors to the reception desk. The clerk on duty looks at the two drowned rats and smiles a welcome. "Sir, Madam. How nice to see you again." Now that is the mark of a really good hotel.

Meters at Sea

After the Second World War, efforts to measure 'g' on moving ships continued but, with the Cold War at its height, the relaxed pre-war attitudes of navies to handing their submarines over to geophysicists were not going to be repeated, and in any case there were simply not enough submarines to go round. Surface ships had to be used and, although the Eötvös effect was by this time well understood, answers had to be found to a range of other problems encountered on vessels exposed to the full force of waves and currents.

The first marine gravity meters, which came to market in the late 1950s, were modifications of the existing land instruments. The LaCoste sea meter, in which the lever-arm was heavily damped to make it incapable of moving fast enough to record short-period wave-generated accelerations, was typical. Even with the damping, the beam was never still, and its average rate of movement became the important parameter. The instrument was kept vertical either by mounting the whole system on a free-swinging platform or by gyro-stabilisation. Errors due to tilt were still present, but were reduced to about a milligal by careful design. In the end, the error that gave the most problem was the one caused by the asymmetry that is a feature all meters that use the zero-length spring. If the mass in the meter shown diagrammatically in Fig. 9.4c experiences a horizontal acceleration in the plane of the movement of the beam, then the beam will move, up if the weight is above the hinge and down if the weight is below it. This 'cross-coupling' of horizontal accelerations into the measurement of vertical acceleration, and hence of gravity, proved especially troublesome because the timescales involved were rather similar, and errors of up to 40 milligal could be produced.

For Lucien LaCoste the answer to cross-coupling was to incorporate horizontal accelerometers into the system, and use an analogue computer to calculate the correction. All seemed to go well, and thousands of line-kilometres of data were collected using his instruments. It was not until more than ten years had passed that questions began to be asked, and then only because of the testing of a marine meter based on a completely different principle.

As Galileo and his father knew well, the natural frequency of vibration of a metal rod depends on tension, and accelerometers based on that principle had been in common use for many years, so much so that by 1970 some were being sold as surplus by the US government. It was almost inevitable that someone would try and use one to build a gravity meter, and a team of scientists at MIT did just that, and persuaded the oceanographers at Woods Hole to put it on one of their ships.

The first results looked good, but when checks were run against a standard LaCoste marine meter it was immediately obvious that something was wrong, because differences of up to fifteen milligal were being recorded. Naturally, it was assumed that it was the new, experimental, meter that was at fault, but no-one could work out why. Moreover, the new meter produced much more consistent readings when the ship re-crossed its own tracks. The problem was eventually traced to a saturating amplifier in the cross-coupling computer in the LaCoste. The VSA (vibrating string accelerometer) system, being symmetrical, required no such correction. What was worse, similar faults were found in several other LaCoste meters. A validity of a whole decade of marine gravity measurements was thrown into question.[5]

---

[5]The story is told, in some detail, in Bowin et al. (1972), which also provides an insight into the days of analogue computers and paper-chart recorders.

Once identified, the problems were quickly dealt with, and the vibrating-string instrument that had caused all the trouble relapsed into obscurity. Other symmetrical meters were designed later, and some are in use today, competing directly with the LaCostes.

Meters in the Air

Measuring 'g' from an aircraft was never going to be easy. One problem, which will be obvious to anyone who has ever travelled by air, is that even the largest and most stable aircraft, flying at altitudes of tens of thousands of metres, can experience accelerations due to air turbulence that are equivalent to significant fractions of 'g'. Survey aircraft are much smaller than commercial jets (many are single-engined) and may be flown only one or two hundred metres above the ground. The problems are at their worst in hot climates and over rough terrain, where the turbulence and manoeuvre 'noise' can be thousands of times greater than the gravity effects of the geology. Much can be done by filtering and by designing instruments with inherently slow responses, since the changes in 'g' arrive more slowly than most of the noise, but this is less effective than it is on ships, which move more slowly.

Airborne gravimetry did not get off to a good start. In 1962 the Cambridge Research Laboratory of the US Air Force installed modified LaCoste and Graf Askania marine meters in a KC135 transport aircraft, along with an Air Profile Recorder, a vertical camera, an N-1 compass, an APN-81 Doppler navigation system and an ASN-7 navigational computer. They got as far as concluding that stabilised platforms performed better than gimbal suspensions, but their attempts to go further came to grief. Because of repeated fuel leaks, no actual flight testing was done before the aircraft was reassigned to a higher priority project.

Undeterred, they tried again, this time in a C130 aircraft fitted with even more auxiliary instruments, including a modified ART-25 stabilized platform on which the gravity meters were mounted. The installation evidently took some time, and by the time it was finished, and before the aircraft could actually get into the air, it too had to be released to another project. Under the circumstances, to entitle the paper in which these 'results' were presented as '*Advances in aerial gravity, 1963–64*' (Thompson and Hawkins 1966) might be seen as an early example of the now-perfected political art of the alternative fact.

A later attempt by the US Government to put modified LaCoste marine gravity meters into helicopters also initially came to grief, but at least there was actual flight testing, with a rather curious, although poorly documented, outcome. The story goes that after the project had been terminated, one of the scientists involved told Frank Carson, who owned the helicopter, that there had simply been too much vibration for the measurements to be reliable. Carson's response was that he had been under the impression that manoeuvrability had been the main requirement and that if he had known what was really wanted, he would have rigged the rotor blades of the aircraft (a twin-rotor Sikorsky S61) very differently. Some confirmation of the truth of the tale comes from the fact that in 1984 Carson was awarded US Patent 4,435,981 for airborne gravity work in which it was noted that *if a helicopter is employed the rotor blades are precisely tracked and aligned for smoothness of flight*. By then it was too late, for the government at least, but the company began airborne gravity on its own account, and dominated the field during the 1980s and early 1990s.

When encountered at trade fairs and conferences, Carson representatives were generally found to know a great deal about helicopters and very little about gravity, but they received what should have been a major boost when Sigmund Hammer, then president of the Society of Exploration Geophysicists and the Grand Old Man of gravity exploration, published a paper in *Geophysics* entitled *Airborne gravity is here* (Hammer 1983). He began by saying that

> 33 years ago, in a public debate with Canadian geophysicist Hans Lundberg, I said that airborne gravity was an impossible dream. It gives me great personal satisfaction to have the opportunity during my lifetime to proclaim that I was wrong.

Poor Hammer! The paper brought a torrent of invective down on his head, and he had to later admit that he had made two significant errors, the more important being an '*overstatement of the precision and resolving power of the Carson airborne gravity method based mainly on 1981 test survey data*'. The letters that brought these errors to the fore and which were published in *Geophysics* were remarkable for the quite unnecessary 'rancour' (Hammer's own word) that characterised them.[6] It can scarcely have been a coincidence

---

[6]The whole story, including Hammer's original paper, the comments and his replies, can be followed in Volume 49 of *Geophysics, v9* (1983).

that they came entirely from representatives of companies in the gravity survey business, who might have had their own reasons for wishing to see Carson fail. Nelson Steenland, of the Geophysical Exploration Company, ended his own diatribe by asking

> Is this discussion trite, concerned with semantics and personalities? No, the misrepresentations relating to the accuracy of the airborne meter and the role of 'residual gravity' in interpretation have inhibited and do inhibit gravity from achieving its proper proportionate place in prospecting. That is the reason for presenting this discussion (Steenland 1984).

Which might have been true, but Steenland's own analysis had its flaws, and it is also true that success for Carson would have been very bad for the employees of the Geophysical Exploration Company. They need not, however, have worried. Carson Helicopters gradually faded out of the gravity survey business, to be replaced by more knowledgeable competitors, and eventually collapsed in 2011 after the jailing of one of its vice-presidents for a fraud on the Forest Service of the US Department of Agriculture (which had nothing to do with 'g').

Despite the doubters, efforts continued to be made to improve airborne gravimetry by reducing manoeuvre noise, by the use of gyro-stabilised platforms, by damping and by post-flight processing, but it was never going to be a serious mapping tool until the navigation issues had been resolved. In a moving aircraft the changes in velocity and heading take place far too rapidly for Eötvös errors to be reliably estimated with the instruments available in the 1980s. It was the introduction of GPS navigation that ultimately made airborne gravimetry work, but by that time none of the US companies were very interested. Today's state-of-the-art airborne gravimeters are manufactured by the 'Joint Stock Company Gravimetric Technologies' of the Russian Federation and the data are processed using software developed by the Lomonosov Moscow State University's Department of Mechanics and Mathematics.

In the United States, on the other hand, the way ahead was seen to lie in a return to gravity gradiometry.

References

Bowin C, Aldrich T, Folinsbee RA (1972) VSA gravity meter system: tests and recent developments. J Geophys Res 77:2019–2033

de La Condamine C-M (1751) Journal du voyage fait par ordre du Roi a l'équateur, servant d'introduction historique a la mesure des trois premiers degres du méridien, Imprimerie Royale, Paris

Ewing M (1938) Marine gravimetric methods and surveys. Proc Amer Phil Soc 79:47–70

Hammer S (1983) Airborne gravity is here! Geophysics 48:213–223

Hecker O (1910) Bestimmung der Schwerkraft auf dem Schwarzen Meere und, an dessen Kuste sowie neue Ausgleichung der Schwerkraftmessungen auf dem Atlantischen, Indischen, und Groszen Ozean. Zentralbureau International Erdmessung, Berlin

Heiskänen W, Vening Meinesz FA (1958) The Earth and its gravity field. McGraw-Hill, New York

Mason RG (1958) A magnetic survey off the west coast of the United States: Geophys. Jour. 1:320–329

Mason RG, Raff AD (1961) Magnetic survey off the west coast of North America, 32°N latitude to 42°N latitude. Bull Geol Soc Amer 72:1259–1266

Menard HW (1986) The Ocean of Truth. Princeton University Press, Princeton

Raff AD, Mason RG (1961) Magnetic survey off the west coast of North America, 40°N latitude to 52°N latitude. Bull Geol Soc Amer 72:1267–1270

Siemens CW (1876) On determining the depth of the sea without the use of the sounding line. Phil Trans Roy Soc 166:672

Steenland NC (1984) On: "Airborne gravity is here" by S. I. Hammer. Geophysics 49:310

Thompson L, Hawkins C (1966) Advances in aerial gravity, 1963–64 In: Orlin H (ed) Gravity Anomalies: Unsurveyed Areas, American Geophysical Union Monograph 9:28-30

Vening Meinesz FA (1931) By submarine through the Netherlands East Indies. Geographical J 77:338–348

Vening Meinesz FA (1932) Gravity expeditions at sea 1923–1930. Vol. I. The expeditions, the computations and the results. Nederlandse Commissie voor Geodesie, Delft

Vine FJ, Matthews DH (1965) Magnetic anomalies over oceanic ridges. Nature 183:882–883

11

The Return of the Gradient

The rise of the gravity meter meant the end of the torsion balance and, seemingly, an end to the use of gravity gradients in exploration. And yet, seventy years later, measuring gradients is again in fashion. Why?

Red October

Gravity gradients decrease with the cube, not the square, of their distance from the masses that cause them, so are even more sensitive to the separation of sources from detectors than is 'g' itself. Gradiometry prioritises the effects of masses that are close to the sensor over the effects of those that are further away, and this can be an advantage. In 1990 the release of the film '*The Hunt for Red October*' reportedly caused something of a stir in the US intelligence world because the crew of the (fictional) American submarine USS *Dallas* could be heard discussing 'milligal anomalies' during some tricky manoeuvring in a submarine canyon, supposedly revealing to the world that the US Navy was using gravity as a navigation tool. As if to confirm the story, and as an admission that the game was now up, the supposedly up-to-then very secret Lockheed-Martin gravity gradiometer ceased to be a secret only a short time later.

I have watched the film several times, on DVD and television, searching for this moment and have never found it. However, it must be there somewhere, because I have it on the best possible authority that it is. Who would dare disbelieve the Central Intelligence Agency? In the summer of 2009 the

J. Milsom, *The Hunt for Earth Gravity*,
https://doi.org/10.1007/978-3-319-74959-4_11

CIA house journal '*Studies in Intelligence*' [1] published an article describing the 'leak' but failing to make clear whether or not it was authorised. All that was said, in true unattributable spook-fashion, was that a gradiometer system was 'allegedly' installed on only a few Trident submarines but was declassified a few months after the film was released (Hadley 2009).

If true, this went far beyond the previous use of gravity in navigation, which had been as the reference for gyroscope-based inertial navigation instruments. If gradiometers were being used in this way, it would not be for navigation in the normal sense but to reduce the chance of the submarine colliding with the sea floor. Submarine sonar had already been developed for that purpose but, with anti-submarine technology at the level it had reached in the late 20th Century, sending out sound waves while trying to remain undetected underwater was not a good idea. Making use of the rapid increase in gravity field close to a massive obstacle was potentially a much better way, since no signal need be sent from the submarine, and measurements of gradient would be better than direct measurements of 'g' because they would change much more rapidly than 'g' itself. Moreover, knowing only that 'g' was increasing would not necessarily be helpful. The gradients could be expected to show the direction in which it was increasing, which would be much more useful.

Despite all this, the idea that the film had inadvertently revealed something that the US Navy wanted to keep secret is not very plausible. For one thing, if it was gradiometry that was being used, the crew should have been talking about Eötvös units, not milligals. For another, the basic instrument had been described two years earlier, in a paper by Chris Jekeli published in the very public pages of the American Geophysical Union's newsletter, *Eos* (Jekeli 1988). And, as a final nail in the coffin, it was Tom Clancy's original book, first published in 1984, and not the film based on it, that introduced into cold-war fiction the idea of using gravity gradients in undersea warfare. In the book it was not the Americans but the Russians who were measuring gravity gradients, with a gradiometer consisting of two large lead weights separated by a distance of '*one hundred yards*' (Clancy 1984). Their relative positions were measured by laser to better than a billionth of a metre, and any changes in distance or orientation were converted into estimates of gradient.

This is really odd. The instrument described bears no resemblance whatsoever to the gradiometers that were developed in the US and, although something similar was later to be done out in space with the GRACE twin

---

[1] The journal is published quarterly primarily for use by US government officials, and most complete issues are classified. However, some unclassified extracts can be found on https://www.cia.gov/library/center-for-thestudy-of-intelligence.

satellites, there are many, many reasons for doubting that it would ever work in a submarine. It may be significant that the book was first published by the US Naval Institute which, while not an official part of the US Navy, has very close links to it. Could it have been a deliberate exercise in disinformation, designed to lure the Russians into a technological dead-end, or convince them that the Americans were already trapped in one? Clancy died in 2013, so it is unlikely that we will ever know.

A History

The fictional sinking of *Red October* marked the beginning of modern gravity gradiometry, but it was the special requirements of airborne survey that led to its development. And, while it was the oil industry that commercialised the torsion balance, when gradiometry again became fashionable it was the mining industry that made it so. The story begins with decisions made by the US Navy and, for different reasons, by the US Air Force to fund the building of a new generation of instruments.

There is little doubt that as early as 1977 the US Navy had become interested in using gradiometers for avoiding collisions and, faced with choosing between three possible designs, they decided to put their money behind Bell Aerospace's 3-D Gravity Gradiometry Survey System (GGSS). In 1983 the Air Force followed suit, handing the responsibility for oversight and testing to their Geophysics Research Laboratory. The Air Force instruments were mounted in a modified camper-van for initial road testing and were carried into the air by the simple expedient of driving the van into a C130 heavy transport aircraft and taking off (Fig. 11.1). By the middle of 1987 the data collection phase of the test programme had been completed and a year later Jekeli's overview was published in *Eos*.

After that things went downhill, and by 1993 the Air Force scientists, evidently taking a lead from their predecessors who had worked with airborne gravimetry, had terminated the programme, even though the main problems seem to have been not with the gradiometer but with the GPS, which performed poorly on well over half the test flights. GPS navigation was itself in its infancy at the time, with only a limited number of satellites in orbit, giving only partial and intermittent coverage. Jekeli was forced to conclude that:

> To rejuvenate the GGSS program with significant payback would require several millions of dollars now and exponentially more in the future. There is,

Fig. 11.1 A different approach to airborne installation. The camper-van containing the US Air Force prototype gradiometer being driven into a C130 for a test flight in 1987. Jekeli (1988) noted that amongst the equipment required was 'optionally a reporter for Ripley's Believe-It-Or-Not'. Photo provided by Christopher Jekeli

unfortunately, no sign of any interest to restore the program, even on the distant horizon (Jekeli 1993).

The Navy took a different view, and by the time that comment was written, and with several years start, they had built several systems and installed them in submarines. The results they obtained in places such as the Persian Gulf were unexpected, and to explain them they turned to Maurice Ewing's successors at Columbia, who concluded that the instruments were not only seeing sea floor topography but also geology, and especially mobile salt. The investigations were still ongoing when, in 1989, the Berlin Wall came down and the Cold War ended almost overnight. The following year saw not only the screening of '*The Hunt for Red October*' but also lobbying of the Pentagon by Columbia's Roger Anderson for gradiometer declassification.

Anderson was lucky. The government committee with the final say was chaired by Anita Jones of the Defense Advanced Research Projects Agency, who informed the audience of admirals that she knew the technology would be useful because her father had been an oil service company field engineer (Anderson 2000). Declassification quickly followed, and in 1994 Bell Geospace was formed by a group of Columbia University oceanographers and Bell Aerospace managers as a company independent of both with exclusive licenses to use the system in the oil industry. Declassification also allowed US Patent 5,357,802 to be granted to Glen Hofmayer and Clive Affleck for the rotating disc 'Gravity Gradient Instrument (GGI)' in October 1994. There must have been some special dispensation from normal patent law, since the system patented was essentially the one described by Jekeli six years earlier. By 1997 Bell Geospace had two ships on long-term leases operating their three-axis Full Tensor Gradiometer (FTG) system commercially, mainly in the North Sea and the Gulf of Mexico, but the following year the oil price collapsed and the company faced bankruptcy. It went into a 'Chapter 11' arrangement, which allowed it to continue operating but at a much reduced level.

Geophysicists involved in exploration have always (or at least until the oil-price crash of 2015) been able to rely on one feature of the resource cycle. When oil has been expensive there have been jobs in the oil industry, but when oil has been cheap other industries have boomed and for lucky geophysicists there have been jobs in the mining industry. Oil was certainly cheap in 1999 when the Minerals Discovery Group of the mining giant BHP took to the air with two single-engined Cessna Caravan aircraft fitted out with the 'Falcon' system that it had developed using the ideas described in the Hofmayer-Affleck patent, Bell accelerometers and a gyro-stabilised platform that had been built by Lockheed-Martin for the U.S. Navy. The new gradiometer was much more sensitive than the naval version but it took eighteen months of testing on a shake-table before the team was satisfied that something useful had been created.

It is probable that Falcon was seen in its early days as a tool that would be kept in-house, to give BHP an exploration edge. While it might be licensed out where there were deals to be done, that would be the extent of its use by outsiders. All that changed soon after serious systematic flying began, thanks probably to conflicting views within the much larger company formed by the BHP-Billiton merger of 2001. Falcon was licensed soon afterwards to a company that specialised in looking for minerals on behalf of other people rather than in extracting them for itself. The company was Fugro, a Dutch conglomerate that was at the time making a bid to become the globally-dominant force in geophysical exploration.

The oil price recovered gradually in the new century and the Bell Geospace that emerged from its near-death experience of 1999 was fitter, leaner, and willing to abandon some of its most cherished ideas, among which was the idea that a surface ship was the most suitable platform for a gradiometer. By 2002 their airborne instrument, which they christened Air-FTG, had completed its tests and was flying commercial surveys. It was still, however, basically the instrument that had been declassified ten years earlier, mounted on a Lockheed-Martin gyro-stabilised platform. The same company, formed in 1995 by the merger of the Lockheed Corporation and Martin Marietta, had also by this time taken over GGI manufacture from Bell Aerospace.

Another company, ARKeX, also entered the scene. It was formed in 2004 as a joint venture between Oxford Instruments, which had already built a superconducting gravity gradiometer for the European Space Agency, and ARK Geophysics, which knew a great deal about airborne geophysical surveys. The eventual aim was to develop a new gradiometer to exploit what seemed to be an expanding and lucrative market, but under the restraints of Chapter 11 Bell had been unable to pay for an additional FTG system that was being built for them by Lockheed Martin, and lost their exclusivity. ARKeX stepped into the breach, bought the instrument and began flying a system virtually identical to Air-FTG but processing the results in different ways. Using technology which was not only twenty-five years old but still shrouded in secrecy was not ideal, but the aim was to generate funds to pay for the development of a new and different commercially-viable system.

The result of this rather tangled history was that in the early 'noughties' there were two systems and three companies competing for gradiometer business, with the FTG operators boasting that theirs was the original system and measured the full tensor described in Chap. 14, Coda 1, while Fugro made much of the fact that Falcon was designed from the outset to be put in an aircraft and not in a ship. Bell Geospace also began using a much larger aircraft, the Basler re-engined modification of the wartime Dakota (C-47 or DC-3) for some of its surveys, providing a much more stable platform, far greater range, and the ability to engulf those geophysicists old enough to have flown survey in the days when the DC-3 was the maid of all survey work in waves of nostalgia. Demand for the services of all three companies was fuelled by oil discoveries in East Africa for which gravity gradiometry was given much of the credit, and on some occasions client companies found that instead of being in the position of choosing between competing contractors, they were desperately trying to get just one to offer to do the work in the time-frame required. Gradiometry became fashionable, to sometimes absurd extents, and consultants who pointed out to their clients that, in their lease area, there was already quite adequate gravity coverage and that the exploration dollars might be better spent on something other than this new and often poorly understood 'silver bullet' were looked at askance.

In 2015 the oil price crashed once more. Falcon (under new licensees) and Bell Geospace struggled on but ARKeX went bankrupt. Bridgeporth, a company that had been formed some years earlier by ex-employees of

ARKeX, stepped into this newly opened breach and took over an improved version of Air-FTG that had been built for ARKeX by Lockheed-Martin and which combined the virtues of its two predecessors.

A 'Black Box?'

So much for the history. The instruments themselves are still hedged around with restrictions, to which all users must agree. They can only be taken to approved countries and their outputs can only be released to approved clients. The details of the systems are still partly classified, and aircraft are prepared for flight simply by persuading voltages measured at unknown points to fall within pre-specified limits. Much of what is happening has to be taken on trust.

So what is actually going on inside each contractually inviolate 'black box'?

The individual sensor is the heart of every gravity instrument, but it takes two of them, a fixed 'baseline' distance apart, to measure a gradient. This is as true of the most recent versions as it was of the torsion balance developed by Lóránd Eötvös, in which the sensors were the masses at the ends of the suspended beam. The sensor described by Jekeli (1988) was the Bell 'pendulous force rebalance accelerometer', in which the position of a suspended mass was sensed by changes in the electrical forces between conducting rings positioned on either side (Fig. 11.2). The electrical signals were amplified and fed back into the system to produce torques that forced the mass back to its null position, and the current required was a measure of the acceleration. Compared to modern micro-electrical (MEMS) devices, the instrument seems clumsily archaic, but information on any modifications or improvements that might have been made since 1988 is sketchy in the extreme. It would have been possible to re-engineer the whole system to take advantage of new technology, but it might also have been decided that well should be let alone.

The accelerometer pair is far from being the end of the gradiometer story, especially if the instrument is to be mounted in an aircraft. Measuring gradients by differencing the outputs of paired accelerometers cancels out the rectilinear (fore-and-aft, up-down and sideways) accelerations but not the accelerations associated with yaw, pitch and, most seriously, roll, which are different at every sensor. To deal with these, airborne gradiometers are mounted on platforms that are gyroscopically maintained in their orientations relative to the Earth's spin axis despite changes in aircraft attitude.

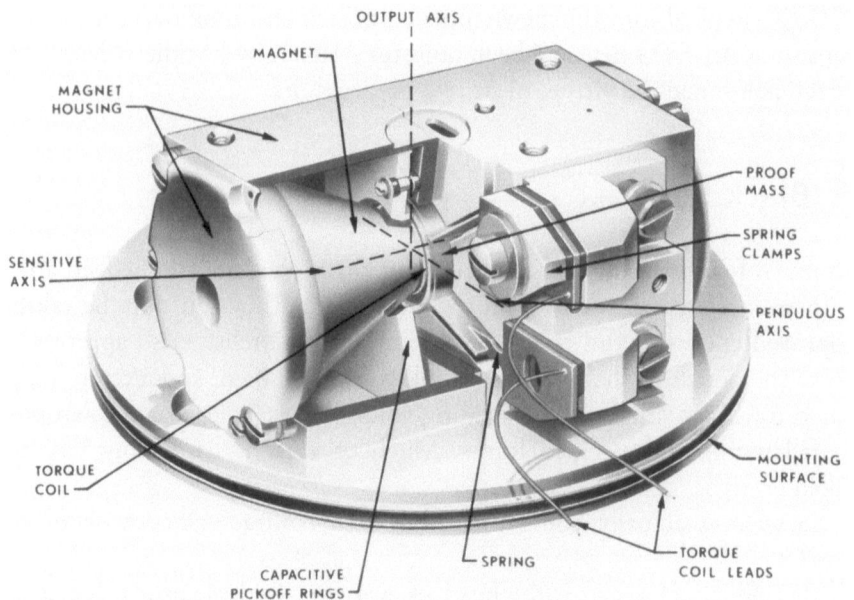

Fig. 11.2 The Bell Aerospace Mk VII accelerometer. It seems astonishing that such an apparently crude device could be capable of the extraordinary measurement precision required for airborne gradiometry. Photo provided by Christopher Jekeli

Differencing outputs also emphasises the inevitable built-in differences between accelerometers, and it was for this reason, amongst others, that the rotating disc system known as the GGI was developed. The initials stand, rather non-specifically, for 'Gravity Gradient Instrument' but in practice what is generally known as a GGI consists of a disc a few tens of centimetres across on which are mounted an even number of equally-spaced and as near as possible identical accelerometers. A full rotation is typically completed every few seconds, which introduces yet another acceleration, the 'centrifugal force'. Because the accelerometers are oriented at right angles to the disc radius they should fail to notice this, but because alignments are never perfect there is yet another effect added to the accelerometer-dependent part of the signal. This is acceptable because the spin causes the whole of this part to be directed alternately in opposite directions, giving it a frequency equal to the spin frequency. Modern digital filtering allows signals or parts of signals with known frequencies to be very efficiently removed or enhanced.

The Hofmayer-Affleck patent provided a generic description of a GGI, but the working systems use it in different ways. Falcon relies on a single horizontal disc with a diameter of about 30 cm and with eight accelerometers mounted at 45° intervals around the circumference. In 2014 a modification called Falcon Plus was introduced for which significant, but unspecified, improvements were claimed. The original Full Tensor Gravity Gradiometry (FTG) systems had discs that were only 15 cm in diameter and only four accelerometers per disc, but in each assembly there were three discs with rotation axes at right angles to each other (Fig. 11.3). In 2015 Lockheed Martin developed the 'Enhanced FTG' system with 30 cm discs that is now being flown by Bridgeporth.

No doubt the colleagues of Lóránd Eötvös would have considered the ability to measure gravity gradients from the air and over such tiny baselines even more '*fast unglaublich*' than the original torsion balance, but it is now routine. In emphasising just how sensitive these instruments could be, Edwin van Leeuwen, the BHP manager responsible for the Falcon project, anticipated the later developers of superconducting gravity meters in using family members as test masses (see Chap. 9) and stated that Falcon could measure the gravitational field of a three-year-old child one metre away (Leeuwen 2000). It can be assumed that he had a three-year old child available for testing.

Measuring gravity gradients became popular after more than half a century of being ignored because it could be done from the air, but for some tasks it has been brought back down to Earth. The aim in these cases has been to see how gravity is changing with time, when magma is moving within volcanoes or when oil reservoirs or aquifers are being depleted or oil recovery is being enhanced by steam flooding, water flooding or gas injection. Measurements may be made months, rather than days or weeks, apart and although ideally the instruments should be left in place during

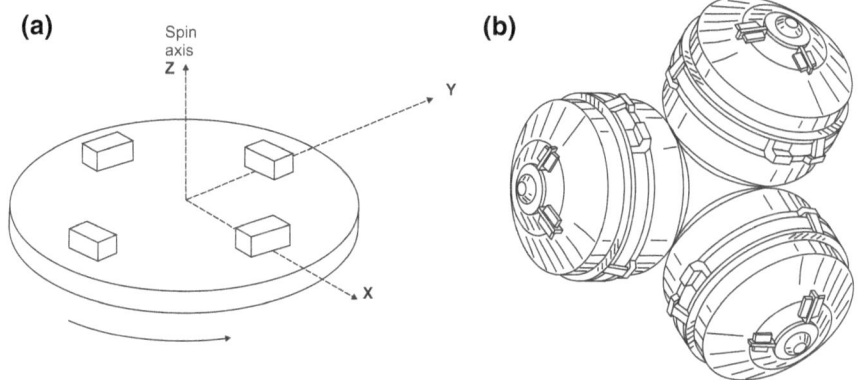

Fig. 11.3 GGI basics. **a** A single GGI, with two pairs of accelerometers mounted on the rotating disc. **b** The three-GGI assembly as used in FTG systems

the intervening times, this would be far too expensive for all but the short-est-term projects.

The End Product

It is hard enough to measure the Earth's gravity field, but a modern gradi-ometer has to measure its gradient, to better than ten Eötvös units. Since one Eötvös unit represents a change of just one milligal over a distance of 10 km, or one fifty-thousandth of a milligal over the 15 or 30 cm that sep-arate the sensors in standard airborne gradiometers, this sounds impossi-ble and in one sense it is. The probable error in a single gradient reading is likely to amount to several hundred Eötvös but becomes almost irrelevant if enough measurements are made. It is then usual to refer to the errors as *noise* and to characterise their relationship to the *signal* by a single signal-noise ratio based on the average values.

A low signal to noise ratio is self-evidently a bad thing, and Fig. 11.4 shows a plot of measurements made along one line of a typical 2010-vintage FTG survey. The noise was about ten times as strong as the signal, which would at the time have been considered quite good, but the dominant frequencies of noise and signal were quite different. Measurements were made every tenth of a second, and this time interval determined the dominant frequency of any genuinely random noise. The survey aircraft, however, was flown at about 200 km/h and took twenty seconds to cross even the narrowest real features of the gravity field. Filters that rejected the high frequencies were very effec-tive in separating the noise from this signal, which is another way of saying that in gradiometry the software is as important as the hardware.

Software comes in two parts, from the manufacturer and the contractor. In the manufacturer's software, those parts of the outputs of the accelerom-eter pairs that are synchronised to the rotation frequency are removed and

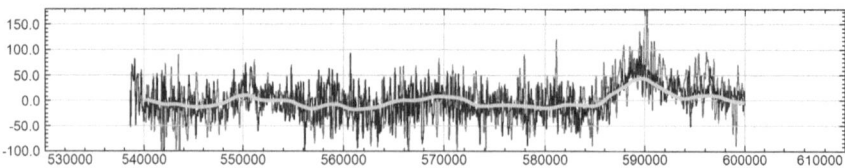

Fig. 11.4 FTG vertical gradient 'raw' data (signal plus noise). The grey continu-ous curve shows the vertical-gradient signal that was eventually extracted with the help of information from other lines

those due to things such as temperature changes are corrected for. Only after this has been done are the signals from all the pairs brought together, to be further manipulated. The programs that do all this are as secret as the instruments themselves, but they produce a usable output. It is then time for the noise to be separated from the signal on the basis of frequency and for the results to be processed still further, using programs that are proprietary to the contractor flying the system and which are also secret. Finally, filtering is applied not merely along line but across lines. There can be few other products where the end-users have so little information about the production process. The results can be validated only by seeing if they look plausible, and for that to be done the availability of some old-fashioned conventional gravity readings can be vital. In Fig. 11.5 FTG results are compared with those obtained from overlapping reconnaissance surveys using gravimeters. The (a) and (b) images are both of gravity rather than gravity gradient, and should ideally be identical, but in this case the differences are very large, mainly because of the large gaps between the traverses of the various ground surveys. The flight lines of the FTG survey were generally only half a kilometre to one kilometre apart and provided much more detail, but residual noise is clearly also contributing to the differences between the two maps.

Of all the possible gradient options, it is the vertical gradient that is mapped in Fig. 11.5c, because it is this that has proven to be the most useful. Although the exploration literature is full of clever things that can be done with the other gradient components, they are not used to any great extent, even in Falcon surveys where only horizontal gradients are measured and vertical gradients have to be calculated. Falcon cannot use the trick, common in FTG surveys, of combining the horizontal and vertical components of the gradient to provide estimates of noise levels, but its supporters claim that the greater distance between the sensors more than compensates.

Airborne Surveys

Working with airborne gradiometers needs careful planning. Line directions should ideally be determined by geology but in all but the flattest areas there is an easier and a more difficult direction in which to fly. As the terrain becomes more rugged these categories can very quickly translate into safe and unsafe.

In this respect, as in all others, the choice of flying height is critical. In each survey, flight paths are planned on an imaginary 'drape' surface that is a smoothed version of the topography set at a constant average height above

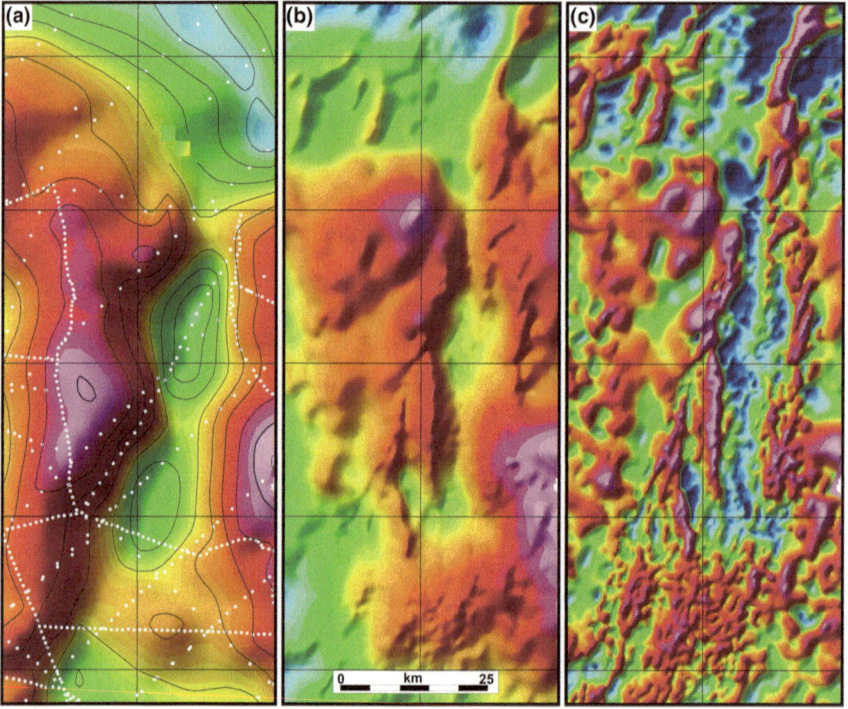

Fig. 11.5 Conventional gravity and FTG. **a** Bouguer gravity map produced by computer-contouring measurements of 'g' made at the points marked by the white circles. **b** equivalent map obtained from an FTG survey along lines that were generally either half a kilometre or one kilometre apart. **c** Topographically-corrected vertical gradient of gravity, from the FTG survey. Publication approved by Bell Aerospace

it which, in very flat areas, can be less than 100 metres. The spacing between lines is dictated by this height, and must be small enough for features seen on one line to be confidently matched to features seen on the next. Curves of the sort shown in Fig. 11.6, which record the changes (anomalies) in 'g' and its gradients due to a spherical mass, are then used to decide this spacing. In the figure, dashed lines are used for the gradient anomalies and continuous lines for the gravity anomalies, and for each there are two curves, for flights 100 and 300 metres above the centre of the mass. The units for 'g' and gradient are, of course, different (milligals and Eötvös respectively), but to make comparisons easy the vertical scales have been chosen so that the peaks of the 100 metre curves coincide. The actual values depend on the mass.

The differing widths of the anomalies of 'g' and its gradient are very obvious, as are the very different rates at which the peak values decrease with

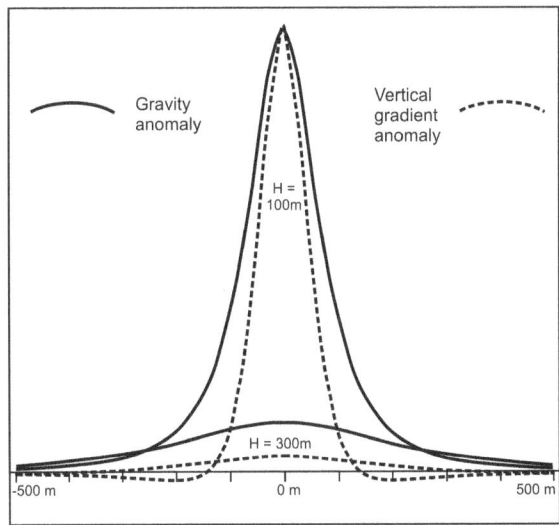

Fig. 11.6 Anomalies in 'g' and its vertical gradient. The vertical scales are arbitrary, since the anomaly amplitude depends on the mass of the source body, but have been adjusted to give similar apparent amplitudes for the 100 m gravity anomaly, in milligal, and the 100 m gradient anomaly, in Eötvös units

flying height. Because gradients decrease with distances from sources much more rapidly than does 'g' itself, gradiometry demands lines that are even closer together than those of 'ordinary' airborne gravity. The curves suggest that lines flown at 100 metres above ground level should be no more than 200 metres apart to be sure of 'capturing' every feature of the gravity field, and no more than about 150 metres apart to be sure of capturing every gradient feature. However, the anomalies defined in this way will include those produced by masses that are at or very close to the ground surface, which may not be interesting. In oil industry surveys, which are typically concerned with deep sources, much greater separations may be acceptable.

There is, of course, a point beyond which no stabilising gyro can do its job, and rapid manoeuvres can 'topple' the platform, so the turns linking the end of one flight line to the start of the next have to be made very gently. Even when 'on-line' the practical acceleration limits for a successful survey can very quickly be reached in the turbulent air characteristic of hot deserts, and in such areas useful work can often be done for only two or three hours in the very early morning and possibly one or two hours in the evening. Most people who have been involved in airborne geophysics have memories of getting up before dawn for a miserable breakfast and then sitting in the dark in an aircraft, waiting for the first glimmer of light that will allow take-off. That was how things were even when only magnetic fields were being measured. Gravity gradiometry is different, and worse, and the people who sign contracts for such work have to look very carefully at the conditions and costs relating to the 'standby' periods when conditions are so bad that little or any useful flying can be done. Airborne gradiometry may be the quickest way of obtaining gravity information over large areas but it is painfully slow compared to most other airborne geophysics.

Most clients still also want the old familiar Bouguer maps to which they have become accustomed, but although FTG operators generally offer the option of directly measuring 'g' with a gravity measuring assembly (GMA), even they seem to have little faith in this device and prefer to extract 'g' from the gradient measurements. With Falcon this is the only option, and is routine. If there is existing conventional gravity information, it provides the ultimate validity check on the Bouguer maps obtained from gradient measurements, and hence on the gradients themselves. The users can then decide whether the whole tortuous sequence has produced something that resembles what they already have, and can demand an explanation if it has not.

References

Anderson R (2000) Technical innovation: an E&P business perspective. Lead Edge 12:632–635

Clancy T (1984) The hunt for red october. Naval Institute Press, Annapolis

Hadley T (2009) The hunt for red october. Stud Intell 53:23–26

Jekeli C (1988) The gravity gradiometer survey system (GGSS): Eos 69:116–117

Jekeli C (1993) A review of gravity gradiometer survey system data analysis. Geophysics 58:505–514

Leeuwen EH (2000) BHP develops airborne gravity gradiometer for mineral exploration. Lead Edge:1296–1297

12

A Map of the World

Understandably, questions concerning the global variations in 'g' were not uppermost in most people's minds during the 1930s. Even the tiny minority of people who were interested in such things could not have imagined that a mere two decades later the armed forces of the major powers would be pouring money into measuring gravity, or that a country such as Sweden, desperate to preserve its neutrality, would have made it illegal for a foreigner to do this within its borders. For these developments geophysicists have to thank the Third Reich and its pioneering use of both cruise missiles and long-range ballistic rockets.

The Reasons Why

One of the earliest of the American Geophysical Union's Monographs (Monograph 7, in a series that is now well past the hundred mark) is entitled '*Gravity Anomalies—Unsurveyed Areas*'. Why, it might be asked, should its twenty-five authors have concerned themselves with guessing 'g' in localities so obscure that no-one had bothered to make the measurements? The answer is that in 1966, when the Monograph appeared, the Cold War was well into its stride and few things that governments did were without military significance.

The German V1 cruise missiles and V2 rockets that bombarded London towards the end of the Second World War (Fig. 12.1) were sophisticated for their time, but navigationally crude. The V1 'buzz-bombs' or 'doodlebugs' kept going until they ran out of fuel and then stopped buzzing and fell on to whatever was underneath them at the time. One removed the end of the

© Springer International Publishing AG, part of Springer Nature 2018
J. Milsom, *The Hunt for Earth Gravity*,
https://doi.org/10.1007/978-3-319-74959-4_12

Fig. 12.1 The German V1 cruise missile and V2 rocket were both notoriously random in their selection of targets. Following the end of the Second World War the major military powers set about remedying this deficiency, and became interested in the Earth's gravity field. (*Photos* Bundesarchiv, Bild 146-1975-117-26 and 141-1875A)

school where my father worked, fortunately in the middle of the night when no-one was there. Soon after that, one of the first V2s landed in woods half a mile from our home and destroyed a few trees. It was the very unpredictability of these weapons that made them terrifying and my parents, having kept us with them in London throughout the blitz, decided that enough was enough and packed us back off to Wales.

Unpredictability is an advantage for a terror weapon, but as the Cold War threatened to become a hot one the need for better targeting was recognised. For the rockets, which were now inter-continental and which, once on their way, followed a set course, detailed information on the exact shape of the Earth was needed, and geodesy became fashionable. Different considerations applied to the successors of the V1. Their launch platforms were mobile, and some form of navigation was essential. In the 1950s and 1960s the available options relied either on the Earth's notoriously ill-behaved and variable magnetic field or on inertial navigation systems in which direction was measured gyroscopically and accelerations (from which the velocities were derived) were measured by accelerometers. Since not even Einstein had been able to tell the difference between an acceleration and a gravity field, 'g' had to be known in advance, in more detail than anybody outside the oil industry had ever before required.

Guessing having proved less than satisfactory, in the 1950s and 1960s generously funded teams fanned out from Washington, London, Paris and Moscow to collect gravity data. The world's geophysicists suddenly discovered that there was money available for what they wanted to do anyway, and they went out and did it.[1] Some worked directly for the military (the US Defense Mapping Agency or DMA being a prime example), some worked for national geological surveys or mapping agencies, and some worked for universities. University geology departments found it rather easy to get funding for gravity surveys, and the French and the British hurried to collect data from their rapidly shrinking empires. The oil industry and, to a lesser extent, the mining industry, looked on and gratefully accepted these free additions to their exploration databases. In 1963 *International Gravity Measurements* (Woollard and Rose 1963) was published, a book so comprehensive that fifty years later I was using it to make two surveys agree across a politically inaccessible national border in the Middle East. This was an impressive testament to work well done, even though the values recorded had to be corrected for the Potsdam error before they could be used.

The Figure of the Earth

The shape of the Earth is determined by its gravity field, and *vice versa*. Any change in the estimate of the one inevitably leads to a change in the other. By the beginning of the 20th Century geodesists were producing quite accurate formulae describing the way in which the gravity field changed with latitude, based on what they then knew of the shape of the Earth but with numerical constants that ultimately relied on the flawed pendulum observations in Potsdam. The best pre-war version appeared in 1930, and so, inevitably, came to be known as the 1930 International Gravity Formula. It had one term that defined the 'ideal' sea level value of 'g' at the equator, another that defined how this value varied with the square of the sine of the latitude angle and another that defined how it varied with the fourth power of the sine of the latitude angle.[2] It was, of course, wrong. Not only was the Potsdam datum wrong, but later geodesists were able to show that the Earth was not quite the shape that their predecessors had supposed. By 1967 a new Earth model described by a new International Gravity Formula had been accepted as standard, and the Potsdam datum was being replaced by the International Gravity Standardization Net, eventually formalised in

---

[1]Given the possible end-uses of the results, this activity might be considered ethically suspect. However, the people concerned would almost certainly have argued that the better the navigation system the greater the chance that the missile would hit a military target and avoid hitting civilians.

[2]Because of the relationships between sines and cosines and their squares, the formula can also be written in ways that involve the cosines of the latitude angles.

1971 and universally known as IGSN71. The idea was that as organisations moved over to the 1967 equation, they would also move over to IGSN71 base-station values. To the final processed maps this change (and the later change to GRS80) made very little difference. Since the 1930 formula had been compatible with the Potsdam base, and the 1967 formula was compatible with IGSN71 values, the discrepancies almost disappeared when the latitude corrections were made.

In modern practice, a national network fans out from a single national reference point where 'g' has been measured by pendulum or weight drop or by gravity meter links to another country, but the networks within each country are usually based wholly on gravity meter ties. Where, as at international borders, two or more networks collide, any differences are likely to be due more to accumulated errors along the networks than to errors in the national base station values. Normally the mismatches are small, and undetectable on Bouguer gravity maps. Where the maps differ, it is almost always because the formulae, rather than the readings, have been incompatible. Major problems can occur when one of the changes has been made but not the other. In those cases, a 14 milligal discrepancy appears. Where different changes had been made in adjacent areas, the mismatch can rise to 28 milligal.

1967: The Papua New Guinea Isogal Survey

Few countries have gone about the business of establishing a national gravity database more systematically than Australia. This smallest of the Earth's continents is occupied by a single country, which helps enormously. There is a single datum to which all gravity values are referenced and in support of which a network of well described base stations and calibration ranges is maintained. This happy state of affairs is largely due to the work of one man, an unsung hero of the gravity world called Brian Barlow.

Brian was one of the most relaxed and amiable people that it was possible to meet. For many a young geophysicist fresh out from Europe, the first introduction to the Australian way of life was a barbecue at the Barlow house in Melbourne or, later, Canberra, but few of those who attended these events were able to remember much about them when they woke the following morning. Brian was, however, in one respect a fanatic. He was obsessed with accuracy in gravity measurement. Not for him the casual attitudes of his colleagues in the Sedimentary Gravity division, who cared little if their measurements were a few tenths of a milligal adrift because their aim was to produce Bouguer maps in which the uncertainties in the elevations of their gravity stations produced errors measured in milligals. For Brian, an uncertainty of a hundredth of a milligal was only just about acceptable, and in pursuit of his ideals he relentlessly detailed the irregularities in the calibrations of all the meters that passed through his hands, and ensured that every Australian capital city had a well-described calibration range on which they could be tested.

The Australian National Gravity Network was built around one of his best ideas. It was ultimately referred to the National Gravity Base Station at the BMR laboratories at Footscray near Melbourne, but covered the entire continent. At first it just linked Footscray to fifty-nine widely-scattered base stations with relative pendulums, but Brian added many more stations and used gravity meters. Each link was made with at least three instruments, but in the early days only the LaCoste had a range of more than about two hundred milligal. This mattered less than it might, because Australia is mostly very flat, and latitude is the principal cause of variation in 'g'. Changes along east-west lines are generally small, and Brian set up his base stations at airfields along such lines. Larger changes were encountered when linking the lines, but were kept below 200 milligal by judicious choice of line spacing. One station from each line was included in the Australian Calibration Line. Because the east-west lines roughly followed contours of constant 'g', the whole network was known as the Australian Isogal project. It was completed in 1965.

Since Papua New Guinea was still an Australian colony, Brian's next project was to extend his network there, and the existence of the Eastern Papua project, with a gravity component that required base stations, gave him his chance. Almost inevitably, the Eastern Papua geophysicist became the New Guinea Isogal geophysicist, and I exchanged my helicopter for a light aircraft.

One immediate problem with planning a PNG Isogal survey was that the lines of constant gravity could no longer be even roughly straight and would certainly not run east to west (Fig. 12.2). There were large gravity anomalies but, even more importantly, some of the airstrips were more than two thousand metres above sea level. Variations in elevation being the main reason for variations in 'g', the flight pattern looked weird, but it was practicable and included some interesting places. Landings at most airstrips in the highlands were made uphill, and take-offs were made downhill, regardless of wind direction. At Keglsugl, the highest of all, pilots were officially instructed to maintain power after landing in order to taxi up to the parking area, because if the throttles were cut right back after landing (the normal practice in sensible places) the aircraft might roll backwards down the steeply sloping strip. At Tapini it was possible to ignore the usual need to take off upwards and instead run at speed over the cliff at the end of the runway, beyond which there was a thousand foot of clear air in which flying-speed could be

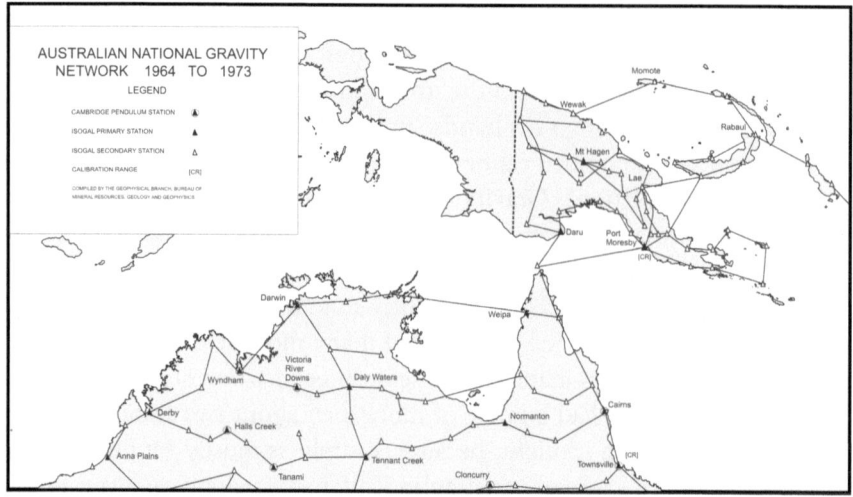

Fig. 12.2 The Papua New Guinea 'Isogal' network, together with the northern part of the Australian net

reached. The pilots who flew potatoes down to Port Moresby in little Cessna 'push-pulls' were reputed to make use of this fact when calculating the allowable load, to the terror of any passengers. At Mt Hagen, the 'hub' for the Highlands, we followed the commercial airliners (the inevitable DC3s) in through the morning clouds, but only after they had used their standard technique, in a region without landing aids, of diving just into and then out of the cloud layer until a hole had been blown big enough for them to fly through. At Kavieng an intact Mitsubishi Zero fighter, left over from the war, was parked next to the terminal building as a sort of impromptu war memorial. I have never before or since sat in the cockpit of an aircraft that so felt that it wanted to fly. A Japanese group eventually came to Kavieng and took it home (it seems there were none left in Japan). We went east to Bougainville in the PNG part of the Solomon Islands, and beyond, to Honiara on Guadalcanal, to link up with the gravity net that the DMA had set up in the southwest Pacific.

Every link was at least an A-B-A-B. The procedure was simple. Read the gravity meters (all three of them) in the cold morning light at the overnight stopping point (A), fly immediately to B, race around to find a good spot for the base station, read the meters, fly back to A, read the meters, fly on again to B, read the meters, calculate the drift, and if it was low enough on all three meters, relax. If not, make another round trip to A. At B there was more work to be done. A second base had to be chosen and occupied, in case the site for the first became unrecognisable at some time in the future, because landmarks around airports seldom survive for long. In principle each site would be marked by a copper disc that could be glued to a concrete floor or (using a version pre-fitted with a long spike) banged into the ground. Neither method guaranteed permanence, and the important part of the job was the station description. A site had to be found that was unlikely to be changed, a sketch map had to be drawn, distances from walls and steps had to be measured, and then a description had to be written that would allow the next user to occupy the exact point, within half a metre horizontally and five centimetres vertically. Churches, war memorials and statues were preferred locations in Australia, but Papua New Guinea was poorly supplied with any of these.

The brass discs suffered various fates. Some years later, landing at Madang, I saw the one I had placed there in 1967 through the open door of the airport manager's office before I had even left the aircraft. It was glinting in the sun and had, I was assured, been polished every week since I put it there. Disappearance was more usual, and one of my colleagues insisted that he had seen one of the local alpha-males in the highlands wearing one as an addition to his normal finery of feathers and paint. I am not sure that he was telling the truth, but it sounded terribly possible.

Sometimes a disc was not necessary. At the hotel in Losuia, on Malinowski's 'Isles of Love', there was an image of one of the local beauties engraved into the concrete floor by the reception desk. The picture included a perfect circle with a dot in the middle. Describable, unmistakable and unforgettable. Adding a brass disc would have been sacrilege.

The survey should have been a great deal safer than the helicopter work, but it came much closer to ending my career than anything else that happened to me in New Guinea. The absolutely final leg of the entire network was a simple trip back to Port Moresby, but the checks before take-off revealed a rather flat tire. We pumped it up, and watched it for an hour, and it didn't seem to be losing very much air, so we decided to give it a final pump and go. It had been a long project, and we were keen to reach the end. Once in the air, the pilot had a bright idea. The aircraft was a Cessna 310, with fuel tanks at the wing tips. Normally fuel was fed from both equally, but they could be drained individually. He decided that if he drained the fuel from the tank on the side with the suspect tyre, it wouldn't matter if it had lost a bit of air by the time we landed.

He definitely overdid it. When the flaps went down for landing, the heavier wing just dropped. We landed on one wheel and, very nearly, on a wing-tip tank full of fuel. The crash investigators might have found it challenging to work out what had happened, but I don't think we would have been around to help them.

For Brian Barlow there was a sequel to the BMR's geological and gravity work in Papua New Guinea. The country achieved full internal self-rule in 1974 and became independent a year later. There was room for a few expatriate geologists in the new Geological Survey, but those places were taken mainly by the people already in post at what had been the Port Moresby Geological Office. The remaining PNG specialists, based in Canberra, were told that from then on they would have to map Australia.

With a few exceptions, they hated it. The old, eroded continent, where nothing much had happened for four hundred million years, failed to excite them. They pined for the active tectonics of the island arcs, and eventually some unknown genius had a brilliant idea. Australia was committed to aid programmes in its nearest neighbour, Indonesia, and what better aid could there be than to send a group of malcontent geologists north to map the western, Indonesian, half of New Guinea using the techniques they had perfected when mapping the eastern half? While they were about it, they could train young Indonesian geologists to do the same. And, since gravity had been a part of the programmes in PNG, it became a part of the aid programme. Brian went north, taking his ferociously high standards with him, and enforced them mercilessly. They continued to be respected in Bandung long after the project ended, but everything fell apart when the politicians in Jakarta decided that, as part of the celebrations of the 50th anniversary of Indonesia's independence, a gravity map should be issued covering the entire country. Quality control was abandoned in the rush to fill in the remaining blanks, and never returned.

1973: The Australian Calibration Line

As the 'Net' in its name suggests, IGSN71 was based not on a single measurement, assumed to be good, at a single place but on a network (Fig. 12.3). Gravity was measured at 473 primary stations using nine invariant pendulums and three weight drop instruments, and these were linked to more than a hundred thousand auxiliary bases (ex-centres) by gravity meter ties (Morelli et al. 1972). The sheer volume of data shows just how far the technology had moved in the seventy years since Potsdam. The entire data set was statistically analysed to minimise errors and, statistics being the sort of science that it is, it was decided that several different teams should do the work independently and then compare results. Eventually they agreed. The final product is not perfect, because only a few of the measurements of absolute field were accurate to bet-

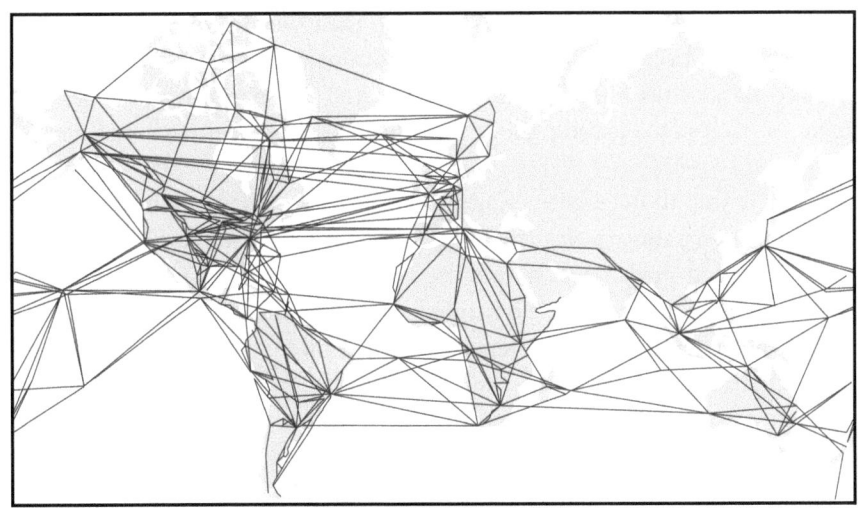

Fig. 12.3 The primary links used in establishing the IGSN71 base station network. It was the lack of information from Russia and China that prompted much of the work described in AGU Monograph 7

ter than 1 milligal, but it will be a very long time, if ever, before it is worth making another change. The use of gravity meters added an extra level of confidence, since even in 1967 they were routinely measuring gravity differences to a few hundredths of a milligal.

The West Pacific Calibration Line (WPCL) established in 1964/65 could be considered typical of the work done to create the network. A chartered Boeing 707 was flown from Anchorage to Hobart via Tokyo, Taipei, Manila, Hong Kong, Singapore, Darwin and Sydney, filled with geophysicists from all the countries *en route*, each with at least one and usually several gravity meters. At each stop, all the geophysicist got out and read their meters. They then all piled back into the aircraft and, if this had been the first visit to that particular stopping point, returned to the previous point to repeat the operation and check the drift of the instruments. In this way they leap-frogged their way down the western edge of the Pacific in a series of A-B-A-B ties.

The Russians, however, hadn't been invited. There was no WPCL station in the Russian Far East or, indeed, any IGSN71 station in the Soviet Union, and the Australians felt that was wrong. In 1972 the Sydney gravity base was linked to Moscow using five Russian relative pendulums and in 1973 this was followed by collaboration between Australia and Russia on the Australian Calibration Line, with Australia contributing four LaCoste G-meters and five Wordens and the Russians bringing nine of their GAG-2 meters. The Russian meters could only be used within a few tens of meters of the aircraft, because they needed continuous power from the aircraft supply, and those sites had later to be connected by special ties to the airport bases. At each place it took about 40 min for the full suite of readings to be completed

When it was all over, there was a party in Canberra to celebrate a job well done and, inevitably, it was Brian Barlow who first saw the funny side. Less than a hundred yards away, across the broad sweep of Anzac Parade, was the twin of the office block then occupied by the BMR, and this was the home of Australia's Ministry of Defence. Brian pointed this out to his Russian counterpart, noting that with the new data the Warsaw Pact would be able to land a missile precisely on the building.

The Russian, it turned out, was also a humourist.

That, we could already do. Now, we can choose the room.

A Global Data Base

Australia, thanks to Brian, has shown the world how a gravity data base should be run. Elsewhere, things are less well organised. In North America, every state and province has its own data repository, which can be a nuisance. It is not that any of them have got anything wrong, just that it may be necessary to deal with an awful lot of people to get the full picture. There is, however, a very adequate 4-km grid available for the conterminous United States from the US Geological Survey, and an even better 2-km grid for Canada from Natural Resources Canada.

In South America, Asia and Europe things are sometimes much less well organised, and the situation in UK universities, which have sent their students to every continent with gravity meters, would be particularly scandalous had it been in anyone's interest to be scandalised. Ph.D. theses, which contain much original data (and not just gravity data) are public documents, but it is rare for more than three or four copies to be produced. Once accepted, they are stored in the dustiest recesses of university libraries, and can be read only by visitors in person. In principle, the work should lead to publication in the wider scientific press, but cash-strapped ex-Ph.D. students are usually in too much of a hurry to find a job and earn some money to engage in the time-consuming and poorly rewarded task of writing papers. Only those who remain in the academic world have any incentive to do so. Their supervisors, who are at least nominally supposed to have been involved in their research, could take on the task, but are probably sitting on too many committees to spare the time. So it is that a vast store of good scientific work is consigned to oblivion, having been looked at only by the student, the supervisor (perhaps) and two examiners.

They do things better in many other countries. In some it would be unusual for a Ph.D. to be awarded without at least one accepted publication, and in others fifty or more copies of theses are routinely produced, and the work is defended in crowded arenas reminiscent of scenes from 'Gladiator'. But in Britain, successive governments of all colours, while bitterly complaining about the cost of universities and doing their best to minimise these by whatever destructive means come most ideologically to hand, have miserably failed to ensure a decent return on their investment.

For a global gravity map to be produced, data have to be assembled, and there have been three important organisations doing just this. The first in the field was the DMA, now rebranded as the National Geospatial Data Authority or NGA, which not only acquires its own data but solicits more, from anyone prepared to hand it over. The United Nations, through UNESCO, followed with the Bureau Gravimetrique International, based in Toulouse. Finally, in the late 1980s (even the founders seem a little vague as to the exact date), a group led by Derek Fairhead at the University of Leeds expanded their original focus on the Cameroon Volcanic Line and began to compile gravity and magnetic data from all over Africa. Ten years later they had gone global, and shortly afterwards a company, Getech, was formed that has become probably the largest of all commercial repositories of gravity and magnetic data. Getech certainly holds far more gravity data than the BGI. It may even hold more than the NGA, but that possibility is shrouded in mystery, because no outsider knows what the NGA holds.

Gravity data have commercial value, to the mining and oil industries. Even governments and universities have noticed this, and many prefer to sell their data rather than provide it free of charge. Private companies are even less willing to part with information that could not only give them an exploration advantage but might have a cash value. The NGA, the BGI and Getech have all had to promise some of their data providers that their data would be held confidential, and in the case of the NGA that confidentiality extends even to the locations of the data points.

If data cannot be used, what use are they? The answer is that most providers are prepared to allow the release of maps based upon their data, but only at scales at which the commercial value is negligible. In the case of Getech, these 'public domain' products are sold, the standard being a map based on a 5-km grid. The NGA and BGI have wider ambitions. They want to map the world, and in 1996 and again in 2008 they made very reasonable attempts to do so and released their resulting Earth Gravity Models as grids and map images. They used all the information that they had, and where they had no information from conventional sources, they used satellites.

Gravity from Satellites

Search the internet and you can find quite decent gravity maps of the moon, but it took much more than the single 26-station gravity traverse of Apollo 17 to produce them. More cunningly, but less excitingly, the maps were made by recording in great detail the slight changes in the paths of orbiting spacecraft, manned and unmanned, caused by density changes below the Moon's surface. Very clever computer programmes were then used to calculate the changes in gravity. The same trick can be played with the Earth, using the myriads of artificial satellites now in orbit, and the first attempts to do so were made only a few years after the first Sputnik had been launched, in 1957 (Kaula 1963).

As time passed, the estimates improved, and the remarkable precision in global positioning and navigation now achievable using GPS is possible only because the variations in 'g' along the satellite orbits are now known in minute detail. Conversely, 'g' is known to that level of detail only because the orbits have been studied so minutely. The first decade of the 21st Century saw the launch of several satellites that had amongst their objectives (and, in some cases, as their only objective) a better mapping of the Earth's gravity field. To do this effectively, they had to be placed in the lowest possible orbits compatible with the atmospheric drag that causes orbits to decay and satellites to burn. Drag is significant below about 500 km and takes complete control below about 200 km.

One of the first tasks to be undertaken at the start of each new satellite programme is the formation of a committee to decide on a suitable acronym. The earliest of the (externally almost identical) specialised gravity satellites was CHAMP (CHallenging Minisatellite Payload). The project was managed by the GeoForschungsZentrum (GFZ), the direct descendant of the Royal Prussian Geodetic Institute that still occupies a part of the Telegrafenberg where Kühnen and Furtwängler created the Potsdam gravity datum. NASA were partners, and the satellite was carried into a near-polar orbit in the summer of 2000 on a Russian COSMOS rocket from Plesetsk, near the White Sea. Initially placed at 454 km above the Earth's surface, it lost height at something over a kilometre a month until it burned up in September 2010. It carried sensors to measure magnetic field and solar activity as well as gravity, and the multiple objectives inevitably led to compromise. Ideally for gravity and magnetics it should have been lower but for the solar data it should have been higher. It was its own gravity-field sensor, with orbital irregularities tracked by GPS satellites, but the instrumentation included a three-axis accelerometer. The model of the geoid produced after just eighty-eight days was widely thought to be better than all the information previously obtained, from all sources. However, it was rapidly made obsolete by the flood of data from later missions.

CHAMP was complemented in 2002 by GRACE (Gravity Recovery And Climate Experiment). This time NASA took the lead, but the launch site, for two identical satellites in polar orbits 500 km above the surface of the Earth, was still Plesetsk. Precision radar measured their 220 km separation more accurately than could be done by GPS alone. Effectively, and for the first time, gravity gradiometry was being done in space, but over a very long baseline. The mission ended in October 2017.

The baseline length was reduced drastically in March 2009, when the GOCE (Gravity field and steady-state Ocean Circulation Explorer) satellite was launched by the European Space Agency. The main instrument was an electrostatic gravity gradiometer consisting of six 3-axis accelerometers with each accelerometer pair separated by a distance of half a metre. The three pairs were mounted at right angles, recording along-track, cross-track and vertical gradients. It is hard enough to accept that quantities as tiny as gravity gradients can be reliably measured over distances of a few centimetres at the Earth's surface, and still harder to believe that this can be done in the weaker fields with much lower gradients at orbital altitudes. But they can, as the results have demonstrated. Positioning was again dependent entirely on GPS, but the number of GPS satellites was increasing as Russia, Europe and China decided that they no longer felt safe relying wholly on the United States for what has become an absolutely crucial element in the organisation of a modern state.

Another distinguishing feature of GOCE was its very low altitude, of only 255 km. Atmospheric drag is significant at this height and occasional boosts were needed to keep the satellite in orbit, but smaller-than-expected fuel consumption allowed it to be brought even lower, to only 224 km, in 2012. It then remained operational for more than a year before finally burning up in October 2013.

None of these missions could match the resolution obtained by gravity measurements on land, at sea or from aircraft, but they did have one significant advantage in addition to providing uniform coverage. They provided snapshots of the variation of the gravity field in time. Originally this was intended as a way of monitoring ocean circulation, but unexpectedly large mass changes were found to be taking place within the continents due to the draining of aquifers by the ever increasing use of ground water in agriculture. It is not a comfortable thought that so much water is being removed from the subsurface that the gravity effects can be detected from space.

The Geoid

Although it is not now much remembered, Archdeacon Pratt did not just write about the Himalayas. He published two papers in the *Philosophical Transactions* in 1859, and it was in the first of these that he set out his ideas about the support of the mountains. In the second he considered the likely effect of the mass deficit represented by the Indian Ocean, and in its second paragraph he said that

> ... the attraction of the mountains northwards, and the deficiency of attraction southwards, which last is in fact equivalent to a repulsive force northwards, combine to produce another effect on the measures of the Survey besides the deflection of the plumb-line, an effect of some importance. They have a sensible influence in changing the sea-level, so as to make the level of Karachi – to which a great longitudinal chain of triangles is brought from Kalianpur in the centre of India – many feet higher than the level at Punnoe, near Cape Comorin, the southern extremity of the Great Arc of the meridian. (Platt 1859)

This was a remarkable insight for its time, but it also begs a question. Historically it has been sea level that has been used as the base level for mapping (at least in countries that have sea coasts), but Pratt's remarks only make sense if there is another, independent, reference. This is now taken to be the ellipsoid that best fits the real Earth, and by 1830 measurements of the shape of the Earth had advanced far enough for Pratt's future sparring partner, George Airy, to define the ellipsoid that is still used today by the UK Ordnance Survey. Other versions exist, in embarrassing profusion, some, like that produced by Friedrich Bessel, that are avowedly global in intent and others designed to provide the best possible match to some chosen part of the Earth's surface.[3]

---

[3]A few years ago, Kazakhstan was plunged into deep gloom because Khan Tengri, their second highest mountain, had been resurveyed and had come out just slightly short of the magic 7000 m of which everyone had been so proud. I was there soon afterwards and did suggest that they must be able to find some acceptable reference ellipsoid that would reinstate the lost metres, but I was not talking to the right people.

Inevitably, there is a geographical bias to all ellipsoids, even those intended to be global, and a perfect fit cannot be achieved even for the oceans. Gravitational potential, however, is almost constant over the surfaces of seas, oceans and other large bodies of water. Were that not so, the water in the areas above the mean level would flow down to infill the areas below the mean level, taking any shipping with it. To a limited extent this does happen, with tides and ocean currents, but these are merely variations about a mean. However, as Pratt realised, sea level is affected by changes in 'g', and 'g' changes over the surface of the Earth for all sorts of local reasons. Geodesists take this idea and define mean sea level as a surface of constant potential (which is not, it must be noted, a surface of constant 'g') which they call the geoid. In some places it is more than 50 metres above the ellipsoid and in others it is more than 50 metres below it and, as Pratt noted, it is higher at Karachi than at Comorin (now Kanyakumari) at the southern tip of India (Fig. 12.4). Geodesists spend much of their lives trying to locate it, especially in land areas where the sea is not directly accessible. Bizarrely (or so it seems to non-geodesists) the original definition of its position in terms of the height of the water surface in imaginary sea level canals cut through the land masses still seems to hold sway. More practically, its position can be calculated from measurements of 'g', which is its gradient and therefore closely linked to its curvature.

To measure 'g' in inaccessible areas, the mapmakers turned increasingly to airborne gravity. As part of this effort, surveys have been flown over entire countries such as Ethiopia and Mongolia where 'g' had previously been poorly known, using equipment similar to that used for resource-oriented surveys but in very different ways.[4] The aircraft were flown hundreds and sometimes thousands of metres above ground level, airspeeds might be more than twice those used in conventional surveys and flight paths might be separated by distances of the order of 20 km. After it was all over, the results were often not processed beyond the free-air correction. It is free-air gravity that geodesists crave, because it is that which determines the orientation of the vertical. In more accessible areas more traditional methods may be used, and sometimes the geodesists collaborate with the geophysicists but more often they do not. In Sumatra there is a sedimentary basin high up in the Barisan Mountains that has been peppered with gravity measurements not only by the Indonesian Geological Survey and by the petroleum research institute, LEMIGAS, but also by BAKOSURTANAL, the national mapping

---

[4]Forsberg et al. (2012) provides a summary of such a survey, atypical only in that it was largely offshore.

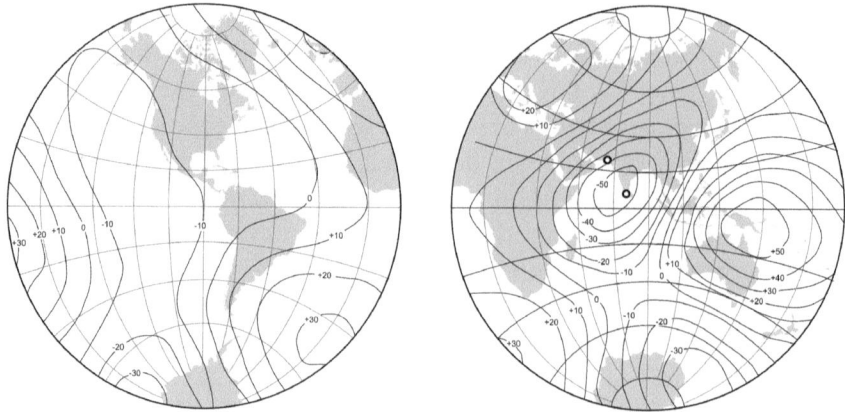

Fig. 12.4 The 'Kaula 1961' geoid, based on early results from tracking artificial satellites. The contour interval is 10 metres, and negative values indicate that sea level is below the ellipsoid. The most prominent high is centred over Eastern Papua, and more recent models, based on vastly more data, have moved the peak very little. The two open circles identify the points on the Indian sub-continent cited by Pratt, but the difference in geoid height between them is only partly due to the contrast between the mass of the Himalayas and the mass-deficit of the Indian Ocean. In the main, the regional anomalies are due to much deeper sources

organisation. There may also have been surveys by oil and coal companies, but none of these people are talking to any of the others.

A Different 'Satellite Gravity'

As gravity maps of the land areas improved during the second half of the 20th Century, mapping of the much larger water-covered parts of the globe lagged behind. Ship tracks along which 'g' had been measured were few and far between, and not all the measurements were reliable. Observations made at satellite altitudes helped expand the picture, but were lacking in detail. Then, almost literally out of the blue, came a technique that changed everything.

Imagine a flat ocean floor, with just one isolated sea-mount. Its mass will attract the water towards it, forming a hump in the sea-surface. The hump is a measure of the gravity effect of the mass, and from its shape the gravity field can be calculated, with an accuracy limited only by the detail with which sea-level can be measured. This is, of course, a very important factor but for the deep oceans it is now possible to map changes in 'g' with wavelengths of only a few kilometres and with magnitudes of only a few milligal, using the sea surface itself as the measuring device and a radar satellite to record its shape. As a result, satellite technology is now producing some of the most spectacular images of our planet ever obtained.

The story began in 1978, when the United States Navy launched a two and a half tonne satellite they called Seasat. Its objectives have been variously reported as the monitoring of sea-surface winds, sea-surface temperatures, wave heights, internal waves, atmospheric water, sea ice features and ocean topography, and it is not entirely clear which of these was considered the most important, or even if that was ever decided. The mission was described at the time as a feasibility study and it is, of course, possible that its real purpose has yet to be revealed, but amongst the most important instruments mounted was a radar altimeter capable of measuring the distance from the satellite to the radar beam's reflection 'footprint' on the sea surface with a precision of 10 cm.

The mission was not an unqualified success. The power supply failed after just over a hundred days and, according to some reports, useful data were collected for only forty-two hours. That was, however, enough there to keep the analysts happy for many years, and eventually the Alaska Space Facility made the files available to everyone in a user-friendly form. Measuring 'g' had not been on the list of Seasat objectives, but the radar altimeter data defined the shape of the sea surface and from this free-air gravity could be calculated. By 1983 Bill Haxby, a marine scientist working at the Lamont-Doherty Earth Observatory, was presenting the world with maps showing the variations in gravity over much of the 70% of the globe that is covered by water (Haxby et al. 1983). Oceanographers were by this time well aware of the existence and distribution of mid-ocean ridges, and that these were often cut by axial rift valleys and offset by fracture zones, but until Haxby no-one had any real idea of the intricacies of these systems and just how much could be learned about sea-floor spreading from them. Figure 12.5, which shows the patterns in the North Atlantic, is based on much more data than was available to Haxby, but the main features were all visible on his earliest maps.

Of course, there were limitations. The Arctic and Antarctic Oceans were omitted, not because of ice cover but because Seasat did not pass over the highest latitudes. Even in latitudes below the cut-off, measurements were limited to single passes along tracks that were many tens of kilometres apart, and this satisfied no-one. In some cases the gaps were as much as 200 km. Marine scientists demanded that more satellites be sent up, and in 1992

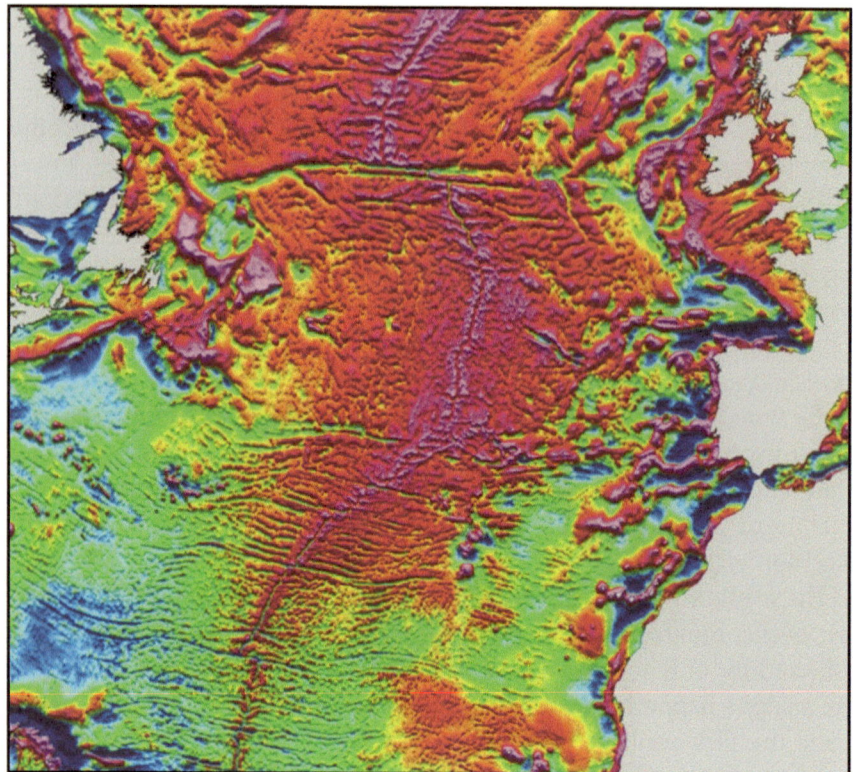

Fig. 12.5 Mercator projection of free-air gravity patterns in the North Atlantic, from satellite altimetry. The purple area in the centre of the ocean is the crest of Mid-Atlantic Ridge. The central rift is clearly visible, as are its offsets along fracture zones. The most prominent fracture, at roughly the same latitude as the south-west tip of Ireland, is the Charlie Gibbs Fracture Zone. There never was a Charlie Gibbs, the fracture being named after Weather Station Charlie, located almost on it, and the USNS Josiah Willard Gibbs, the research vessel that first surveyed it

NASA and the French CNES jointly launched the dual-altimeter Topex/Poseidon which went on providing data, claimed to give the average height of the sea surface to within almost 3 cm, for almost ten years. Soon oceanographers all over the world were trying to merge Topex and Seasat data to improve on the Haxby map, but this was not easy, because of compatibility problems between the two data types. New processing methods had to be developed, and it turned out to be easier to estimate sea-surface slopes (the gradients of the geoid) than to merge the estimates of geoid height from the two very different systems.

In one sense Topex/Poseidon was redundant before it had even been launched. In 1985 the US Navy had followed Seasat with GEOSAT, which orbited the Earth for almost a year and a half but, at least as far as the academic and commercial worlds were concerned, it orbited it silently. The Navy kept the data to itself. Late in 1986 they placed the satellite into a new orbital pattern known as the Exact Repeat Mission, the idea being to eliminate the effects of winds and tides by repeating a limited number of tracks again and again and again. Every seventeen days, the flight pattern was begun again. The features that were unchanged defined the geoid and those that were different defined tides and currents.

There must have been some lively meetings within the US Navy at that time, with the classifiers at daggers-drawn with the de-classifiers. As some sort of a compromise (or perhaps just to annoy scientists who didn't have security clearances), data from the not-terribly-exciting oceanic areas south of 60°S and north of 72°S were declassified in 1990. In 1992 the northern limit was moved up to 30°S and some really interesting areas were included and at last, in 1995, the full data set was released. Dave Sandwell and Walter Smith of Scripps took over where Haxby had left off (Sandwell et al. 2013) and soon hundreds of people were producing maps from their grids. The data release was probably due less to the de-classifiers having won the argument and more to it being pointless to keep the secrets any longer, because the European Space Agency was flooding the world with data from its ERS-1 satellite. It did more than release the processed data, it released the radar signals on which the results were based, and it was when examining these that a young London University scientist called Seymour Laxon made the next major advance.

What Seymour realised was that usable results could be obtained from areas of the Arctic and Antarctic Oceans even when they were largely covered by ice, because water reflects radar beams so strongly that most of the signal comes from any areas of open water, however small. By analysing the shapes of the returning radar pulses (a process he called re-tracking), Seymour was able to distinguish between reflections from the ice and reflections from the water surface between ice-floes and, working in collaboration with Dave McAdoo of the US National Oceanographic and Atmospheric Administration (NOAA), he was able to produce the sort of maps of 'g' over the polar regions that had previously been thought impossible (Laxon and McAdoo 1994). In a few short years, from about 1995 onwards, the shape of the sea-surface was defined with extraordinary accuracy over almost all of the water-covered parts of the globe. It is, of course, a surface that is continually being disturbed by wave action, but even the tightest of the radar beams has a 'footprint' covering several square kilometres and wave disturbances are simply averaged out. Astonishingly, the averages are being measured with an accuracy of a few centimetres.

Seymour's method had wider applications. One of the major defects of satellite altimetry was (and is) that it becomes increasingly unreliable as a coast is approached. Tidal fluctuations create very obvious problems, and the bits of land that start emerging above the surface as the water shallows also confuse the signal. In shallow tropical seas this can be particularly important because of the abundant coral reefs that are sometimes wet and sometimes dry and always sources of interference. When only Topex/Poseidon and Seasat data had been available, the image interpretation specialists of Nigel Press Associates, the UK's leading imagery interpreters, had been working on a study of Indonesia, and while their programmers had successfully developed their own ways of merging the two types of data, they had not solved the coral reef problem. It seemed possible that Seymour's re-tracking technique could distinguish the genuine sea-surface signals from the signals from any drying rock within the radar footprint and, since his office was only a few yards away from mine, I went to see him. The map of Indonesian waters that he and his student produced fitted the measurements that had been made by gravity meter along the coasts of the islands far better than anything that had gone before, and his technique went on to become almost routine amongst the people who were trying to turn radar altimetry into gravity.

Tragically, Seymour was killed in a freak accident a few years later, and the UK lost one of its brightest and best climate scientists. His monument was CryoSat, a satellite launched in 2010 that was designed and built to his specifications to monitor the ice-covered regions of the Earth.[5]

Bathymetry from Gravity

Maurice Ewing, in describing one of the earliest attempts to measure gravity at sea, wrote of his 'shock' at realizing that Siemens had been trying to do this in order to estimate water depth. It is not clear whether the shock came from recognising a man far ahead of his time or because he knew that isostasy would have made what Siemens wanted to do even more difficult than he could ever have imagined. Siemens had mentioned both Airy and Pratt

---

[5]This was CryoSat-2. The very similar CryoSat-1 crashed soon after launch in 2005 due to errors in the control programs.

in his discussion, but thought that problems would arise only in estuaries or at the shores of mountainous continents. We now know he was very wrong, and in 1938 not even Maurice would have predicted that regional bathymetric maps would eventually be routinely derived from free-air gravity data. When he wrote his review (Ewing 1938) making bathymetric maps was very hard work indeed.

In 1899 the General Bathymetric Chart of the Oceans (GEBCO) was presented to the world at the Seventh International Geographical Congress in Berlin, and in 1910 Prince Albert of Monaco decided to further his ambition of establishing a globally important Musée Océanographique in Monaco by sponsoring a second edition. The world was so impressed that when the International Hydrographic Bureau was established in 1928 to act as a clearing house for bathymetric soundings from all over the world, it was sited in Monaco. From these soundings successive generations of GEBCO maps were drawn, but it became a truly onerous task only after the Second World War. Before then, deep-water soundings had been very few in number. Most had been made by cable companies as they established their global networks, and where there was no need for cables there were no soundings.

After the war, the semi-redundant hydrographic vessels of the navies of the victorious powers were unleashed on the world to map its oceans. The measurements they made were not only vastly greater in number than those made before the war, but vastly more accurate. By the 1980s it had become obvious that the painstaking drawing of charts by hand was no longer an adequate way of dealing with the data flood, and the British Oceanographic Data Centre (BODC) was given the job of producing a digital atlas. The first edition appeared in 1992, to much jubilation, but for the traditionalists a disturbing new development was already on the horizon. Bathymetric maps were being produced from the satellite-derived gravity maps.

This was not, of course, possible from satellite data alone, because isostasy implies that free-air gravity values average to zero over sufficiently large areas. It proved, however, to be perfectly feasible to merge relatively small amounts of reliable bathymetric data with free-air gravity to produce maps from which the effects of isostasy had been removed, and sea-floor features too small to be isostatically compensated began to be mapped with increasing accuracy. Fast high-capacity computers were needed to do this, but by the time the satellite data started arriving in large quantities almost every geophysicist and geodesist in the world had such machines on their desks.

One consequence was that the deficiencies in the conventional maps became apparent. Until GPS navigation became a reality, the technology for measuring water depth was far better than the technology for determining position when out of sight of land. Charts such as those produced by the British Admiralty were a major source of information for GEBCO, but their main purpose had always been to prevent shipwrecks, and if a ship's captain reported the presence of an island or reef, it went on the chart. If a second captain then reported a sighting of the same reef in a slightly different location, that went on the chart too. Over time, a single reef 'drying at low tide' could be converted into an entire archipelago. Satellite data, however, are extremely accurate where position is concerned, and they showed that many small islands, reefs and sea mounts were misplaced on GEBCO maps, sometimes by several kilometres. The technique advanced from being a useful way of filling the gaps where there were no conventional soundings to being the standard regional source of bathymetric data, and conventional soundings came to be seen merely as providing control information for satellite-derived maps.

It was a hard thing for some of the old-style oceanographers to accept. In the official *History of GEBCO, 1903–2003* (Carpine-Lancre et al. 2003) there is only one mention of satellite gravity. It comes on p. 134, and says that:

> The satellite gravity fields … provide unique insights into the distribution of mass below the sea surface (which is closely associated with bathymetry) – although of limited application in sedimented areas, they nevertheless provide key information on trends and structures in sediment free areas. They have also been used to good effect to predict the bathymetry in areas where the sounding coverage is sparse.

Which seems a very grudging acknowledgement of a technique that was, even in 2003, completely revolutionising the mapping of the global sea floor.

Maps of the World

By combining the results of painstaking gravity-meter measurements on land with satellite altimetry at sea, adding in a little airborne data and topping these up with observations of satellite orbits and direct measurements by GRACE of gravity at satellite altitudes, it has now become possible to produce maps of the geoid and gravity field of the entire world. Much of the work in poorly mapped areas has been done by the Danish National Space Institute (DTU Space), but it was the US National Geospatial-Intelligence Agency (NGA) that took the lead in making whole-Earth models, including the most recent 2008 Earth Gravity Model (EGM2008). The mathematicians who produced it describe it as a model in spherical harmonics to degree 2159, but for almost anybody else it is more useful to know that it consists of grids of geoidal heights and free-air gravity for every 5 arc-minute by 5 arc-minute block over the entire surface of the Earth. At the equator these blocks are approximately nine kilometres square, but the east-west width decreases to north and south, becoming zero at the poles.

EGM2008 was a remarkable achievement, especially considering that even where conventional regional gravity data do exist the measurement points are often much more than nine kilometres apart. Just how much of an advance it was over previous efforts can be judged from the fact that its predecessor, EGM96, was defined only for blocks 30 arc-minutes across. The accuracy of the geoid model, which was NGA's primary objective, increased from approximately one metre to just 15 centimetres. Even so, this massive accumulation of gravity data would have been all but useless without the provision by the Shuttle Radar Topographic Mission (SRTM) of consistent and accurate estimates of ground height within thirty meter squares for almost all of the global land surface. In the highest latitudes, beyond the region covered by SRTM, data from the ICESat satellite was used.

The NGA stopped short of going beyond geoid heights and free-air gravity to Bouguer and isostatic gravity and, not surprisingly, there have been people willing fill this gap. In many cases this has been for local studies using only sub-sets of the data, but the Bureau Gravimetrique has risen to the challenge in full. They took the EGM2008 free-air values and, not content with following the conventional route of correcting for topographic effects via the flat 'Bouguer' plate and subsequent terrain corrections, they went right back to the method that Bouguer had suggested (and rejected) three centuries before and began by calculating the gravity fields of spherical shells with the Earth at their centre. For each 5-minute square there was a shell with a thickness equal to the mean height, and the effects of the deviations from it of the real topography around the entire globe were then calculated. Finally, the effects of the assumed isostatic compensation were added. With all this done, the BGI were able to announce and make available their own global, topographically-corrected Earth gravity model, which they called WGM2012 (Balmino et al. 2011)

It was a massive effort and it deserved to succeed but, unhappily, there is a fundamental problem with the input data to the onshore parts of EGM2008. On land, gravity measurements have seldom been evenly distributed. In regional surveys their locations have always been dictated by accessibility and they have tended to be made along roads or occasionally, as in parts of Papua New Guinea, along rivers. This has introduced not only a locational bias but also, because of the ways in which roads are constructed and rivers flow, a topographic bias. The rugged terrains of the upper slopes of mountains have generally been shunned, and where the free-air gravity value assigned to a 5-minute block has been based on gravity meter measurements, it has been an average biased towards its low-lying parts. Free-air gravity depends strongly on elevation, and if the average free-air gravity for a block based on the average measured values is converted to Bouguer gravity using the average height of the block, the result can be almost meaningless.

The obvious solution would be to use the actual heights at the individual observation points, but these are not part of the EGM2008 data set. The coordinates of the points on which the gravity grids are based are kept confidential to protect the commercial value of some of the data, and the result has been a sub-optimum product. To repeat the work done in developing EGM2008 for Bouguer gravity beginning from the raw data from all the various sources would be a massive undertaking, but it is only way in which a reliable global Bouguer gravity map will ever be created. In some cases WGM2012 was almost certainly based solely on the point-data that the BGI held in its own files and could have used without restriction.

The Nature of the Map

Geoid maps, free-air maps, Bouguer maps, isostatic maps. To this bewildering assortment another form of map has been added, and is now very common. It shows terrain-corrected Bouguer gravity on land and free-air gravity at sea, and is for many purposes more useful than any of the others. Unlike the isostatic map, which requires the effect of an assumed model of the isostatic compensation of the topography to be subtracted, it makes few assumptions about what is going on beneath the Earth's surface, yet it provides a vivid picture of the effects of any changes.

The power of this form of presentation is demonstrated by the map of Sumatra and the surrounding seas in Fig. 12.6. Offshore, free-air gravity dramatically picks out deep lows corresponding to the Sunda Trench, where ocean crust is subsiding into the mantle, and to the deep sediment-filled forearc basin which is underlain by relatively light material scraped from the upper layers of the ocean as it plunges beneath Sumatra. These features are oriented at roughly 45° to the spreading fabrics in the Indian Ocean and are separated from them by a belt of relatively higher gravity along the forearc ridge. Onshore, Bouguer gravity emphasises the area surrounding the site of the Toba super-volcano, which erupted some 80,000 years ago, perhaps with critical and almost terminal, consequences for human evolution. The presence of an even deeper low further north suggests that there might be

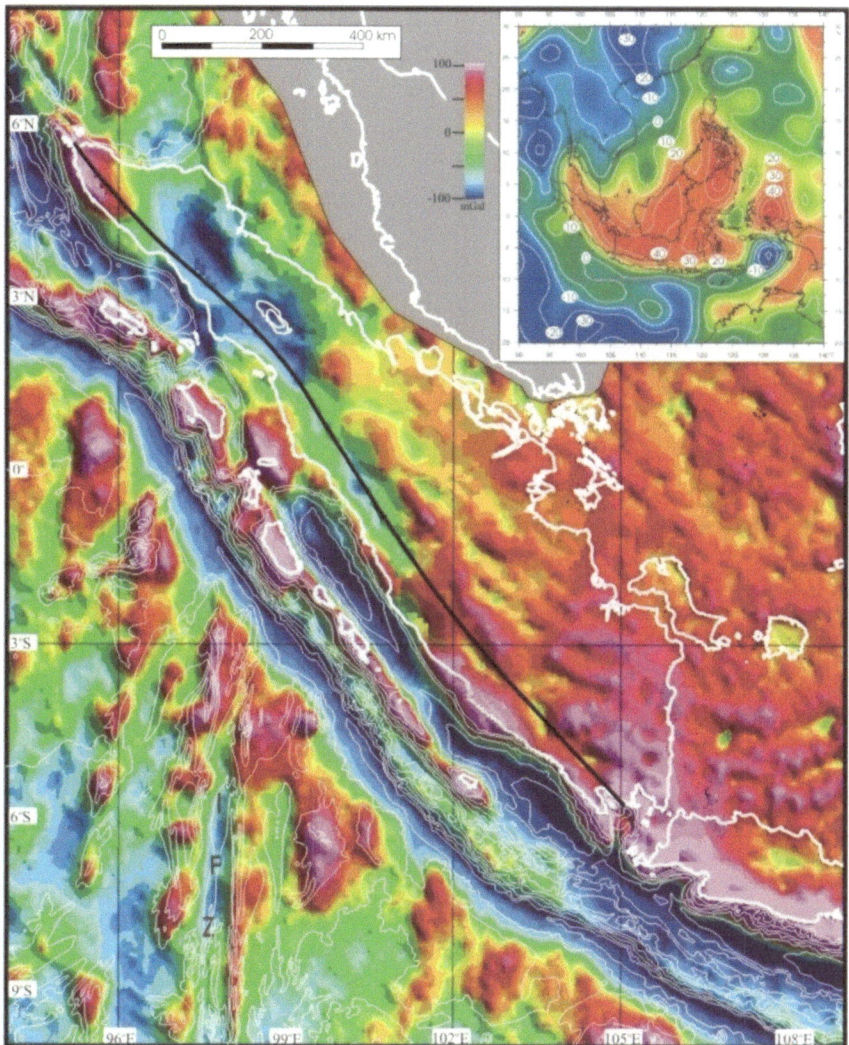

Fig. 12.6 Sumatra. Bouguer gravity onshore, free-air gravity offshore. The large lake in the northern part of the island marks the site of the Toba super-volcano. The thick black line is the Sumatra Fault Zone, which absorbs the trench-parallel component of the motion of the Indian Ocean relative to South-East Asia. *Inset* regional field; GEM-T3 Earth model, 1992

another similar magma pool forming, for future generations to worry about. The division of the island into two distinct geological provinces, with high gravity dominating the south and east and low gravity in the north and west (with the exception of the extreme northernmost tip) is very clear. Less obviously at this scale, Bouguer gravity features also mark the sites of important hydrocarbon-producing sedimentary basins.

In the offshore areas the contours are based on satellite-defined sea-surface topography, and off the east coast of Sumatra they are dominated by arcuate highs and lows that cut across the coastline at high angles. One of my brighter research students speculated that these features might not be gravity at all but artefacts produced by standing waves in the seas occupying the corner-region east of Sumatra and north of Java. It sounded terribly plausible and, if true, would have cast doubt on the whole idea of obtaining 'g' in this way. He was busy trying to find a way of demonstrating the possibility mathematically when the onshore data became available, and showed that the trends continued through much of eastern Sumatra, covering roughly half the island. They are now thought to be created by strain patterns associated with the suggested rotation of Borneo.

The inset to Fig. 12.6 shows a version of the gravity field obtained from observations of satellite orbits. At these long wavelengths the image is dominated by the high gravity fields produced by the slabs of dense oceanic lithosphere that have descended back into the more fluid asthenosphere, in compliance with the principles of plate tectonics.

References

Balmino G, Vales N, Bonvalot S, Briais A (2011) Spherical harmonic modelling to ultra-high degree of Bouguer and isostatic anomalies. J Geodesy. https://doi.org/10.1007/s00190-011-0533-4

Carpine-Lancre J et al (2003) The history of GEBCO, 1903–2003. GITC, Lemmer b.v

Ewing M (1938) Marine gravimetric methods and surveys. Proc Amer Phil Soc 79:47–70

Forsberg R, Olesen AV, Alshamsi A, Gidskehaug Ses S, Kadir M, Peter B (2012) Airborne gravimetry survey for the marine area of the United Arab Emirates. Mar Geodesy 35:221–232

Haxby WF, Karner GD, Labrecque JL, Weissel JK (1983) Digital images of combined oceanic and continental data sets and their use in tectonic studies. EOS 64:995–1004

Kaula WM (1963) Determination of the Earth's gravitational field. Rev Geophys 1:507–551

Laxon SW, McAdoo DC (1994) Arctic Ocean gravity field derived from ERS-1 satellite altimetry. Science 265:621–624

Morelli C et al (1972) The international gravity standardization network 1971. International Association of Geodesy, Paris

Pratt JH (1859) On the influence of the ocean on the plumb-line in India. Phil Trans Roy Soc 149:779–796

Sandwell DT, Garcia E, Soofi K, Wessel P, Smith WHF (2013) Towards 1 mGal global marine gravity from CryoSat-2, Envisat, and Jason-1. Lead Edge 32:892–899

Woollard GP, Rose JC (1963) International gravity measurements. Society of Exploration Geophysicists, George Banta, Wisconsin

13

Epilogue

This is a book about how 'g' came to be measured and how it then came to be used to investigate hidden geology. It is all too easy for the people who have this as their day job to forget that gravity also drives geology. To purer geophysicists, the shape of the Earth, even down to the detail of its topography, is something that has been dictated by processes that have varied in importance with time but which have all had as their underlying imperative the reduction of gravitational potential energy. These are the processes that shape the lives of all the inhabitants of the biosphere.

Gravity and Geology

At the most superficial level, gravity is the force that distributes the products of erosion downwards, sometimes in spectacular fashion in rock-falls, landslides and submarine slides offshore. The largest marine slides determine the shape of entire continental margins and produce the chaotic debris flows known as olistostromes that are the despair of geologists who try to map them where later uplifts have exposed them on land.

The steady rise of Finland is a gravitational response to the disappearance of the ice sheets that covered Scandinavia a mere ten thousand years ago. Gravity also drives the upwards surge of mobile salt into the domes, pillows and walls that were the main targets of torsion balance surveys in the 1930s and which remain important to this day. Any low-density and mobile rock can respond to gravity in similar ways, and shale diapirs ooze mud out

© Springer International Publishing AG, part of Springer Nature 2018
J. Milsom, *The Hunt for Earth Gravity*,
https://doi.org/10.1007/978-3-319-74959-4_13

over the land surface or sea floor on the landward side of many subduction zones and in the oilfields of Azerbaijan, where entrained gas provides extra lift. Volcanoes also are examples of light rocks (in these cases liquid magmas) concentrating to produce unstoppable, upwardly mobile, forces.

Gravity is also the driving force of Plate Tectonics (Fig. 13.1). The plates that move horizontally are the lids of massive convection cells that carry heat away from the interior of the Earth. The special feature of this type of convection is that the material that has been cooled and sinks differs from the material that originally rose up, because the lighter components are stripped away in two stages of partial melting. The first takes place beneath the mid-ocean ridges and creates oceanic crust, the second at depths of hundreds of kilometres beneath volcanic arcs and forms the precursors to new continents in which oxygen and silicon are concentrated at the expense of denser 'mafic' minerals rich in iron and manganese. The mantle plumes that produce islands such as Iceland on the mid-ocean ridges and chains such as the Hawaiian Islands in the deep ocean basins are further, although slightly different, manifestations of gravity in action through large scale convection.

There is yet one more effect of gravity on the geology and topography of the Earth, and it is one that has only recently been recognised.

Orogeny

In the Greek of Plato and Socrates, the word 'oros' meant a mountain, although it has now taken on wider and different meanings. In the 19th Century, it was first the French and then the British who used it to coin new names for mountain building. The British called the process 'orogeny' and a

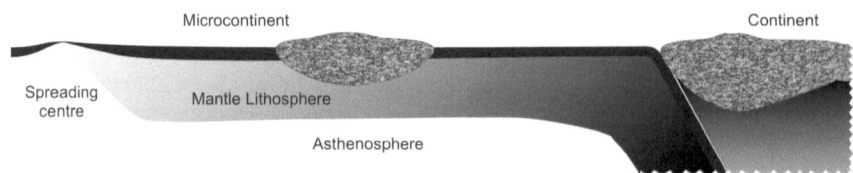

Fig. 13.1 Simple Plate Tectonics. The 'plates' consist of the crust and the underlying, rigid, part of the mantle. As temperature increases with depth, the mantle becomes warm enough to act as a viscous fluid (the asthenosphere). The relative importance of the three forces that can drive the plates (push from the elevated mid-ocean ridge, pull from the dense downgoing slab and drag by asthenosphere currents on the base of the lithosphere) is still being debated, but all three rely on gravity. If horizontal movement is resisted, as when a continent or microcontinent reaches a subduction zone, a fold mountain belt will form

mountain chain became an orogenic belt or, more briefly, an orogen. Such a word is needed because a single mountain chain can have been constructed during more than one orogeny and quite separate mountain chains can be products of a single orogenic event. The term has also proved useful in distinguishing between different types of mountain belts. Some, as in Japan and the Philippines, are almost purely volcanic and are the surface expressions of hot melted rock at depth. Others, such as the Alps, were formed almost entirely by the folding and thrusting of solid rock, and some, like the North American Rockies, involve both folding and volcanism. The word itself is restricted by many geologists to the fold mountains, but it took Plate Tectonics to provide a satisfactory explanation of how these formed.

The new understanding was reached almost as soon as Fred Vine and Drummond Matthews had formulated their hypothesis, and the rapidity with which their ideas were transferred from the oceans to the continents was partly due to the presence close to their study area of the world's greatest and youngest orogen, the Himalaya. It was while studying the ridge along which India had begun its most recent phase of drift that the pair found the evidence they needed to explain the origin of oceanic basins. The site of their survey owed much to pure chance, but the idea that the rise of the Himalayas and the Tibetan Plateau was due to collision between India and Asia followed almost inevitably. Ocean crust and the underlying cold and rigid part of the mantle combine to make up an oceanic lithosphere that is dense and sinks at the deep oceanic trenches, but continental crust is thick enough and light enough to resist the pull of its attached mantle and will not go down. The small continent in Fig. 13.1 will eventually reach the subduction zone, the crust will thicken and a collision orogen will be created.

Collapse

The North Sea is a failed ocean, with a central rift that never managed to split the continent. The Norwegian mountains can be traced across it into Scotland and Ireland, and then, beyond a rift that did produce an ocean, to Newfoundland and the Appalachians. The North Atlantic is only the most recent of a series of oceans in roughly this same location, each closed by an orogeny. The worldwide occurrence of this pattern of repeated ocean opening and closure led John Dewey, charismatic head of the Geology Department at Oxford University, to speculate on the ultimate fates of mountain chains. In late 1987 he sat down to write a seminal paper describing the *extensional collapse of orogens* (Dewey 1988). His starting point was

the idea that the highlands formed when continents collide are gravitationally unstable and, being composed of weak continental material rather than
strong oceanic material, are destined to eventually fall apart and provide sites
for ambitious new oceans. The mechanism he invoked focussed on an inadequacy in simple isostatic theory when applied to very high mountains.

In Airy isostasy (Pratt isostasy is largely restricted to the deep oceans) a perfectly strong and rigid crust floats on a perfect incompressible fluid, but even
on geological time scales an orogen is not like that. It is made up of layers of
widely differing strengths, and because pressure is equalised only at the level of
the deepest part of the Moho, there is instability at higher levels. In collision the
crust may double in thickness, forcing the Earth's surface to rise. The pressure
at sea level below the elevated block of Fig. 13.2 is greater than at sea level elsewhere, because there is a greater mass above sea level. Unless it is relieved by flow
in a weak layer, that overpressure is maintained throughout the underlying crust
and only begins to reduce at the top of the downward-projecting root. Even if
there is a weak layer, the overpressure can still only produce changes if the wider
surroundings allow material to escape, and Dewey argued that the properties of
the rocks in the Himalayan region suggested that this would happen only if the
block reached to more than 3000 metres above sea level. More than simple collision may be needed to attain these extreme elevations, but there is at least one
possible mechanism available.

It is not just the crust that is thickened by collision but the whole of the
lithosphere, which includes the brittle uppermost part of the mantle. The
lower boundary of the lithosphere is determined not by a change in rock

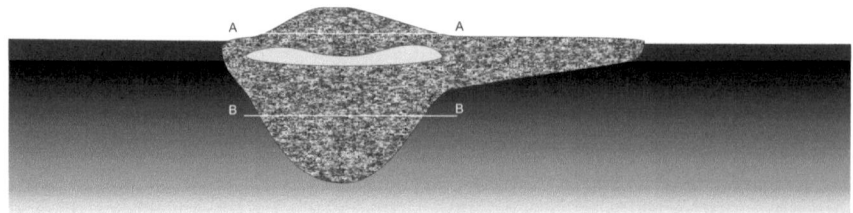

Fig. 13.2 The problem with isostasy. The pressure at any point along the line
AA passing beneath the peak of the collision orogen will depend on the weight
of material above that point, and the sides will tend to expand out over the surrounding areas of 'normal' crust. However, the material at this level is cold and
relatively strong. The overpressure will be maintained below sea level and will
only start to reduce at levels where, as along the line BB, there is dense mantle
rock beneath the surrounding areas of 'normal' crust. Above these levels a weak
layer (light grey) which was thickened along with the rest of the crust during
collision may be squeezed out sideways if free to do so

type but by a gradual decrease in strength due to increasing temperature. Cold lithosphere that projects down into the hotter, more fluid parts of the mantle will gradually warm up and expand, and the crust above it will rise. This is a slow process that may not even keep pace with erosion, but increasing temperature reduces strength, and a part of the heavy root may detach and sink. Freed of that load, the block as a whole can rise rapidly, increasing overpressure to the point where it can no longer be contained. This is the catastrophic phase of orogenic collapse. In the Himalayas, where this stage has been reached, there are no strong constraints to the east and the flows of the upper crust to the east and then to the south have been tracked by repeated GPS measurements, even though they amount to only a few millimetres a year (Maurin et al. 2010).

The India-Asia collision may have been an unusual, because the converging continental margins were almost straight and almost parallel, and collision would have been almost simultaneous over the whole length of the contact zone. In many other orogens there would have been collision in one place while significant widths of unsubducted ocean remained in others. Moreover, continental margins are not always simple and the quite large fragments that are often separated from them can create small collision orogens well ahead of the arrival of the main continental mass. Size, as John Dewey pointed out, is no bar to the development of an orogen that is capable of collapse. The only requirements are high elevation and freedom of movement, and a small orogen may have more options for expansion than a larger one.

Oroclines

There have been long periods in Earth history when nothing much has been happening, but geologists have arrived on the scene at a time of what is, by the leisurely standards of Earth processes, frenzied activity. As well as the highly visible head-on crash between India and Asia, many less simple collisions are on display elsewhere. When Africa first encountered Asia's European promontory the two margins were probably very irregular, and several microcontinents may have arrived before the main collision took place. A network of mountains and basins now stretches from the Straits of Gibraltar to the Black Sea, and many of the mountain chains are strongly curved and wrap around young internal basins. This is the pattern seen in the Rif-Betic Arc, the Alps, the Balkans, the Carpathians and the Tyrrhenian and Aegean seas (Fig. 13.3), but it is not restricted to Europe. Sam Carey, mapping in New Guinea in the 1930s, noticed it there and, in comment-

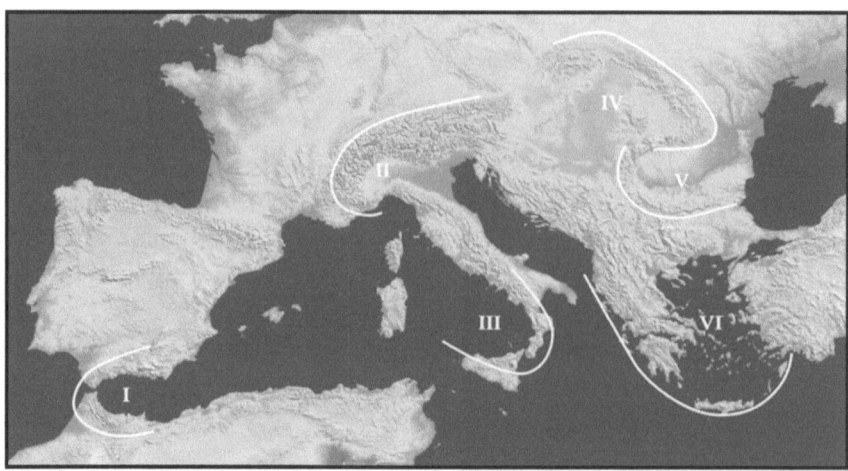

Fig. 13.3 The European oroclines. I Rif-Betic, enclosing the Alboran Sea; II Alpine, enclosing the Po Basin; III Aeolian, enclosing the Tyrrhenian Sea; IV Carpathian, enclosing the Pannonian Plain; V Balkan, enclosing the Moesian Plain; VI Hellenic, enclosing the Aegean Sea

ing on its worldwide repetition, gave these features a name. He called them oroclines, or bent mountains. John Dewey, pondering the consequence of orogenic collapse, recognised the near inevitability of orocline development where a free margin existed and listed many examples, but he somehow neglected to include amongst them Vening Meinesz's former playground in eastern Indonesia (Fig. 13.4).

The Banda Sea

The Himalayas provide geologists with an extraordinary natural laboratory, but they have their drawbacks. They are hard to reach and difficult to work in. They are criss-crossed by borders across which hostile armies eye each other with deep suspicion. Moreover, they provide insights into just one stage in the orogenic process. Other laboratories are needed, and few parts of the world provide more of these than Indonesia. Volcanoes on the large islands of Java and Sumatra testify to the subduction of the Indian Ocean at the Sunda Trench, while further east, where the Australian continent is just entering the subduction zone, the trench transforms itself into the much shallower Timor Trough. The great paired belts of positive and negative free-air gravity discovered by Vening Meinesz continue across this boundary with

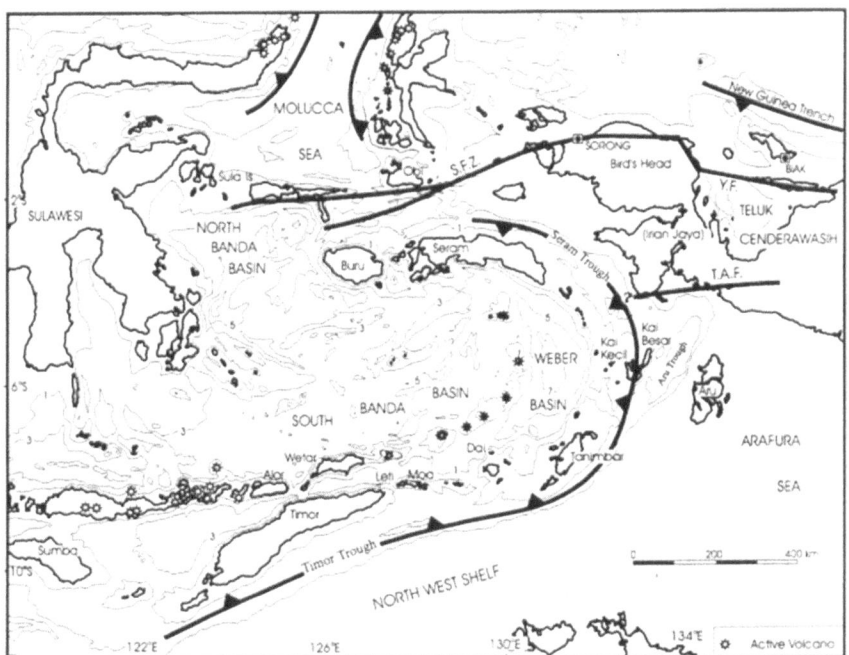

Fig. 13.4 Eastern Indonesia. Stars indicate active volcanoes and thick black lines indicate major faults

only the most minor of hesitations, cutting between Timor and Australia to enclose the Banda Sea, Indonesia's internal ocean, in a gigantic loop.

Before Plate Tectonics the Banda Sea was not a problem. Like all other global features, it was simply 'there'. By the mid-1960s it had become a problem. The origins of Java and Sumatra in subduction-related volcanism were easily understood, and the blocking of the trench further east by the Australian continent was entirely predictable in plate-tectonic terms. Active and recently extinct volcanoes north of Timor were evidence for continuing, if waning, subduction but the active volcanoes of the Banda Islands seem be explicable only by subduction of oceanic crust that was no longer visible but which must have lain to the east, not the south. Even worse, the large island of Scram, which is very similar to Timor, seems, with its attendant extinct volcanic islands, to define a subduction system that faces north, not south. In the late 20th Century a variety of solutions were being offered to explain these observations, but as earthquake patterns became better and better defined opinion coalesced around the idea that there is, hidden beneath the Banda Sea, a shovel-shaped slab of the oceanic lithosphere that had previously occupied the region (Milsom 2001; Pownall et al. 2013). The process invoked is called roll-back, and it takes the idea of subduction zones as gravity-driven repositories for unwanted oceans one step further. Gravity can make lithosphere slide down dipping subduction zones but it can also cause the lithosphere ahead of the trench to sink vertically if it does not arrive at the trench fast enough. Effectively, the trench rolls forward to meet the ocean.

Seram

In the 'pure' Plate Tectonics of the 1970s, the boundaries represented by subduction zones were supposed to be absolute. The geology on one side could have no relation to the geology on the other, because they would have been separated for almost all of their history by hundreds, and possibly thousands, of kilometres of ocean. When, in the early 1960s, a young geologist called Mike Audley-Charles, working deep within one of Vening Meinesz's gravity lows on the island of Timor, recognised rocks containing distinctive Australian fossils, it was not a problem. It became a problem with the advent of Plate Tectonics, because those rocks were on the Asian side of the plate boundary, as expressed by the Timor Trough. Eventually, and reluctantly, the purists admitted that material could be transferred from one plate to another but insisted it could only happen in a narrow chaotic belt above the subduction zone. As Warren Hamilton, who had just published his massive *Geology of Indonesia* (Hamilton 1979), memorably said at one of the conferences where such things were being discussed *It's all spaghetti, Mike. You can't map it. It doesn't mean anything.*

But Mike, who had mapped it, and was by then teaching in London University, was unmoved. It wasn't spaghetti (or melange, as geologists prefer to call it), but regular, mappable units, and he was determined to prove it. However, his field area in East Timor had just been invaded by Indonesia and was no longer a good place to visit, and he decided to head for Seram, on the opposite side of the Banda Sea, where he expected to find a mirror image of Timor geology. He took with him Martin Norvick, a BP geologist who had been one of the BMR geologists who had mapped Papua New Guinea in the 1970s, Dave Carter, an expert on microfossils, and Tony Barber, who was later to found London University's South East Asia Consortium. He also decided, thanks to the Vening Meinesz map, that it would be a good idea to have some measurements of 'g'.

Eastern Indonesia in 1975 was very different from the place it is today. Communications were sketchy, and information was hard to come by. The only way to work there was to equip for all possible circumstances and then go. This was Mike's method, and so it was that the team found itself in the customs hall in the old Jakarta airport surrounded by a truly monumental pile of baggage. There was no way that any customs' officer was going to ignore that, and we were honoured with the personal attention of a very senior one indeed, complete with military style uniform and sunglasses. It was close to midnight, the customs hall was not well lit and he might not have been able to see too well, but his English was extremely good and his approach was unusually direct:

Senior customs' officer: '*What is all this for?*'

Audley-Charles (blustering): '*An international co-operation project. For the benefit of all Indonesians*'.

SCO (with a truly charming smile): '*I am myself very interested in international co-operation. I co-operate for about 10,000 rupiah*'.

'Administration fees' being inescapable in Indonesia in those days, we were through customs very swiftly, and two days later were back in the same airport, waiting for a flight to the city of Ambon, the capital of Maluku Province, on the island of the same name just south of Seram. BP (still, in those days, British Petroleum) had provided much of the funding and their representative in Jakarta came to see us off. '*I admire your enthusiasm*' he told us '*but I think you will be able to do very little*'.

That evening in Ambon, it seemed all too possible that he had been right. The staff in the office of the small Australian company that operated the oil field at Bula on the north-east coast of Seram obviously did not want to see us. The promises of help that had been made on their behalf were very definitely not going to be fulfilled. Nobody else in the town seemed to know anything about Seram, except that it was inhabited by tribal people with attitude problems. It was raining torrents and Mike had just walked into a deep monsoon drain, ripped his pants and lacerated his shins. We were sitting on the verandah of the crumbling ex-Dutch hotel wondering how long we would have to stay before we could decently admit defeat and get on a flight home when a small dark figure appeared out of the even darker night.

You want help? I have references.

We certainly needed help, but it wasn't with much hope that we took the folded and refolded scrap of paper. It was a reference written by no less a person than Maurice Ewing and was one of the most enthusiastic we had ever read. Lamont, it appeared, would have achieved nothing in the Banda Sea without the note-holder, who liked to be known as Danny. Our lives changed.

There was no point in my going inland with the geologists, because my only way of estimating the heights above sea level at the places where 'g' was being measured would be by looking at sea level, so only coastal sites were useful. What I needed was a boat, and Danny was the man to find one. A day later I was the proud lessee of about 70 ft of fully-found, diesel powered, trading-cum-fishing boat. The geologists were then organised on to the next flight to Bula, from where they could get a boat to Wahai, the jumping-off point for their trek into the interior. They had a three day wait, and at the last minute Danny decided that theirs would be the easy bit and that he should come with me. We got our gear together, waved good-by to Ambon and headed north into the sheltered waters of Piru Bay. And to Piru itself, capital of Western Seram.

As a district capital, Piru was the headquarters of what in Papua we would have called the District Officer or DO. In Indonesia this is the camat (pronounced with a 'ch'). There was also a chief of police and a commander of the Army base, and each had to be visited. Each had to be shown my Surat Jalan, the Letter of Travel, without which no foreigner could move in Indonesia outside the designated tourist areas. Each had to add his impressive stamp to it, each had to discuss the state of the world over coffee and each had to be paid an administration fee. It took most of the day, and when it was over, they all decided that they would like to see 'g' being measured.

It was an impressive group that headed back to the boat, with each chief followed by his retinue in their various uniforms. The one thing that such an escort should have prevented was an ambush, but an ambush is what happened. Out of an office near the jetty leapt yet another uniformed figure (in Indonesia, thanks to the Dutch, every civil servant has a uniform), complaining he had not yet had the joy of providing me with coffee. This was the harbour master. He had no power over me, but he could prevent the boat sailing, on health and safety grounds (it didn't look healthy, and it certainly wasn't safe). Another administration fee had to be paid.

It was at this point that the police chief had his idea, which he put to Danny. My interpretation of their conversation may be a little rough, but it went something like this

If you don't do something about it, you are going to get this sort of nonsense all the way round the island. You don't want that, do you?

No

I will hire you one of my policemen (who happens to be the brother of the captain of your boat) to inform all the officials you meet that the formalities have been observed. He will come fully equipped with a Lee-Enfield .303 rifle, as evidence of this

This last was no idle statement. When, a week later, we by-passed Amahai on the south coast in search of a quieter beach, the local police came after us, brandishing their Lee-Enfield and demanding that we stop, drink coffee and pay an administration fee. Our policeman produced his Lee-Enfield, as did our captain. Outgunned, the Amahai police returned to base.

The armament was also useful because Seram is famous for deer. Alfred Russel Wallace, who placed the island on the Australian side of his famous floral and faunal dividing line, would have been shocked to find them there, but dried deer meat fetched a good price in Ambon. So did parrots, which the crew bought whenever they could, and Danny decided to buy a goat for his children. It was clear that 'my' charter was widely regarded as a great opportunity for a trading voyage at someone else's expense. With the meat of two deer drying on the top deck, a loose-bowelled goat occupying the prow and half a dozen equally loose-bowelled parrots perched on the 'bridge', it was no wonder that we had not been able to sneak past Amahai. Their noses must have told the police that we were coming.

All that, however, was in the future, and we had a long way to go. The first stop was on the Kaibobo Peninsula, only a short distance south of Piru. Get ashore, set up the meter, make a measurement. Simple.

Except that something was clearly wrong. Obstinately, the needle that was supposed to be somewhere in the middle of the field of view stayed firmly to the left. It took seventy turns of the adjustment dial to persuade it to move to the reading line, implying an increase in 'g' that was almost impossibly high. Not wanting to admit that I'd come so far with a malfunctioning gravity meter, and certainly not wanting to go back to Piru, I ordered the boat west across the bay to the other shore. The reading was almost the same as at Piru. Back to another point on the peninsula. Monstrously high. South down the bay, and the readings fell back again.

Parts of Kaibobo are made up of the same sort of rocks that produce high 'g' in Eastern Papua, so high values were to be expected, but even in Papua I had seen nothing like this. The results were exciting, and already publishable (Fig. 13.5; see Milsom 1977), but getting them home might be easier said than done. It was getting dark as we reached the southern end of the bay, and time to moor somewhere for the night. Ahead and to the west lay the headland that we had to round to continue up the west coast of the island,

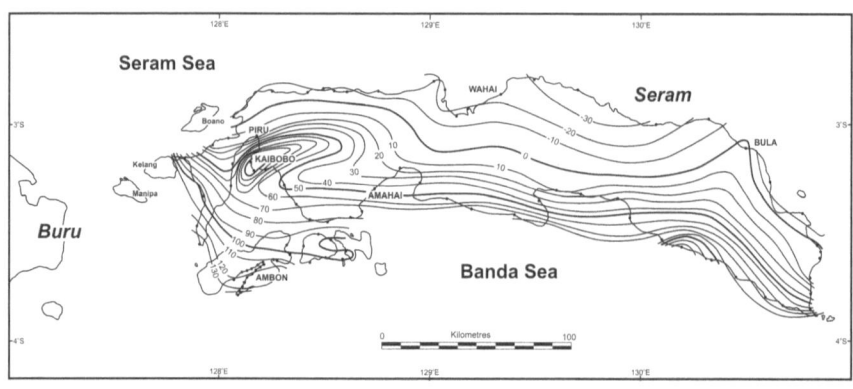

Fig. 13.5 The 1975 Bouguer gravity map of Seram. Contour interval 10 milligal. Black filled circles indicate locations of gravity stations. Shortly after the survey was completed, a joint Woods Hole Oceanographic Institute—Indonesian Marine Geology Institute expedition visited the area and identified a gravity high similar to that at Kaibobo on Kelang Island. After Milsom (1977), modified by the addition of place names mentioned in the text

and it ended in a series of jagged rocks and hidden reefs. Ignoring their presence, our captain kept going into the darkness. To add to the excitement, the kerosene stove in the galley erupted in a sheet of flame, destroying any night vision that anyone might have had. I sat on the deck, with my passport and precious notebook clutched to me, ready to abandon ship as soon as we hit anything or the flames really took hold.

Miraculously, neither thing happened, and at dawn we slid into the boat's home harbour on the west coast. It turned out that we had been in a hurry to get there so that the expedition could be blessed by the matriarch of the port. Water was sprinkled on the boat, and on me, and I relaxed. Not only were we now blessed but it seemed that the captain had known what he was doing after all (I was wrong about that. A few years later he drowned, attempting something equally insane). The next three days passed without major problems. The boat stopped, I went ashore, I measured 'g', I got back on the boat. Each evening we headed for a sheltered anchorage for the night, and I sat in the bows, watching the green forest grow closer, while Danny handed me a cup of hot cocoa. Life was good.

On the fourth day there was a hitch. It was only mid-afternoon, but there was solid resistance from the crew to going any further. When forced, they went on, but with bad grace and, instead of mooring after the last measurement, they turned the boat round and headed it back the way it had come. Eventually Danny told me what was going on. There was, it seemed, to be dancing at the village we had just passed, and no-one wanted to miss it. Well, I didn't want to miss it either. Seram was still the haunt of indigenous tribes who had only recently been described by Robin Hanbury Tennyson as almost untouched by the modern world, so it promised to be an interesting night.

It was, but not in the way expected. The modern world had arrived, or at least the Beatles had, and I, being from the land of the Beatles, was expected to be an expert. To my everlasting shame, I introduced the twist to Seram.

If it hadn't been for the dancing, we might have been more use to the geologists, because we caught up with them the next evening in Wahai. They should have already been well inland, but had only just arrived. They had flown to Bula as planned but the only boat available to take them on the next stage of their journey had been powered by sail only. Within sight of Wahai the wind had dropped and they spent most of the day becalmed, with almost no shelter. If we had come a few hours earlier we could have towed them in, but as it was they had all had too much sun, and Mike was in a particularly bad way. The cuts from the storm-water drain in Ambon hadn't healed, and he now had sunburn as well. He should have gone straight home, or at least come with us to Bula and medical care, but he was insanely determined to see some rocks so we left him to it. He eventually had to be carried back to the coast in a hammock because he could no longer walk, still slashing with his hammer at any rocks that came within reach. When he did get back to London, it took his wife, a trained nurse, some time to sort him out, but she was used to it. They had met when he was admitted to the Hospital for Tropical Diseases in Bloomsbury after a previous field trip.

I got home several weeks before him, because the rest of the Seram circumnavigation (apart from the incident at Amahai) was uneventful. I didn't see BP's man in Jakarta again but I had a satisfying chat with him on the phone. He was, when he heard my voice so unexpectedly soon, immensely sympathetic. He could imagine, he said, how difficult it must have been, and he was not surprised that we had been unsuccessful. It was nice to be able to tell him that, having been all the way round Seram and having been driven along all the available roads on Ambon, there was little useful left for me to do.

And, having come quite close to losing them, I was anxious to get the results from Kaibobo into print. They would add something to the map of 'g' that not even Vening Meinesz had anticipated.

Kaibobo

Where does the extraordinary change in 'g' at Kaibobo fit into the picture of gravity as a driving force? This is an even more complicated story and one that, even now, is only partly understood.

In the end, it turned out that Mike had been right. Rocks painstakingly collected on Timor, Seram and Buru in the Banda Arc and also in eastern Sulawesi contain fossils typical of the southern side of the wide ocean that once separated Australia from Asia. A plausible explanation is that a microcontinental precursor of Australia collided with the outer edge of Southeast Asia, where Sulawesi is today, forming a collision orogen that then collapsed to form the Banda Sea. That a rather minor collision led to the making of a new ocean may seem surprising, but John Dewey had already made the point that it is not the size of a collision orogen but its elevation that determines whether or not it will undergo catastrophic collapse. If there is a free margin to be exploited, the size of the successor basin is dictated by the size of the space into which expansion is possible.

There may have been an extra mechanism fostering the creation of the Banda Sea. Immediately to the north is another oddity of Plate Tectonics, represented by the convergence of the opposed subduction zones on either side of the Molucca Sea (see Fig. 13.4). There can be little doubt that this is happening, because the dipping slabs beneath the arc in the west (Sangihe Arc) and the arc in the east (Halmahera Arc) are both clearly defined by earthquakes. The space between them must be contracting, so the intervening mantle must be being forced out (Fig. 13.6). If some of this material

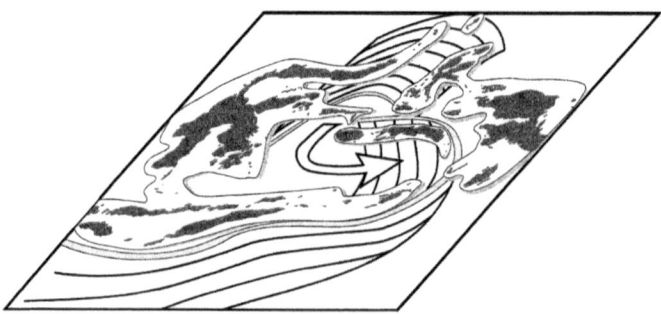

Fig. 13.6 The creation of the Banda Sea (Milsom 2001). Once the Sulawesi orogen had begun to disintegrate, the eastward retreat of the subduction zone may have been encouraged by the escape of mantle material from the decreasing space beneath the opposed (Halmahera and Sangihe) subduction zones lying to the north

is escaping to the south, then it could be ending up within the scoop of the Banda subduction zone, forcing it eastwards.[1] Whether or not this is the case, it is clear that, uniquely, we are witnessing in the Banda area not only the very early stages of one continental collision (between Australia and Asia), but the final consequences of another.

As for Kaibobo, it has turned out that the extraordinary gravity pattern there is not unique. Shortly after we left Seram, Peter Jezek of the Woods Hole Oceanographic Institute took another LaCoste gravity meter to eastern Indonesia and measured 'g' on many of its islands, including a few places around the coast of Seram. He visited both Piru and Kaibobo, and made four measurements on beaches on Kelang, a small island west of Seram that I had passed by. There he measured gradients and maximum values of 'g' even greater than those at Kaibobo. Also, many years later, working on Buru, the large island between Seram and Sulawesi, I found an equally strong, very localised gradient that I had no time to define.

The reasons for these gradients are in one sense obvious. Some of the rocks that are exposed on Kaibobo and Kelang are very dense, and if present beneath the surface in sufficient quantity they are certainly capable of causing the observed changes in 'g'. The difficulty lies in understanding how they got there. Very strong forces would be required to bring such large masses close to the surface and hold them there against gravity, and it seems likely that the journey from the 25 to 30 km depths recorded in some of the minerals from which they are formed must have been made in at least two stages. The first of these would have involved thinning of the crust as the Banda orogen collapsed, stretched and fractured. Faults recording this phase have been mapped on Seram (Pownall et al. 2013) but the process cannot account for the steep gradients, and not all the high density rocks associated with these faults are also associated with high 'g'.

Very different conditions would have existed during the second phase, when a recognisable Seram would have had roughly its present orientation, having been rotated and then stranded as the subduction zone rolled eastwards. The main movements would then have been of great blocks of crust moving past each other, and for that to go smoothly, the faults between them would have to be absolutely straight. They never are. Where there are kinks there is either stretching or compression, and where there is compression there can be forces strong enough to overcome gravity and bring very

---

[1]Faccenna and Becker (2010) similarly emphasise the importance of mantle flow in the formation of the European oroclines.

dense material to the surface. Remarkably, it seems that gravity, the force that makes heavy things sink, can indirectly and in some circumstances make some heavy things rise.

In general terms this idea works, but without more measurements to define more exactly the locations and excess masses of the bodies involved, there is little chance of it being taken further. On Seram, as in many other parts of the world, there will still be a need for more measurements of 'g' for many years to come.

References

Dewey J (1988) Extensional collapse of orogens. Tectonics 7:1123–1129

Faccenna C, Becker T (2010) Shaping mobile belts by small-scale convection. Nature 465:602–605

Hamilton WB (1979) Geology of Indonesia. USGS Professional Paper 1078. US Government Printing Office, Washington, DC

Maurin T, Masson M, Rangin C, Min UT, Collard P (2010) First global positioning system results in northern Myanmar: constant and localized slip rate along the Sagaing fault. Geology 38:591–594

Milsom J (1977) Preliminary gravity map of Seram, eastern Indonesia. Geology 5:61–643

Milsom J (2001) Subduction in eastern Indonesia: how many slabs? Tectonophysics 338:167–178

Pownall J, Hall R, Watkinson I (2013) Extreme extension across Seram and Ambon, eastern Indonesia: evidence for Banda slab rollback. Solid Earth 4:277–314

14

The Codas

Coda 1—The Vector 'g'

Maskelyne on Schiehallion and Cavendish on Clapham Common were both trying to 'weigh the Earth', but they measured different things. Cavendish compared the magnitudes of gravitational forces, while Maskelyne measured changes in their directions. Gravitational forces and accelerations have directions as well as magnitudes. They are vectors.

Vectors

The easiest route to understanding vectors is by beginning with the ones that are most familiar, the distances between places.

Figure 14.1a shows a triangle OPQ. A journey might be accomplished by first going the distance *A* from O to P and then the distance *B* from P to Q. The end result would be to reach Q from O, which might have been achieved more directly by travelling the distance *R* (known in mathematical jargon as the *resultant*) directly from O. Almost miraculously, the same construction, known as the triangle rule or, sometimes, the parallelogram rule, can be used for adding together vectors of any other type, including accelerations or forces due to gravity. It can be extended into three-dimensions and can also be used for disassembling (*resolving*) single vectors into component parts. This can be done trigonometrically as well as geometrically. The resolved part of *R* in the direction of *A* is *R*.cos S, and is equal to *A* if the angle *OPQ* is a right angle.

© Springer International Publishing AG, part of Springer Nature 2018
J. Milsom, *The Hunt for Earth Gravity*,
https://doi.org/10.1007/978-3-319-74959-4_14

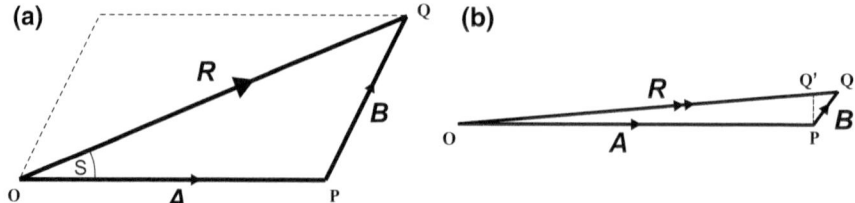

Fig. 14.1 The vector combination rule and its application to gravity fields at the Earth's surface. In **b** the magnitude of the smaller field **B** is still very much larger in relation to **A** than it would be in any real case in which **A** is the Earth's gravity field. The approximations discussed in the text would then be very much closer than they appear in this diagram

It is only in rather special circumstances that vectors need to be resolved into components that are not at right angles, but very common for vectors that are not at right angles to have to be combined into a single resultant.

At every point on the surface of the Earth 'g' is a composite of the gravity effects of the main mass of the Earth, of all the masses associated with its geological and topographic irregularities, and of the Sun, the Moon and even the planets. Fortunately, most of these can be ignored for all practical purposes and, also fortunately, the combination of vectors according to the triangle rule is not too difficult to visualise. This is especially true in the case of the Earth, where one very large vector is being combined with a large number of very much smaller ones (which can be combined into a single vector that will still be, relatively, very small).

Figure 14.1b shows that the effect of adding a small vector **B** to a much larger vector **A** is very similar to the effect of adding to **A** the resolved part (component) of **B** in the direction of **A**. Even when (as in the figure) **B** is about 10% of **A**, the approximation looks reasonable. When dealing with the Earth, the effects of topography and changes in rock density never amount to even one thousandth of the main field and the approximation is very good indeed. Because there is also a component of **B** at right angles to **A**, the direction of **R** is slightly different from the direction of **A**, and it was this difference that Maskelyne measured.

Potential Fields

There is energy in gravity, as Galileo showed when he found that the speeds reached by balls rolling down slopes depended on the height differences between the start and end points and on very little else. When (if) he climbed the steps of the Leaning Tower carrying a cannon ball and a mus-ket ball, he was increasing what is known as their potential energy. When he let them fall, he was allowing that potential energy to be converted into energy of motion (kinetic energy) and when they hit the ground this was transformed into ground vibrations and heat. The amount of stored poten-tial energy at the dropping point would have been no different had he taken a different route up the tower, and the subsequent events would have been the same. These ideas lead to the definition of a quantity known as the grav-itational potential, defined for any point in the universe as the potential energy possessed by a unit mass at that point, as compared to its potential energy at a point where it has a known or assigned value. Most real physi-cists place this point at 'infinity', that imaginary realm beyond the edge of the universe and remote from all masses, where the potential energy is zero. Geophysicists tend to use a more local reference, which is often sea level.

Potential energy has no direction, and is therefore a *scalar*. A surface over which it is constant is called an *equipotential*, and bodies that are able to move only on such surfaces, such as ships at sea, have no (gravitational) rea-son to do so because the gradients parallel to the surface are all zero. The maximum gradients are at right-angles to the equipotential surfaces and are constant only if a surface is perfectly flat.

Tensors

The direction of 'g', the gradient of the gravitational potential, defines the vertical. In most geographic reference systems this is the 'z' direction ('y' being north and 'x' being east), so 'g' could reasonably be written as g_z. This is only done occasionally because there is, by definition, no gravity field at right angles to the vertical, and therefore no g_y or g_x.

As Halley told Hooke, and Bouguer went to some lengths to quantify, 'g' decreases with height, implying the existence of a vertical gradient of this gradient, which might be written g_{zz}. 'g' also varies from place to place, even at sea level, so it must have gradients in the x and y directions. Therefore, and despite the lack of a g_x or a g_y, there must, at every point, be a g_{zx} and a g_{zy}.

This is not the end of the story. The horizontal is not a planar surface but a slightly bumpy ellipsoid (the geoid). The 'g' vectors at two adjacent points are unlikely to be parallel, so there will be a component of 'g' at one of those points that is at right angles to the 'g' vector at the other. If we imagine three adjacent points along a line in the x direction, then at the two outer points there will be components of 'g' that are at right angles to 'g' through the central point. It follows that even though there is no $\mathbf{g_x}$ at this central point, there is a horizontal gradient of $\mathbf{g_x}$, i.e. there is a $\mathbf{g_{xx}}$. Inevitably there is also a $\mathbf{g_{yy}}$.

The splitting of these gradients into 'x' and 'y' components at right angles to each other is, of course, a mathematical convenience. In reality these are the resolved components of a gradient that is in neither the x nor the y direction but in some direction in between (unless one of the components is zero). This direction can also vary, implying that the components can vary, which in turn implies the existence of a $\mathbf{g_{xy}}$ and a $\mathbf{g_{yx}}$. Although in practical terms the vertical is the most important direction, in the formal equations it merely defines one direction in a right-angle (Cartesian) co-ordinate system. There is nothing mathematically special about it, and so there should also be a $\mathbf{g_{yz}}$ and a $\mathbf{g_{xz}}$. There are thus nine separate quantities to worry about, often written in matrix form:

$$
\begin{vmatrix}
g_{xx} & g_{xy} & g_{xz} \\
g_{yx} & g_{yy} & g_{yz} \\
g_{zx} & g_{zy} & g_{zz}
\end{vmatrix}
$$

This matrix is known as a *tensor*, and all of its elements are needed to completely define the gradient of a gradient. Perhaps Aristotle had a point when he rejected even the idea of such a thing! The full tensor is what Eötvös was trying to measure, but not even his instrument could directly measure $\mathbf{g_{zz}}$, which is the most geologically useful of the nine components. However, and long before Eötvös, the French mathematician Laplace had proved that for gravity and similar fields the sum of $\mathbf{g_{zz}}$, $\mathbf{g_{xx}}$ and $\mathbf{g_{yy}}$ (known, because of their positions on the down-to-the-right diagonal of the matrix, as the in-line components) is zero. It can also be shown that $\mathbf{g_{xy}}$ is equal to $\mathbf{g_{yx}}$, $\mathbf{g_{xz}}$ is equal to $\mathbf{g_{zx}}$ and $\mathbf{g_{yz}}$ is equal to $\mathbf{g_{zy}}$. The tensor is therefore fully defined if any two of the in-line components and three of the non-identical cross-line components are known.

Coda 2—The Graphs of Galileo Galilei

In *Two New Science*, Galileo provided geometrical proofs of his theorems concerning Roll and also arguably made the first ever use of graphs to display scientific results. His discussions and diagrams hark back to the experimental results recorded on Folio f107v (Fig. 1.4). In Table 14.1 the three

Table 14.1 The route to the Law of Roll

t^2	Time 't'	Distance 'd'	Ratio (d/32)	Ideal d (t^2 x 33)	'd' - 'Ideal d'	Increment in 'd'
1	1	32	1	33	-1	32
4	2	120	4	132	-12	88
9	3	298	9	297	+1	178
16	4	526	16	528	-2	228
25	5	824	26	825	-1	298
36	6	1192	37	1188	+4	368
49	7	1620	51	1617	+3	428
64	8	2104	66	2112	-8	484

columns of numbers recorded on the folio are shaded. The remaining four columns contain the results of some simple calculations.

The **Ratio** column shows the consequences, to the nearest whole number, of dividing the distances in the **Distance** column by 32, which was the first distance recorded. Galileo must surely have done something like this, either mentally or on some scrap of paper that has since been lost, and could scarcely have failed to see that what he had produced was very close to a listing of the squares of the first eight integers. He even wrote these down (almost certainly some time later) to the left of his original two columns.

An even closer fit between the recorded and theoretical results can be obtained by assuming that the first distance should have been 33 punta, not 32. Multiplying this by 1, 4, 9, 16, 25, 36, 49 and 64 gives the results in the **Ideal d** column, which are what Galileo's measurements would have been had his only observational error been '32' for '33' after just one time interval. The **'d'—'Ideal d'** column, obtained by subtracting the numbers in the **Ideal d** column from the numbers in the **Distance** column, then gives some idea of his likely accuracy.

Stillman Drake devoted several pages of *Pioneer Scientist* to this experiment (Drake 1990). The final, **Increment in 'd'**, column shows the distances travelled in successive time intervals, and Drake thought that Galileo calculated these and immediately noticed that their ratios to the first distance approximately followed the odd-number sequence (1, 3, 5, 7....) but only later recognised in this the operation of a times-squared law. This seems unlikely, because the relationship between the odd-number and integer-squared sequences has been known since the time of Pythagoras and as a lecturer in mathematics Galileo would have been well aware of it. It appears in the *Dialogo* (Galilei 1632) and is prominent in the discussions of experiments on slopes in *Two New Sciences* (Galilei 1638).

The route suggested by Drake is also implausible because only total distances were recorded on the folio and there is no evidence that interval distances were ever calculated. Even if they had been, it would have required all

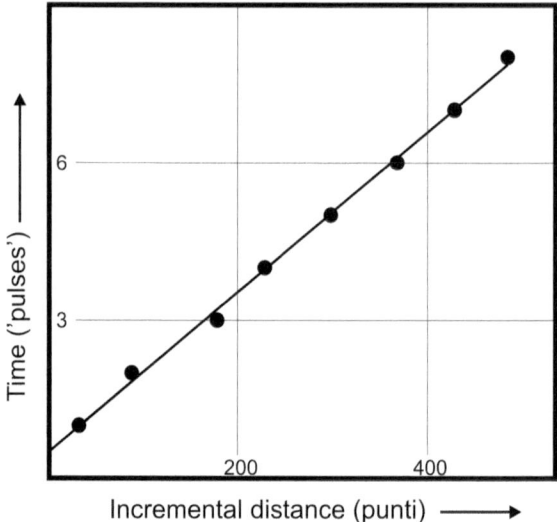

Fig. 14.2 The successive differences in the distances recorded on f107v. The points define a straight line that does not go through the point (0, 0), suggesting a fixed 'reaction time' delay in Galileo's (unknown) timing system

of Galileo's admitted genius to have recognised the odd-number sequence in the numbers listed in the final column of Table 14.1. Only if he drew a graph of the sort shown in Fig. 14.2 would it have been obvious. In Fig. 47 of *Two New Sciences*, which forms the basis of Fig. 14.3a, he came very close to doing this, although for theoretical rather than actual results.

At the time that Galileo was doing his experiments the concept of continuous acceleration was anathema to most philosophers, because Aristotle had rejected the idea of 'change of change', and acceleration is a change of velocity, and velocity is change of place. Galileo was one of the first people to say, out loud and clearly, that if a body started with one velocity (which could be zero) and reached a different velocity some time later, it would have passed through every intermediate velocity on the way. In Fig. 14.3a (a copy of his Fig. 47) the direction **AB** represents time of travel and the direction **AG** represents velocity. In the terms used in describing modern graphs, **AB** is the time axis, **AG** is the velocity axis and the slope of the line **AE** represents the acceleration. Using these ideas, Galileo proved, by extended argument and via the geometry of similar triangles, that the area of the triangle **AEB** was proportional to the distance travelled.

Figure 14.3b takes this idea a step further. At the uppermost level the area of the triangle **ABC**, representing a distance travelled, is equal to the area of the rectangle **ADEC**. At each successive level there are two additional rectangles, so

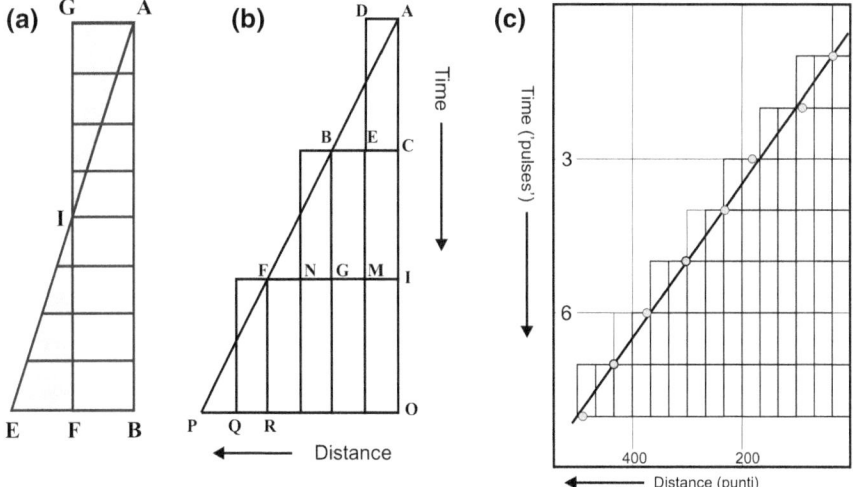

Fig. 14.3   The 'graphs' of Galileo. **a** Redrawn from Fig. 47 of 'Two New Sciences'; **b** Redrawn from Fig. 48 of 'Two New Sciences'; **c** The results of Folio 107v, plotted as Galileo might have done (but didn't)

that their number increases to the next odd integer. Their total area increases to equal the total area of the right-angle triangle extending from **A** down to the base of the level concerned. Galileo used this construction (his Fig. 48) to provide a geometrical illustration of the relationship between the squares-of-integers series and the odd-number series No-one today would bother with such a demonstration, but he lived in an age when mathematical thinking was dominated by Euclidean geometry and people wanted to see geometrical proofs.

Figure 14.3c applies this idea to the results recorded on f107v to produce a diagram that Galileo and his contemporaries would have been comfortable with.

Roll and Fall

Galileo must have been very happy when he discovered that (as far as he could tell) a ball rolling down a slope to the base of a vertical circle took the same time from any point on the circumference of the circle. So happy, in fact, that on pages 221–223 of *Two New Sciences* (pages 188–191 of the Mott translation) he provided no fewer than three distinct proofs. In all probability, although Alexander Koyré would doubtless have argued otherwise, he first discovered the result experimentally and only later worked out the theory.

We also know that he had devised a technique that allowed him to compare the speeds at which brass balls were moving when they reached the ends of sloping groves, and from hints in the folios and discussions in *Two New Sciences* we can deduce how he did this. A grooved surface with an adjustable slope would have been placed on a flat table, with its lower end close to the table edge. After reaching the bottom of the groove the ball would travel the short distance to the edge of the table without losing very much speed and would then fall under gravity. The horizontal distance it covered while falling was a measure of the speed at which it left the groove, because the time of fall was always the same. In this way he was able to show that this speed depended not on the angle of the slope but only on the vertical distance between its top and bottom and, in doing so, he was beginning to define the concept of potential energy that was to become fundamental to so much of physics.

Moving on further, he then argued (correctly) that because the final velocity for Roll through a given height difference was independent of slope, the average velocity would also be independent of slope, and that the times of roll down different slopes all starting and ending at the same two heights would be proportional to their lengths. From this idea and his incorrect (but useful) assumption that free-fall was just a limiting case of Roll, he could derive two independent relationships for Roll down a slope confined within a vertical circle.

In Fig. 14.4a the line DA is a slope with height **h** and length *s*, and the line EA, height and length *H*, is the vertical diameter of the enclosing circle. If t_s is the time taken by the ball to roll down the slope and t_h is the time it would take to free-fall through the same vertical distance, i.e. through the distance **h**, then it follows that, since the average velocities are the same, the ratio of t_h to t_s is the same as the ratio of **h** to *s*.

The second relationship comes from applying the times-squared law to vertical fall along EA, and is that if T is the time taken to fall the distance *H* from E to A, then the ratio of T^2 to t_h^2 is equal to the ration of *H* to **h**.

Galileo also knew his Euclid, and he knew about similar triangles and that an angle inscribed in a semicircle is a right angle, and he was therefore able to deduce a third, purely geometrical, relationship, which is that the ratio of **h** to *s* in the triangle AFD is the same as the ratio of *s* to *H* in the triangle AED.

Juggling these three relationships without the benefit of algebraic notation would have required at the very least a high degree of mathematical skill,

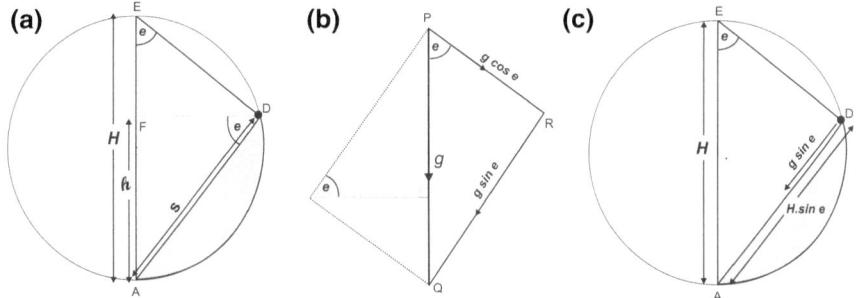

Fig. 14.4 a The geometry of the slope in a circle. **b** Components of 'g', as defined by the parallelogram law. **c** The vector approach to equality of times

and perhaps something that we might fairly call genius.[1] With algebra the problem is a simple one. The three equations are:

$$t_h/t_s = h/s \tag{14.1}$$

$$T^2/t_h^2 = H/h \tag{14.2}$$

and

$$h/s = s/H \tag{14.3}$$

Combining (1.2) and (1.3) to eliminate **h** gives:

$$T^2/t_h^2 = H^2/s^2$$

And combining this with (1.3) gives

$$T^2/t_h^2 = t_s^2/t_h^2 \quad \text{i.e.} \quad T^2 = t_s^2$$

Which is another way of saying that the time of roll is the same for all starting points on the circumference of the circle. That this proof also puts that time equal to T, the time of vertical fall from the highest point on the circle, is a consequence of ignoring rolling resistance, an error that no-one would make today. That Galileo made it is a measure of how far he still had to go,

---

[1] Galileo's own proofs consist of series of propositions of which a typical example (in translation) is that "*since the time of fall along DC is to that along DG as the mean proportional between CD and GD is to GD itself..... it follows that the time of fall along DC is to that along DG as the length FD is to GD*". Mathematical life is easier these days.

but it also shows just how revolutionary his innovations were at the time he was making them.

In modern textbooks very different methods are used, and the starting point is likely to be that 'g' is a vector. The slope, which makes an angle **e** with the horizontal (Fig. 14.4c), prevents the ball from falling vertically and instead it rolls down it. Vector theory then says that it will respond not to 'g' but to $g \sin(\mathbf{e})$, the part of 'g' that is directed down the slope. The distance travelled under this acceleration in any time t is equal to $\frac{1}{2}gt^2 \sin(\mathbf{e})$, while the slope length is equal to $H.\sin(\mathbf{e})$. Putting these two facts together, we find that

$$t^2 = 2H/g\sin(\mathbf{e}) = 2H.\sin(e)/g\sin(\mathbf{e}) = 2H/g$$

This last expression is equal to the square of the time of fall from E to A, and does not depend on the position of the start point on the circle. The Galilean and modern methods thus produce the same answer. They seem very different but actually rely on the same underlying assumptions and also, in the modern approach, on the assumption that 'g' can be handled according to the laws of vector geometry. Instead of taking this successful outcome as a proof of the original proposition, it might be regarded as a validation of vector geometry, which Galileo himself invoked when he referred to the *momentum ponderis* (component of weight) of the ball down the slope.

Missing, of course, from both derivations is any allowance for either air resistance or rolling resistance. Air resistance would probably have been negligible in Galileo's experiments but rolling resistance could have been significant, and variable. The range of slopes used would probably have been too small to show this variation, but by ignoring it Galileo ended up by incorrectly treating vertical Fall (where rolling resistance would not apply) merely as a special case of Roll.

The Asinelli Tower

There is very little in Galileo's writing to suggest extensive experimentation with Fall, and certainly nothing approaching the commitment shown by Giovanni Riccioli. According to the *Almagestum Novum*, he and his colleagues dropped clay balls from a number of buildings, but the results they chose to present to the world were those obtained from the tallest, the Asinelli Tower in Bologna. Riccioli's own sketch of the tower (on the left of Fig. 14.5), does not really do justice to its terrifying reality (Fig. 1.2), but it might not have had, in the middle of the 17th Century, the tilt that it has today. At any rate, he made no mention of it, emphasizing instead that a ball dropped over the parapet surrounding the upper platform would fall directly on to the pavement of the lower platform, 280 (Roman) feet below, and that the experiments could therefore be completed without danger to either the experimenters (if immune from vertigo) or to passers-by in the square below. The actual drop technique is not described, but presumably the balls were lowered to the chosen heights on threads, either from convenient windows or from the upper platform, and then released.

Riccioli's tabulation (on the right of Fig. 14.5) records the results of three sets of experiments. The first stage in the first of these was to find the drop height corresponding to five swings of his very short pendulum, which beat six times to the second. Conveniently, this came out as five feet, and it might have been this result that led him to use this basic interval in this first experiment, rather than the six swings that would have corresponded to one second. Because he began by disbelieving Galileo's times-squared rule, the ten-swing and fifteen-swing determinations must have involved a fair amount of trial and error, but from then on the pattern would have been clear and the height required would have been known before each drop was made. If this is what he did, then not all his measurements can be considered independent (Graney 2012).

Ordo experimentorum	Vibrationes Simplices Pendiculi alti vnciam 1, 5/12.	Tempus primi Mobilis respondens Vibrationibus.		Numeri Quadrati Vibrationum.	Spatia côfectâ à Globo argillaceo Vnciarû 8. in fine temporû.	Spatia seorsim confectâ singulis temporibus.	Proportio Incrementi Velocitatis Grauium in Aëre nostrate.
	Vibr. Simpl.	Secûda	Tertia	Quadrata	Pedes Romani	Pedes Romani	Numeri minimi
	5	0"	50'''	25	10	10	1
	10	1	40	100	40	30	3
I.	15	2	30	225	90	50	5
	20	3	20	400	160	70	7
	25	4	10	625	250	90	9
	6	1	0	36·	15	15	1
	12	2	0	144	60	45	3
II.	18	3	0	324	135	75	5
	24	4	0	576	240	105	7
	26	4	20	676	280	40	8 1/4
	6 1/2	1	5	42	18	18	1
III.	13 0	2	10	169	72	54	3
	19 1/2	3	15	381	161	90	5
	26 0	4	20	676	280	118	6 7/13

Fig. 14.5 Riccioli's sketch of the Asinelli tower, together with the results recorded on p.385 of Almagestum Novum. The second and third columns record the times of fall measured in, respectively, pendulum half-periods and seconds, and the fourth lists the squares of the times, in half-periods. The fifth column gives the fall distances and the sixth the fall increments, which in the final column are 'normalised' against the first values in each set in order to demonstrate the odd-number sequence

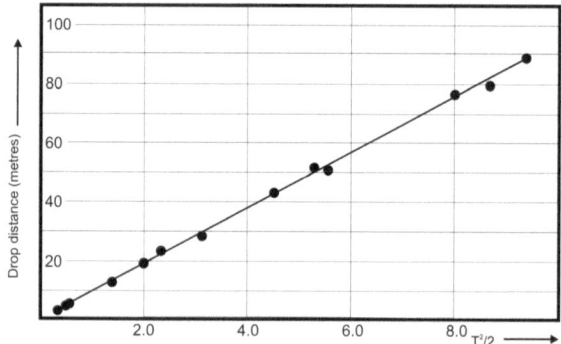

Fig. 14.6 Riccioli's drop distances, in metres (converted at the rate of 0.318 m to one Roman foot) plotted against half the squares of the drop times measured in seconds. The excellence of the fit to a straight line is partly spurious, since many of the experiments must have been merely confirmations of expectation, but the shortest and longest drop measurements would have been truly independent

In the second set of experiments the intervals were multiples of one second, except for the final drop, which was made from the upper platform. For the third set the basic interval was 6½ swings, with the final drop again made from the upper platform, so that for the first and third determinations half-swings had to be measured.

Because the drop distance **L** is related to the drop time **T** by the equation $\mathbf{L} = \frac{1}{2}g\mathbf{T}^2$, the most convenient way of illustrating Riccioli's results is to plot **L** against $\frac{1}{2}\mathbf{T}^2$ (Fig. 14.6). The value of 'g' implied by the gradient of the line of best fit is approximately 940 Gal using the Graney (2012) estimate of a Roman foot as 31.8 cm, which he obtained by comparing his own measurement of the distance from the upper to the lower platform with Riccioli's quoted value. The more usual 29.6 cm gives a less impressive 880 Gals, providing an object lesson in the uncertainties surrounding any assessment of the errors in early experiments.

Coda 3—The Curious Curves of Christiaan Huygens

A vertical spring produces a force proportional to the amount of stretch, and a weight bobbing up and down at its lower end is accelerated back towards its position of static equilibrium. The further the weight is from that position, the greater the force and therefore the greater the acceleration, but on its return it will overshoot because it then has both kinetic energy and momentum. If there are no frictional or other losses, the overshoot will equal the original maximum displacement. This form of motion, known as Simple Harmonic, is easily analysed using the techniques of infinitesimal calculus, and quite difficult to analyse in any other way. In the notation devised by Leibniz, it is defined by the equation

$$d^2x/dt^2 = -k^2x$$

where x is the distance from the equilibrium position at time t and k is a constant depending on the weight and on the strength of the spring. The expression on the left is Leibniz's way (and now almost everybody's way) of writing an acceleration, with the '2's denoting not the squares of any quantities but changes of change, in this case in x with respect to t. The minus sign appears because the acceleration, which half the time is actually a deceleration, always acts to decrease x. The solution of the equation is that $x = \sin(kt)$, which implies that the full period (the time taken by the weight to move from one extreme position to the other and back again) is equal to $2\pi/k$.

If a weight suspended on a thread (a simple pendulum) is pushed sideways, it swings back and forth and has a period, but follows a curved path (Fig. 14.7a). The displacement measured along this path is equal to $L.e$, where e is the angle (in radians) that the thread makes with the vertical and L is its length. The only force that matters in accelerating the weight back to its rest position is the one at right angles to the thread, because that is the only direction in which the weight can move. This force is equal to $mg.\sin(e)$, and the acceleration of the bob is therefore $g.\sin(e)$. The equation of motion is:

$$d^2(L.e)/dt^2 = -g.\sin(e)$$

This is not Simple Harmonic, but when e is small (and measured in radians) it is almost identical to $\sin(e)$. The equation can then be rewritten as

$$d^2e/dt^2 = -g.e/L$$

which is Simple Harmonic, with a period equal to $2\pi\sqrt{(L/g)}$.

The pendulum equation also applies to balls rolling down slopes forming arcs of vertical circles (Fig. 14.7b), and Galileo thought that for this motion, and for that of a pendulum, the period would be independent of the initial displacement (i.e. that the arc would be the *isochrone*), and would also be

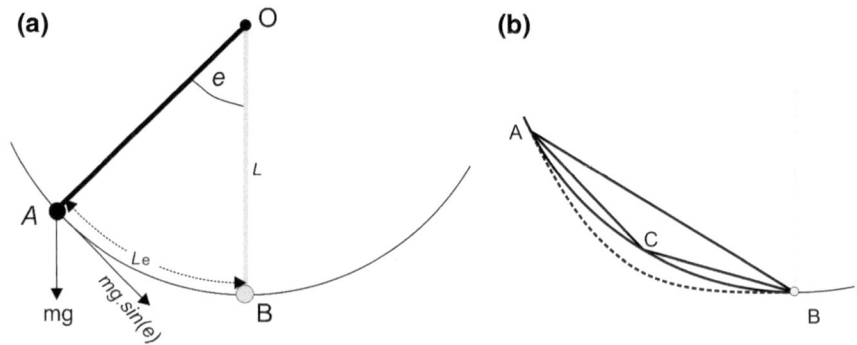

Fig. 14.7 Pendulum paths: **a** The simple pendulum. Although the gravitational force on the weight A is equal to its mass, 'm', multiplied by 'g', only the component of this force at right angles to the string can actually produce movement. **b** The brachistochrone. Galileo showed that, for a rolling ball, the path ACB is quicker than the direct path AB because the greater initial acceleration produced by the steeper initial slope compensates for the longer path length, and he deduced that a slope formed by the circular arc AB would provide a quicker path than any made up of segments contained within it. He did not investigate slopes outside the arc (such as that shown by the dashed line) and so never arrived at the quickest path of all

the shortest possible (sometimes known as the *brachistochrone*). That he was wrong on both counts was quickly demonstrated by, among others, Marin Mersenne.

There is no obvious reason why an isochrone slope should exist at all, or why, if it does, it should also be the brachistochrone. By common consent the first person to show that there is an isochrone was Christiaan Huygens,[2] but how he did it is still something of a mystery because the discussions preserved in *Horologium* and the *Oeuvres Complètes* are strewn with logical lacunae. The editors of the *Oeuvres*, working long after Huygens was dead and writing in French, struggled to bridge these gaps in voluminous footnotes that were often four or five times as long as the original (Latin) texts.[3] For only a few of its Anglophone readers will a path through this linguistic maze be opened up by the footnote noting that one of the editors had already published a full discussion of this aspect of Huygens work. In Dutch.

---

[2]That the same curve is also the brachistochrone was demonstrated fifteen years later by Jakob Bernoulli.

[3]The *Oeuvres* consists of a 22-volume collection of letters and notes prepared by different editors between 1888 and 1950. Mahoney (2000) provides a comprehensive discussion in English of this part of Huygens' work.

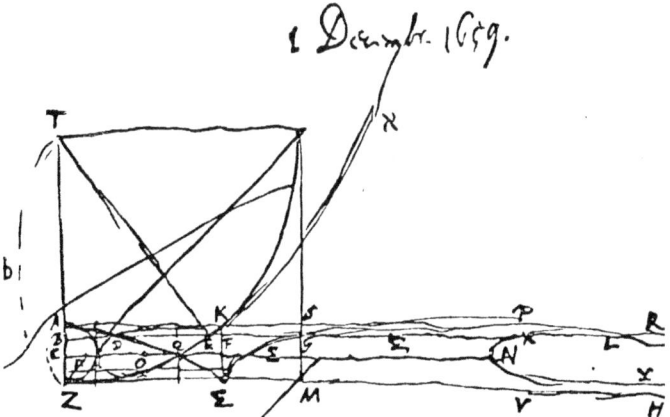

Fig. 14.8 One of Huygen's 'pendulum' sketches, reproduced from Vollgraf (1929; p. 392). It shows both a physical pendulum with its support point at T and geometrical constructs related to the parabola of velocities ZX, and is typical of Huygens' artwork

Writing before the introduction of the symbol π and before the invention of calculus (which he was moving towards but never quite reached), Huygens' work lacks the clarity that these advances would have conferred. He had a fondness for proof by *reductio ad absurdum* which, while perfectly valid as a method, requires its users to have decided at the outset what it is that they wish to prove or disprove, and he seldom explained how he reached such decisions. Moreover, as Fig. 14.8 shows only too clearly, in his notes his arguments were supported by diagrams that might well have been drawn using Banjo Patterson's 'thumbnail dipped in tar'.[4] Despite these obstacles and shortcomings he arrived at the right answer:

> For a cycloid formed about a vertical axis and with its vertex at the bottom, the times of descent for a body, from whatever point it is released, are equal; and this time bears the same relationship to the time of fall along the cycloid axis as does the length of the half-circumference of a circle to its diameter. (Huygens 1673; Proposition XXV).

The cycloid is the curve traced by a point on the circumference of a circle rolling on a flat surface. Figure 14.9 shows how this works but differs slightly from the illustrations in most textbooks because, in order for the

---

[4]*'Clancy of the Overflow'*.

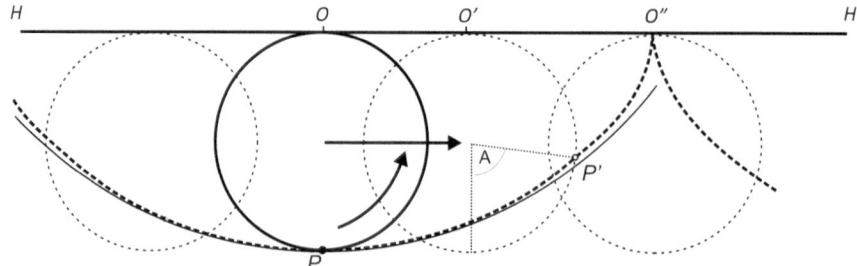

Fig. 14.9 The cycloid. As the circle turns anticlockwise without slippage, its point of contact OO'O" with the plane surface HH moves to the right. The cycloid (dashed line) is traced out by the point P' fixed to its circumference. The thinner continuous line is part of a circle with four times the radius of the cycloid generator. That the cycloid has a steeper slope is only obvious after its generating circle has turned through almost 90°

isochrone slope to be the appropriate way up, the rolling surface *HH* has to be above rather than below the generating circle.

The descriptive term '*the whole axis of the cycloid*' is ambiguous, but the context shows that it refers to the diameter, 2r, of the generating circle, and the time of fall through this distance is equal to $2\sqrt{(r/g)}$. The '*ratio by which the semicircumference of a circle is related to its diameter*' would now be written as $\pi/2$, so Huygens is claiming (correctly) that the time of roll down the slope to P is equal to $\pi\sqrt{(r/g)}$. This is, however, only a quarter period. The time for a full period, corresponding to the $2\pi\sqrt{(L/g)}$ for a pendulum of length L, is $4\pi\sqrt{(r/g)}$, which implies that for a ball to roll on a cycloidal slope with the same period as a pendulum swinging through a very small angle, the pendulum must be four times as long as the radius of the generating circle. The periods diverge for larger angles, as they must if the cycloid is to be the isochrone or the brachistochrone (or both), because the circular arc described by a pendulum is neither.

Armed with prior knowledge of the right answer and the calculus of Newton and, more especially, Leibniz, it is not too difficult to show that the cycloid is the isochrone. Huygens had neither and his journey from circle to cycloid seems an unlikely one. It may have been possible only because he was already interested in the curves, known as evolutes, that are obtained by plotting the positions of the centres of curvature of each small segment of other curves. Another way of looking at them is as the '*envelopes*' of all lines drawn at right angles to the original curves (Fig. 14.10a). They generally look nothing like their generating curves (the evolute of a circle is a single point), but the evolute of a cycloid is an identical cycloid displaced both laterally and vertically. Huygens may well have drawn diagrams such as Fig. 14.10a, in which a cycloid is compared to its tangent circle, before he became interested in finding isochrones and brachistochrones. When he did move on to that problem, he would have known that he was looking for curves with initially steeper descents than circular arcs, and his decision to try cycloids might have been a lucky guess, influenced by just such a sketch.

Once he had proved to his own satisfaction that the cycloid was the curve that he needed, Huygens devoted considerable time and effort to designing pendulum clocks using cycloidal restraints or 'chops', to patenting them and having them built. It is not immediately obvious that such restraints will cause pendulum bobs to describe cycloidal arcs (it is only because cycloids are their own evolutes that they do so), and ultimately they proved to be not worth the effort. They will only work with pendulums that use very flexible, and therefore rather weak, support threads (or, to keep them aligned, double threads such as those shown in Fig. 14.10b), and weak threads will stretch and contract significantly during each swing as the effective force upon them changes. The designers of clocks, and of instruments to measure 'g', soon returned to rigid pendulums swinging in circular arcs.

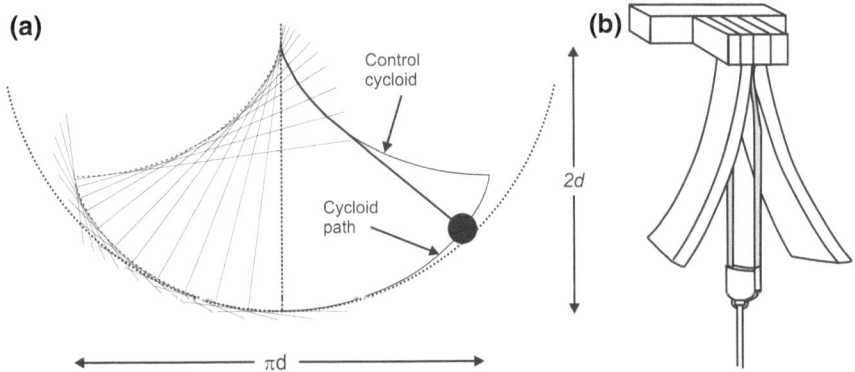

Fig. 14.10 The cycloid pendulum. **a** Pendulum curves. The path of the bob of the cycloidal pendulum is dictated by the 'control' cycloid around which the thread wraps itself and which is defined by the envelope of the lines at right angles to the path, displaced laterally by half a cycle. The outer dashed curve is the path of the equivalent simple pendulum, with radius 2d. **b** The upper part of a pendulum with cycloidal 'chops'

Coda 4—The Three Effects of Pierre Bouguer

In *La Figure de la Terre* Pierre Bouguer discussed the three factors that he thought would cause changes in 'g' on the Earth's surface. They were latitude, distance from the centre of the Earth and the gravitational attraction of the rock masses above sea level. His conclusions are as valid now as they were when he wrote them down, but his reasoning is sometimes hard to follow and many of his most important results were presented without proof.

The Effect of Latitude

The Earth spins, and that has consequences. Because of the spin, observers on its surface who like to think of themselves as being 'at rest' also have to accept that they are subject to a centrifugal acceleration that is proportional to the square of their angular velocity around the Earth's spin axis.[5] The effect on the gravity field decreases with latitude, L, both because the radius of rotation decreases towards the poles and because the angle between 'g' (directed towards the centre of the Earth) and the very much smaller centrifugal acceleration (at right angles to the Earth's spin axis) increases. Both changes involve the cosine of L, so the overall effect depends on the cosine squared (Fig. 14.11).

The Earth makes one rotation in 24 hours and its mean radius is 6370 km, implying a centrifugal acceleration of 0.0337 m/s (3370 milligal) at the equator, and of 3370 $\cos^2 L$ milligal at latitude L. Pythagoras said (and Danny Kaye memorably sang) that in a right angled triangle the square on the hypotenuse is equal to the sum of the squares on the other two sides. The trigonometric consequence is that the squares of the cosine and the sine of any angle add up to one. The centrifugal acceleration can therefore also be written as 3370 $(1 - \sin^2 L)$ milligal, which might seem clumsy but turns out to be useful when it has to be combined with the gravitational acceleration due to the main mass of the Earth.

The 'ideal' Earth is not a sphere but an ellipsoid (the solid body obtained by rotating an ellipse) and an ellipse is a curve made by joining together all

---

[5]Some purists object to any use of the word 'centrifugal' (flying away from the centre), on the grounds that what is actually happening is that a centripetal (towards the centre) force is having to be supplied to keep the body circulating instead of heading off in a straight line. Einstein would disagree, arguing that observers are entitled to use their own frames of reference, and if those require them to be subject to mysterious additional accelerations, so be it.

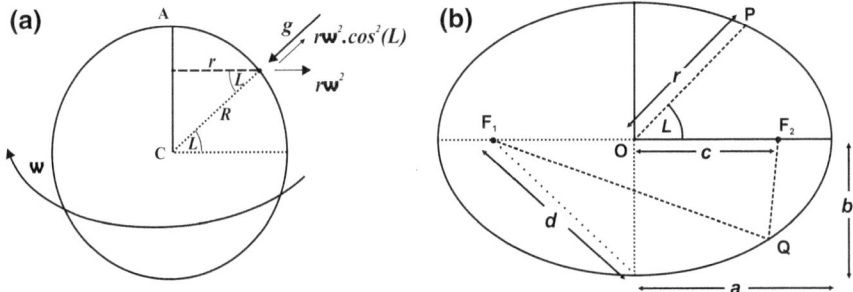

Fig. 14.11   a An ideal spherical Earth, radius R, rotating with angular velocity 'w'. The radius 'r' of the circle around which a point at latitude **L** rotates is equal to R multiplied by the cosine of **L**. The centrifugal force is equal to rw^2 but must be multiplied in its turn by the cosine of L to obtain the component in the direction of the gravity field. **b** The ellipse: The distances 'a' and 'b' are, respectively, the major and minor axes. The foci F_1 and F_2 are at distances 'c' on either side of the centre, and the sum of the distances F_1Q and F_2Q is constant and equal to 2a. The eccentricity, which is always less than 1, is defined as **c/d** or $(1 - b^2/a^2)$

the points for which the sum of the distances from two fixed points (the foci) has a constant value. It has the form of a 'squashed' circle characterised by an equatorial radius, **a**, and a smaller polar radius **b** (Fig. 14.11b). In many treatments of the subject on the internet and even in some textbooks it is assumed that a good approximation to the gravity field at any point on the surface of an ellipsoid can be obtained by using Newton's equation for a sphere, coupled with the local radius. This must be true for very small eccentricities, because the ellipsoid is then effectively a sphere, but is it valid when applied to the real Earth?

At the Earth's poles there is no complicating centrifugal acceleration. For the 'ideal' Earth of the 1980 Geodetic Reference System (GRS80) the polar gravity is 983,218 milligal, but substituting the polar radius and the mass of the Earth into Newton's equation gives 986,049 milligal, which not even an 18th Century observer would have considered close. A similarly poor result is obtained at the equator, where the GRS80 sea-level gravity is 978,032 milligal and Newton's equation gives 976,429 milligal after allowing for centrifugal acceleration. The change from equator to poles calculated using the approximation is thus almost twice as large as the change that actually occurs, showing that the actual Earth is too far from being spherical for the easy approach to be a satisfactory one.

It was left to Claude Clairaut, the youngest of the French scientists who went to Lapland with Maupertuis, to come up with something better, and he did this by going back to fundamentals. Rather than rely on observation, he first calculated the shape that would be adopted by a perfectly fluid rotating mass affected only by its own gravity field, and then calculated the gravitational attraction at points on its surface. The mathematics was not easy, but in 1743 he published an equation in his *Théorie de la Figure de la Terre* that would keep geodesists happy for a hundred and fifty years. It was that:

$$g_L = g_0 + (5c/2 - f. g_0).\sin^2 L$$

where g_L is the gravity field at any latitude L on the surface of the fluid body (or at sea level on an 'ideal' Earth) and g_0 is its value at the equator. 'c' is the centrifugal acceleration at the equator and f is equal to $(a - b)/a$. It was not until the 20th Century that this equation was replaced in common use by variants of an equation due to Carlo Somigliana, written in its 'closed' form as

$$g_L = g_0[(1 + k.\sin^2 L)/\sqrt{(1 - e^2.\sin^2 L)}]$$

where e is the Earth eccentricity and $k = (b.g_p - a.g_0)/a.g_0$. At first sight this equation looks nothing like Clairaut's, but once approximations are introduced, similarities emerge. Until desk-top computers made it easy to use Somigliana directly, calculations were based on expansion of his equation as an infinite series in $\sin^2 L$, taken only as far as the (very small) term in $\sin^4 L$.

Thanks to measurements made from satellites during the last fifty years, the shape of the Earth has been defined in quite astonishing detail. In 1980 this work, which still continues, defined the GRS80 formulae which provide the best current estimates of the constants in the Somigliana equation:

$$g_L = 978032.67715\left(1 + 0.001931851353\sin^2 L\right)/\left(1 - 0.0066943800229\sin^2 L\right)^{1/2}$$

Expressed as a truncated series, this becomes:

$$g_L = 978032.677 + 5163.075\sin^2 L + 22.761\sin^4 L$$

This can be written in several other ways, some of which are more widely quoted, but this is the form in which it is easiest to see what is going on. The first term on the right-hand side is g_0, the sea-level gravity at the equator. The fourth-power term is very small, with a maximum value, at the poles, of only 22.761 milligal.

Table 14.2 Variation of 'g' and centrifugal force with latitude, from Bouguer's experiments

Location	Latitude	Pendulum results		Centrifugal effect (CE)		
		'g' (milligal)	'g' difference from Manta (milligal)	Centrifugal effect (milligal)	Difference from Manta (milligal)	CE difference/'g' difference %
Pello	66° 47′N	982,229	4364	−520	2850	65
Paris	48° 52′N	981,115	3250	−1450	1920	59
Haiti	18° 26′N	978,444	579	−3020	350	60
Panama	9° 32′N	978,065	200	−3280	90	45
Manta	0° 58′S	977,865	0	−3370	0	

Bouguer and Latitude

Since he knew about the measurements made by Jean Richer and others, Bouguer must have been prepared to see 'g' decrease towards the equator, and his sea level measurements at Petit Goave on Haiti and Porto Bello in Panama, as well as at Manta in Ecuador, would have fully confirmed his expectations. Not content with this qualitative result, he went on to consider, at some length, whether the changes he recorded could be entirely explained by changes with latitude in the centrifugal effect. He certainly knew of the existence of Newton's *Principia Mathematica*, but seems not to have known that some of the equations he needed had been derived, with quite un-Newtonian clarity, in its second volume, or that Clairaut was working on the same problem. Instead he credited '*the famous M. Huygens*' with being the first person to consider the relationship between centrifugal acceleration and gravity,[6] and in his calculations he used methods far more laborious than those used by Newton.

In Table 14.2 the values of 'g' implied by the lengths of Bouguer's seconds-pendulums at his sea-level stations are listed in the third column, together with the values obtained in Paris by Mairan and at Pello (in Lapland) by Maupertuis. The fourth column shows these as differences from the value at Manta, which is less than one degree from the equator. The fifth column shows the centrifugal effect at each point, with a minus sign because 'g' is being reduced, and the sixth compares these effects with the centrifugal effect at Manta. The final column (obtained by dividing Column 6 by Column 4) gives the percentages of the observed changes that

---

[6]Schliesser and Smith (1966) discuss Huygen's belief that the change in 'g' with latitude was entirely centrifugal and his rejection of Newton's postulated additional effect due to Earth ellipticity.

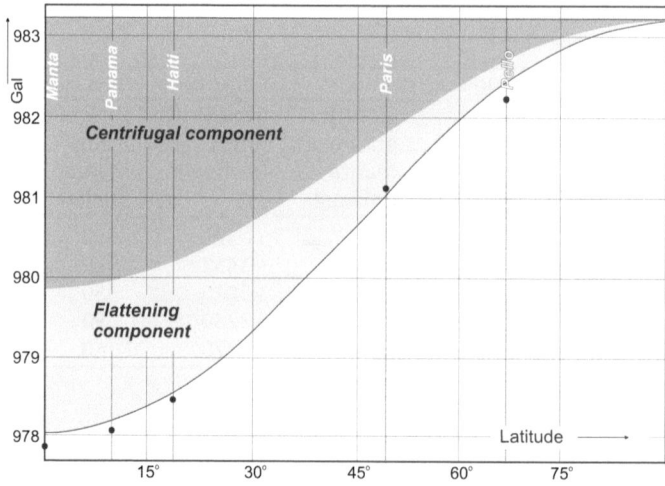

Fig. 14.12 The two contributions to the variation of 'g' with latitude, referenced to the GRS80 polar gravity of just over 983 Gal (represented by the horizontal limit at the top of the graphs). Filled black circles show the results obtained by Bouguer at five points close to sea level

can be explained by the centrifugal effect. What this demonstrates, and what Bouguer realised, is that centrifugal acceleration can only account for, at the most, two-thirds of the variation. This is shown graphically in Fig. 14.12, where the lower curve represents the total change in sea-level gravity with latitude according to the GRS80 Earth model and the upper curve represents the corresponding changes in centrifugal force. The region between these two curves represents the changes due to the changes in Earth radius, which are responsible for roughly a third of the observed effect.

Unlike Newton, Bouguer was unable to explain the discrepancy. This is a little surprising but could be because Newton used only Richer's values in Cayenne and Paris, and a curve with just two unknown constants (g_0 and k in $g = g_0 + k.sin^2L$) can always be made to pass exactly through any two chosen points. Bouguer used the values from Paris and Pello as well his own South American measurements, and, because of experimental errors and real geological effects, none of these plots on the modern theoretical curve (Fig. 14.12). It may have been this scatter that prevented him from identifying the underlying rule.

Using modern global databases the actual values of 'g' at the points listed in Table 14.3 can be estimated to within a few milligal (with minor uncertainties because the locations were not described in any detail), and the accuracies achieved in the 18th Century can therefore be estimated. For the

Table 14.3   Accuracy of French measurements of 'g' in the early 18th Century

Location	Latitude	'g'(original) (milligal)	'g' (corrected) (milligal)	'g'(modern) (milligal)	'g' difference (milligal)
Pello	66° 47'N	?	982,229	982,336	137
Paris	48° 52'N	980,915	981,115	980,930	−185
Haiti	18° 26'N	978,132	978,444	978,680	236
Panama	9° 32'N	977,754	978,065	978,290	225
Manta	0° 58'S	977,553	977,865	978,080	215

The Pello value, which was mentioned by Bouguer but was not listed by him, was obtained by Maupertuis at the northern end of his Lapland transect, and may or may not have been corrected for temperature and other effects. The 'Paris' result was quoted by Bouguer but was probably due to Mairan. The corrected version is actually further from the most likely actual value (the currently accepted value of 980,928 milligal at the Paris observatory) than Christiaan Huygens' fortuitously accurate value of 980,940 milligal (Chap. 1), but the uncorrected value is impressively close

measurements that Bouguer made himself, the values in the *'g'(original)* and *'g'(corrected)* columns are based on the lengths he listed for the seconds pendulum before and after corrections had been made to compensate for known sources of error. Buoyancy, which reduces the effective weight of the bob but not its mass and would be different at different altitudes, worried him more than air resistance; indeed, he convinced himself (although not, perhaps, all of his readers) that the error introduced by air resistance on the downswing would be exactly compensated on the up-swing. He also corrected for changes with temperature in the length of his iron ruler but, rather surprisingly, seems not to have considered the possibility of errors due to the rather crude suspension system. His corrections improved the results from the Americas considerably and the strong systematic element that remains may have been introduced at the upper end of the fibre, where it had to bend rather than swing freely. The consistency is impressive, especially considering the lack of sophistication in the instrumentation, supporting John Smallwood's contention that Bouguer, whatever his defects as a theoretician, was a very, very good experimentalist.

The Effect of Height

Having dealt with latitude, Bouguer went on to discuss, much more briefly, the effect of height. He first considered what is today called the free-air effect, which causes 'g' to decrease with height above sea level. In Fig. 14.13, which is based on one of his illustrations and uses his symbols, A is a point at sea level on an ideal Earth and 'a' is a point at the same latitude at **h** metres

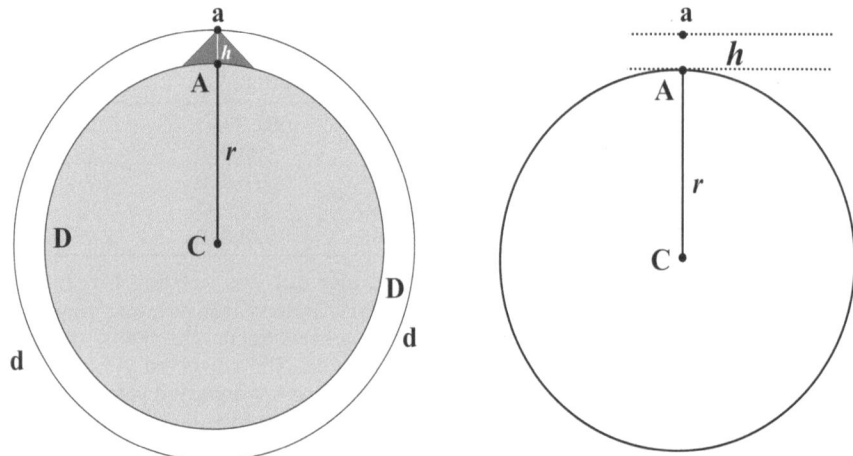

Fig. 14.13 Elevation corrections. The sketch on the left is a re-drawn version of Bouguer's Fig. 43. The simplified version on the right focuses on the free-air effect, calculated on the assumption that there is only air between A and a

above sea level. If there is only air between A and a, then Newton's Law implies that the corresponding gravity fields are GM/r^2 and $GM/(r + h)^2$ respectively. This second term can be written as

$$GM/r^2(1 + h/r)^2$$

which can be expanded as the infinite series $GM(1 - 2h/r + 3h^2/r^2 \ldots)/r^2$.

For measurements made at the surface of the Earth the ratio of h to r is always going to be small, its square is going to be so small as to be barely detectable, and its cube and higher terms can for all practical purposes be ignored. Thus the gravity field at 'a' is approximately equal to $GM(1 - 2h/r)/r^2$ or $GM/r^2 - 2GMh/r^3$. The first term is the gravity field at A, the second, which is directly proportional to height, is the approximate free-air effect, which for most practical purposes can be taken as equal to 0.3086 milligal per metre. The error introduced by ignoring the h^2/r^2 term would be 0.07 milligal for a height of 1000 m and 1.8 milligal for a height of 5000 m. These would certainly be measurable by modern instruments (although not by Bouguer's pendulums) but there are likely to be other and far more important sources of error in surveys involving such extreme topography.

Bouguer's route to this relationship was slightly different, because his ultimate goal was to compare the average densities of the rocks of the Andes and the main bulk of the Earth. This led him to write his expression for sea-level 'g' in terms of density rather than total mass, saying:

> I denote the radius AC by r, and the density of the Earth ..... by Đ. Accordingly, I have the expression rĐ for the gravity field at all the points ... at the Earth's circumference.

This seems a little strange for two reasons, the first being that he had just spent some thirty pages explaining that 'g' was not, in fact, the same at all points on the Earth's circumference. Even today, however, when vastly more sensitive instruments are available, an average value is generally used for this particular calculation. More confusing is the apparent dependence of 'g' on the Earth radius, r, rather than on its inverse square, but this is a consequence of working in terms of density. The mass of a sphere of constant density 'Đ' is $4\pi r^3 Đ/3$ and substituting this for M in the equation $g = GM/r^2$ gives $g = 4\pi GrĐ/3$. For Bouguer's version of the equation to be used for accurate calculations, a density inversely proportional to r^3 would have to be used and the non-sphericity of the Earth would have to be taken into account, but no-one today would take such an approach.

Bouguer wrote rĐ for the mass of the Earth instead of $4\pi GrĐ/3$ because he ignored all constants. Again, this seems odd, but is not very different from what is done today when Newton's Third Law is written as $F = ma$. This works only because the units of force, mass and acceleration have been deliberately chosen to avoid the need for additional constants. A system of units could be constructed that allowed the Law of Gravitation to be written as $F = m_1.m_2/r^2$, but then the Third Law would have to include a constant factor. We cannot have it both ways. To expect Bouguer, working at a time when the ways of writing mathematical equations had yet to be standardised, to present his calculations in forms that are now familiar is not reasonable. He knew that the constant G must exist but he had no idea of its value, and he also ignored π (the symbol becoming popular only after its use by Euler many years later). Whether his equations are actually valid within the system he was using depends on whether the same constant factor, $4\pi G/3$, is ignored in all cases.

For the free-air effect, and following his usual practice of simply stating his mathematical results without proof, Bouguer also said that:

> If I then denote the elevation Aa by h, which will be very small compared to r, the gravity field at a will be less than at A, in proportion to the square of CA to the square of Ca. This decrease will be by 2 *h*, compared to r. That is to say

that the gravity which was $r\text{Đ}$ at A will be $r\ 2\ h\ x\ \text{Đ}$ at a. (Bouguer 1749; caption to Fig. 45)

From the context it is clear that what he wrote as $r - 2\ h\ x\ \text{Đ}$ would now be written with brackets, as $(r - 2\ h)\ x\ \text{Đ}$. His 'free-air factor' is thus equal to 2Đ. The constant term being ignored is $4\pi G/3$, so that within the conventions he was using his equation is valid.

The Effect of Topography

In reality, of course, the rocks between sea level and the point of measurement cannot be ignored, and Bouguer first tried to calculate their effects by imagining the 'sea-level Earth' to be surrounded by a spherical shell made up of rock of constant density. The gravitational attraction of the shell at any point on its outer surface is equal to the difference between the gravitational attractions at such a point of spheres with radii r and r + h. This is equal to $4\pi Gh\text{đ}$, provided that h is so small compared to r that any terms involving higher powers of h/r can be ignored.

Bouguer obtained his equation as an extension of his free-air calculations, for which he had already used a spherical shell. He said that:

> ..... all this will change if we add to the globe a spherical shell with density đ. This new 'shell,' if it has the same density as the rest of the Earth, will increase the gravity field at its surface because the radius is greater. The increase will be in the ratio of r to r + h. Thus the shell does not only restore the 2 h decrease in gravity because of the distance Aa = h above the Earth, but it adds a new element to the gravity, equal to half the decrease, because the gravity which is actually r − 2 h at the point a becomes r + h. It follows from this that the gravity that the spherical shell is capable of producing at its outer surface at a, is greater by 3 h, or three times the elevation.

This is a bit tortuous, and would surely have been set out differently had he not wished to continue by writing that:

> .... however, it is necessary to multiply by the density đ, because we suppose that the densities of the shell and the whole Earth are different.

and by completing his discussion by concluding that:

..... if the Earth is augmented by a spherical shell ...... then the gravity field at a is increased to r–2 $h x D$ + 3 $h\bar{d}$.

Multiplying 3 $h\bar{d}$, the part of this expression due to the shell, by $4\pi G/3$, the factor that converts 'Bouguer' units to modern units, gives the modern $4\pi G h\bar{d}$, so once again his equation is valid. However, having got this far, Bouguer realised that the spherical shell model was completely unrealistic and rejected it. Instead, and because he was thinking of the long chain of the Andes, he decided to calculate the effect of a ridge with a triangular cross-section (Fig. 14.14).

Infinitely long bodies with constant cross-sections are known in modern jargon as 'two-dimensional' (with a third dimension that is infinite, not zero). For the purposes of calculation they can be treated as made up of infinitely-long horizontal line-masses, the gravity effects of which are equal to their masses per unit length multiplied by G and divided by the perpendicular distances from the point of observation, r (and not the r^2 appropriate to point masses). Bouguer evidently knew this, and more, because he stated that if the mass beneath the observation point could be treated as a symmetrical ridge with a height equal to half its base, then the gravitational effect at the apex point would be only a quarter of the effect of a spherical shell of the same thickness. That he was not quite right about

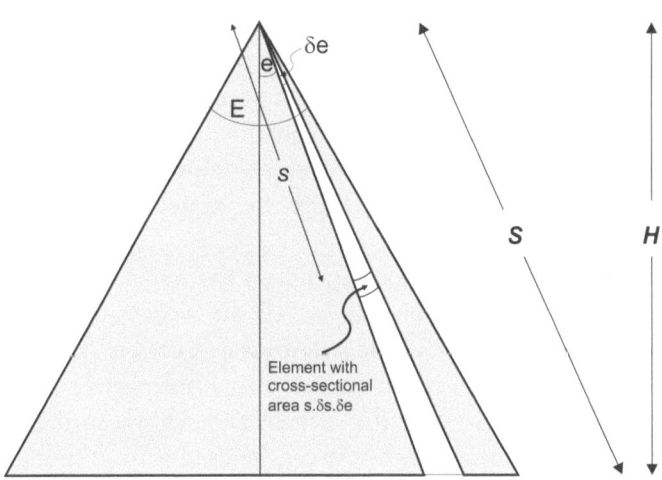

Fig. 14.14 Topographic corrections: the infinite ridge model. In today's physics the Greek letter δ is often used as a prefix to indicate a very small increment in some quantity, e.g. the symbol δs represents a very small increment in the downslope distance s

this scarcely matters, because he then abandoned this model in its turn, saying that in the case of the Andes the full angle at the apex of the triangle would be closer to 170° than 90°, and that this would double the effect he had previously calculated. He then went on to quote a value 3 hđ/2, which he stated, correctly in terms of the conventions he was using, to be half the effect of a spherical shell of the same density and thickness. However, this, when multiplied by the always necessary $4\pi G/3$ to give $2\pi Gđh$, is the effect of a ridge with an apex angle of 180°, not 170°, which is another way of saying that it is the effect of a uniform flat plate extending on all sides to infinity.

As usual, Bouguer stated his result without proof and we do not know how it was obtained. Today the gravity fields of complex bodies are calculated using integral calculus, but at the time this was in its infancy, Newton and Leibniz were still squabbling over the question of priority, and very few people were actually using the technique. It is likely that Bouguer used a forerunner of calculus in which bodies were divided into very small parts or 'elements', the effects of which could be calculated and then added together. For the infinite ridge these elements would be infinite line masses approximating infinite horizontal rods with the shapes in cross-section of segments of a ring (Fig. 14.14). For very small increments δe in the angle **e** and δs in the distance **s** the area is approximately equal to $s.\delta s.\delta e$. The mass per unit length is obtained by multiplying this by the density and is equal to đ.$s.\delta s$. δe. This is also the mass per unit length of the equivalent line mass, and must be divided by **s** and multiplied by the gravitational constant, G, to give the gravity effect at the apex, which is therefore G.đ.$\delta s.\delta e$. Because this does not involve **s** (i.e. the gravity effects of all the elements are the same), the gravity effect of the wedge can be obtained by replacing **s** by **S**, to give G.**S**.đ.δe .This, however, is a force directed along the axis of the wedge, and must be multiplied by cos **e** to get the vertical effect which, since cos **e** = **H/S**, is G.**H**.đ.δe.

Since this expression does not depend on the angle, the full effect of the ridge can be obtained by summing for all the wedges of which it is composed, and is equal to Gđ.**H**.*E*. It is not even necessary for the ridge to be symmetrical, although if it is not there is also a horizontal component to be considered. It seems a pity that this elegant and slightly surprising result has very few practical applications, but it can be used for the special case of the flat Bouguer plate, for which the apex angle is 180° (2π radians) and the field is $2\pi GđH$, as can be proved in many other ways.

Thanks to modern mapping techniques we have a far better idea than did Bouguer of what the Andes actually look like, and Fig. 14.15b shows

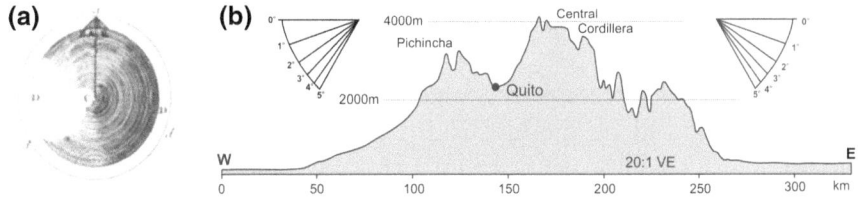

Fig. 14.15 a Bouguer's Fig. 45 from his 'La Figure de la Terre'; **b** Topographic profile across the Andes through Quito and corresponding slope angles (vertical exaggeration 20:1). Bouguer's favoured apex angle of 170° implied slope angles of 5° on either side of the mountain mass which, as an average, is much too large

a cross-section drawn through Quito. There are some slopes at the 5° angle implied by the 170° ridge model, but they are not sustained. The flat-plate model seems equally inappropriate, but remarkably, it works very well, because the parts of the topographic mass that are most important are those closest to and below the measuring point and are shared with the flat plate, and measurements will generally be made in places that are fairly flat, at least locally.

Although Bouguer never specifically made the conceptual leap from the ridge to the plate, he did insist that his rather imprecisely defined model would, '*sans risque de se tromper*', produce the greatest possible gravity effect. The reason he gave, which was that the Andes were a chain of mountains rather than a single peak, was not quite the correct one, but for the flat plate his claim is true. Any topographic masses above it will produce upwards forces, reducing 'g', and any valleys below will substitute air for part of the mass of the plate, also reducing 'g'.

The Bouguer plate is still used today in gravity surveys as the starting point for calculating the corrections that have to be made for rock masses above sea level. Being infinite, it has little resemblance to any actual topography, but that does not matter. What is important is that its gravity effect is usually a reasonable first approximation to the gravity effect of the topography. Roughly 90% of that effect comes from the parts of the plate within a radius equal to five times its thickness, so that even for a 'Quito plate' 3000 m thick, 90% of the gravity effect would come from less than 15 km away. It is now common practice to make additional corrections for all topographic deviations from the Bouguer plate within 167 km of the measuring point, and a 'Quito plate' with this radius would produce more than 99.9% of the effect of an infinite one.

Coda 5—The Compensations of Airy and Pratt

The seismological successors of Mohorovičić refined his techniques, and his interpretations, and by the mid-20th Century geophysicists were satisfied that the Earth could be divided into a crust, a mantle and a core separated by boundaries at which there were major changes in composition. That basic division still holds but has been complicated by the additional idea of a lithosphere consisting of the crust and the relatively cool and rigid uppermost part of the mantle. It is the lithosphere, and not just the crust, that forms the 'plates' of Plate Tectonics. Below it lies the much thicker asthenosphere, which is the part of the mantle that acts as a fluid on geological time scales. In global terms the crust is very, very thin (thinner in proportion to the Earth's radius than the skin of an apple) but the thickness of the lithosphere, while certainly several times greater, cannot be defined exact because its rigidity decreases gradually, not abruptly, with depth. A value of somewhere between one and two hundred kilometres is often assumed, which would not be enough to satisfy William Hopkins but is just about what is needed to make Pratt's isostatic theory compatible with the observed variations in 'g'.

Despite its critical role in some geological processes, the lithosphere only produces important changes in 'g' close to the trenches where it plunges down into the Earth's interior. Elsewhere the density contrasts between air, water, rock and mantle are the main controls on gravity anomalies, but the forms taken depend on the corrections that have been applied.

Figure 14.16 shows an E-W section through the crust and uppermost mantle along latitude 40°S. Beginning on the western slopes of the Andes and ending on the eastern flank of the Mid-Atlantic Ridge, it includes examples of most of the major geological provinces present on today's Earth with the exception of subduction zones. New ocean crust is being generated in a narrow and, at this scale, invisible, rift valley at the crest of the broad mid-Atlantic ridge, beneath which is a zone where it is really not possible to separate crust and mantle, or lithosphere and asthenosphere. The 'ridge' is not actually very ridge-like, since the slopes involved are extremely gentle. Away from its crest the rocks cool and the Moho becomes increasingly well defined, separating the lighter rocks of the crust from the denser rocks of the mantle, and as it does so the main mechanism of isostatic support changes gradually from Archdeacon Pratt's model to George Airy's model. It is Airy's model that best describes the situation beneath the abyssal plain, the continental rise and the continent. In particular, the Andean Cordillera, which is much narrower at this latitude than where it was visited by Bouguer and La Condamine, is supported by thickened crust in the way that Airy supposed.

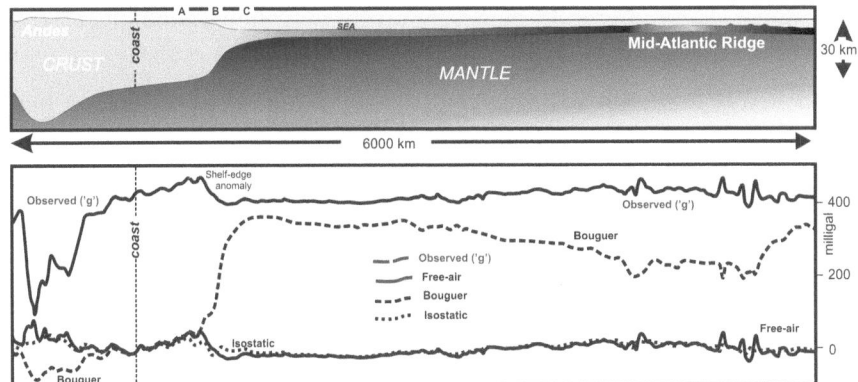

Fig. 14.16   Regional gravity variations, from mountain range to mid-ocean ridge across a passive continental margin. The variable shading of the oceanic part of the crust is intended to show the gradual development of a sharp Moho boundary away from the ridge crest. The vertical exaggeration is approximately 25:1 and even the apparently very steep slope of the sea-floor from the continental shelf down to the abyssal plain involves gradients of only one or two degrees. The gravity profiles are based on the World Gravity Map 2012 (Balmino et al. 2011) and are all to the same vertical scale. A constant 980,000 milligal has been subtracted from the 'observed' gravity profile to allow it to be plotted together with the other profiles but without overlapping onto them

The four profiles show the various forms of 'gravity' calculated from measurements made at the land or sea surface. Raw 'g' (shown by a continuous line) is plotted to the same vertical scale as the other three, but to an arbitrary base. Because the profile runs from west to east, the latitude correction is constant, and because the free-air correction depends on the heights of the observation points, which are at sea level in marine surveys, the free-air gravity and 'g' profiles have the same shapes in detail in the marine area but differ by about 980,000 milligal, which is the effect at this latitude of the total mass of the Earth.

The extreme dependence of raw 'g' on topography is very clear. Because on land the measurements are made on the rock surface, but at sea they are made on the sea surface, the correlation is positive for the sea-floor topography but inverted for the land topography. For free-air gravity the correlation is positive on both land and sea because the land free-air correction, by ignoring the mass of the topography, over-corrects for the extra distance from the centre of the Earth. The Bouguer correction over-corrects for this over-correction in land areas because it takes no account of the mass deficit represent by the low-density crustal root beneath the high topography.

Bouguer gravity becomes massively positive in the oceanic areas because the correction, made offshore by adding the effect of a flat plate with a thickness equal to the water depth and a density equal to the density difference between water and crustal rock, compensates only for the thickening of the water layer, which is light, and not for the underlying rise of the dense mantle. There is a decrease towards the mid-ocean ridge because of the lower density of the hotter mantle beneath it. In Fig. 14.16 the correction has coped well with some of the narrow bathymetric features but less well with others. This may be due, at least in part, to the material forming some of these features, which are isolated sea mounts, differing in density from normal crustal rocks, i.e. real geological effects may be being recorded.

The fourth (dotted) curve represents isostatic gravity, which is corrected for the effect of an assumed model of isostatic compensation, and is very similar to the free-air gravity but is less influenced by small-scale topography because it is computed via the Bouguer correction. In principle the features on this profile should represent anomalous sub-surface (or sub-seafloor) mass distributions but the correction process is never perfect.

An interesting effect occurs at the shelf edge, where on both the 'raw g' and free-air profiles there is high gravity just landward of the shelf break and low gravity over the foot of the continental slope. This pattern occurs because changes in the level of the sea floor and the Moho, even when producing an exact isostatic balance, do not produce exactly compensating gravity fields. In the sketch there is water at C at depths where there is rock at A. Even though there is isostatic balance at both points, the gravity effects do not cancel because at A gravity is already beginning to respond to the high mantle at C. At C, on the other hand, gravity is still affected by the thick crust at A. These high-low free-air gravity couples are characteristic of continental margins everywhere.

The moral of all this is that, when confronted by a gravity map, the first question that has to be asked is 'what sort of gravity map?'

Coda 6—The Coincidences of Captain Kater

Accuracy and precision, which are pretty much the same things in everyday speech, are very different in physics. Accuracy is what common usage suggests it should be, but precision is determined by the number of significant figures used when recording a result. Teachers of sciences wage constant battles against students who take results directly from their computers and present them at precisions much higher than can be justified by the accuracy of their experiments. The story of Huygens' apparently phenomenally accurate value for 'g' in Paris, produced by converting a rather rough estimate of the length of a seconds pendulum in Rhenish feet to metres using a factor defined to a much higher precision, is an object lesson in the care needed when handling, and presenting, error-prone physical measurements.

The Calculation of Errors

In the years since Galileo the methods of analysing errors have been standardised, and the standard method can be applied to estimates of 'g' made using pendulums. Thanks to Huygens, we know that 'g' is related to pendulum length L and period T by the equation $g = \pi^2 L/T^2$. If there is a small error in L equal to δL and a small error in T equal to δT, leading to a small error δg in g, this equation becomes:

$$g + \delta g = \pi^2(L + \delta L)/(T + \delta T)^2$$

Because $g = \pi^2 L/T^2$, the left hand side can be divided by g and the right hand side by $\pi^2 L/T^2$ without affecting the equality. This gives:

$$1 + \delta g/g = (1 + \delta L/L)/(1 + \delta T/T)^2$$

The binomial theorem can then be used to expand $1/(1 + \delta T/T)^2$ as a series, giving:

$$1 + \delta g/g = (1 + \delta L/L)/(1 - 2\delta T/T)$$

if $\delta T/T$ is small enough for terms involving $(\delta T/T)^2$ and higher powers to be ignored. The expression on the right-hand side can be multiplied out and any terms in which a small quantity is being multiplied by another small quantity (e.g. $\delta L/L \times \delta T/T$) can be ignored. We get

$$1 + \delta g/g = 1 + \delta L/L - 2\delta T/T$$

or

$$\delta g/g = \delta L/L + 2\delta T/T$$

The minus sign in the first of these equations is replaced by a plus in the second, because errors can be in either direction and the 'worst case' scenario occurs when they both act in the same direction. The percentage error in '*g*' is thus equal to the percentage error in the length of the pendulum plus twice the percentage error in the period of the pendulum.

Kater's Calculations

Because Kater was so meticulous in measuring the distance between the knife edges of his reversible pendulum, he had to worry about the different standards of length that existed at the time. The official standard was a yard-long brass bar kept in the House of Commons, but there were at least two alternatives, prompting him to quote his length of a seconds pendulum as:

	inches
By Sir G. Shuckburgh's standard	39.13860
By General Roy's scale	39.13717
By Bird's Parliamentary standard	39.13842

The difference between the Parliamentary and Shuckburgh standards amounts to less than five parts in a million, but Kater preferred Shuckburgh because:

> The standard yard made by Bird in 1758, for the House of Commons, better known by the name of BIRD's Parliamentary standard, is little adapted for measurements where great precision is necessary. The yard is determined by two large dots made on gold pins, which are let into a bar of brass. The mean of a number of bisections of these dots gave their distance equal to 36,00016 inches of Sir GEORGE SHUCKBURGH's scale. (Kater 1818; p. 55)

General Roy's yard ruler differed by several tens of parts per million from the other two, but was important because it had been used for the Trigonometric Survey of the British Isles. It might then have been lost for ever, because it was Roy's personal property and was sold after his death along with his other personal effects, but Kater knew the purchaser. It was, almost inevitably, Henry Browne. The sale may not have taken place until some years after Roy's death in 1790, because Browne did not return permanently to England until 1795.

For all three standards Kater quoted his result to a precision of one part in almost four million, which was a little optimistic but not too far out of line with the accuracy with which he measured length. In measuring time he was much less precise, recording only to the nearest second, or approximately one part in 500 of a typical interval. Even so there was some variation, from

Slider 19 divisions Clock gaining 0",18 on mean time.			**Great weight** *above*						Barometer 29,90.
	Temp.	Time of co-incidence	Arc of vibration	Mean arc	Interv. in seconds	No. of vibrats.	Vibrations in 24 hours.	Corr. for arc.	Vibrations in 24 hours.
July 3rd.	o 68,3	m. s. 45. 3 53. 27 1. 51 10. 16 18. 42	o 1,23 1,03 0,87 0,74 0,63	o 1, 13 0,95 0,80 0,68	504 504 505 506	502 502 503 504	86057,16 86057,16 86057,82 86058,49	s 2,08 1.47 1,04 0,75	86059,24 86058,63 86058,86 86059,24
	68,4							Mean Clock	86058,99 + 0,18
	68,4	mean							86059,17
M			**Great weight** *below*						
	68,4	24. 31 32. 54 41. 18 49.42 58. 8	1,24 1, 11 0,99 0,90 0,82	1, 17 1,05 0,94 0,86	503 504 504 506	501 502 502 504	86056,47 86057,16 86057,16 86058,49	2,23 1,80 1,44 1,20	86058,70 86058,96 86058,60 86059,69
	68,4							Mean Clock Temp.	86058,99 + 0,18 + 0,04
	68,5	mean							86059,21

Fig. 14.17 A typical tabulation of the results from one of Kater's experiments, redrawn from his 1818 paper

as little as 503 seconds to as much as 506 seconds on the typical worksheet of Fig. 14.17. The number of 'vibrations' (half-period oscillations) in the seventh column was obtained by deducting two from the number of seconds in the sixth, and not by actual measurement.

Division of each time interval by the number of vibrations in that interval would have given the half-period times directly but Kater did not make that calculation. Instead, he divided the number of vibrations by the number of seconds and multiplied by 86,400 (the number of seconds in a day) to obtain the number of vibrations in a day. He never explained why he took this rather circuitous route, but it was common practice at the time and was also used by de Freycinet. It may have been to make it easier to correct for the clock error, which was measured in seconds per day, but it produced numbers quoted to far greater precision than the times from which they were calculated.

This all seems very odd, especially given that errors in the period would have been twice as important as errors in length, but Kater knew what he was doing. The key to his success lay in the very non-linear relationship between the times and the periods that they implied. Because the number of 'vibrations' was always two fewer than the measured number of seconds ('t'), the period would always be equal to $t/(t-2)$, a quantity that gets closer and closer to one as the distance between the knife edges gets closer to the length of the seconds pendulum. As this point is approached, the time between coincidences increases, but in smaller and smaller steps. It is the size of these steps that matters, and the step from 501/499 to 502/500 is equivalent to about 16 milligal. As Kater himself put it:

> ..... the brass pendulum may arrive at the lowest part of the arc either precisely at this second, or at any portion of the second preceding it. An error might possibly arise from this circumstance amounting to nine-tenths of a second .... and as an error of one second in the interval, occasions a difference of 0.63 in the number of vibrations in 24 hours if 0.55 (the proportional part of 0.63) be divided by 4 (the number of intervals forming each set of experiments) we have 0.14 for the greatest error in defect in the number of vibrations in 24 hours which can arise from this cause.

This extract works as an explanation but also suggests that Kater, although rigorous in his measurements, might have been a little slapdash when it came to calculations. Interval times that differ by just the one second specified appear in the last two rows in the first group of measurements shown in Fig. 14.17, but they produce a difference in the number of vibrations in 24 hours (before the correction for arc i.e. in Column 8) of 0.67, not 0.63. It would have been 0.63 had the times been 525 and 526 seconds, but these were not intervals that were recorded in any of the published experiments. Kater did refer at one point to using intervals of 'about 530 seconds', but for this the corresponding difference would be about 0.62 of a vibration. And, when he reduced 0.63 to its "*proportional part*", i.e. to nine-tenths, he quoted a difference of 0.55, not the almost 0.57 that it should have been.[7]

---

[7]Richard Howarth (pers.com.) has identified similar small errors on some of the other worksheets.

The Merits of Being a Little Bit Sloppy

One of the odder aspects of low precision measurement is that a little inaccuracy can actually improve the final results. If Kater's pendulum achieved a true coincidence after 500.1 swings, i.e. after 502.1 seconds, and his time measurements were always accurate to the nearest second, he would have recorded this as 500 swings in 502 seconds. If his timing method were perfect he would never have recorded anything else, no matter how often the experiment was repeated. If, however, there were some small errors (as inevitably there would have been), he might occasionally have recorded the time as 503 seconds and sometimes, but more rarely, 501 seconds. Averaging the results from a large number of experiments might then produce something closer to the correct answer, although the improvement would be difficult to quantify.

The Effects of Geology

The absolute measurements made by Kater were important as pioneering efforts, but were never actually used in any ways that required high accuracy. The relative measurements made by him, his colleagues and their French competitors and collaborators were much more useful. They contributed to a better understanding of the shape of the Earth and provided the first indications of a link between geology and gravity. This connection was made by assuming (at least in Sabine's case) that 'g' should change with the square of the sine of the latitude angle, L, and seeing how well the actual values of 'g' conformed to this pattern. Sabine would have known that there was also a small variation with the fourth power of the sine of L (because George Airy would have told him so) but he chose to ignore it. He was right to do so, not only because it is trivial compared to the experimental errors but because, thanks to those errors, he could not have calculated the numerical multiplier accurately.

Figure 14.18 shows the results obtained by Kater, Hall, Goldingham, Sabine and de Freycinet, plotted against latitude. Between them these pioneers could have drawn a reasonable approximation to the modern theoretical curve, but even in a plot at this scale there is an obvious scatter.[8] With the global data available today it is possible to make some sort of estimate

---

[8]The curve is based on GRS80, but none of the variants proposed since the 1960s would have produced versions that would have been visibly different.

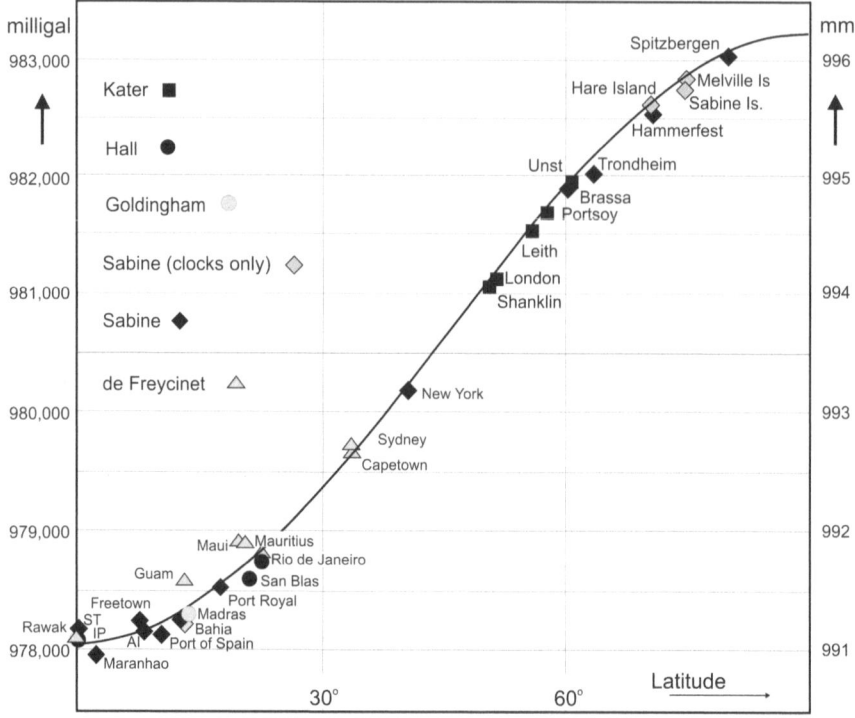

Fig. 14.18 Relative pendulum results, 1818–1823. The continuous line is the theoretical sea-level gravity calculated from GRS80. ST: São Tomé; IP: Isla Pinta (Galapagos); AI: Ascension Island. The scale on the right is of the length of the seconds pendulum, in millimetres

of how much of this is due to geology and how much to errors in measurement, although only approximately because of uncertainties as to locations. Neither the British nor the French recorded longitudes as accurately as they recorded latitudes (Basil Hall, for example, gave longitudes only to the nearest quarter minute, or about half a kilometre) and in some cases they did not record them at all. Sometimes the sites can be identified from the observer's descriptions (Sabine was particularly meticulous in this respect) and for measurements made close to sea level they can be guessed by finding points that are on the coast at the latitudes specified. In many cases, however, even the measurements at 'coastal' sites were made in some convenient building well away from the sea.

In Fig. 14.19a the information given in Fig. 14.18 is replicated but displayed more clearly because it is 'flattened' on to the regional curve. This gives a better idea of the sort of information that would have been available to the people who were trying to estimate the eccentricity of the Earth ellipsoid in the early 19th Century. The standard deviation is 94, and de Freycinet (1826) comes out particularly badly, with differences at Maui, Guam and Mauritius of more than 200 milligal. This is, however, very far from being the whole story.

Figure 14.19b is a plot of the differences between the pendulum results and the values of 'g' that can be extracted from global data sets, with uncertainties of a few milligal due to the lack of positioning information. There are dramatic reductions in the differences at de Freycinet's most problematic stations, and significant, although smaller, reductions at Hall's station on Isla Pinta and Sabine's stations on São Tomé and Ascension. These are all on ocean islands, and Hall was right when he suggested that this might affect 'g'. The islands are the tops of local masses forming isolated and largely submerged mountains surrounded by areas where the dense rocks of the mantle are only about 10 km below sea level. At such stations 'g' is almost always several hundred milligal greater than would be predicted from the 'ideal' curve.

As far as the continental stations are concerned, the agreements with the modern values are quite astonishingly good for Kater's survey of Great Britain and reasonable for Sabine's measurements in the Caribbean and in southern latitudes, and for de Freycinet. Two of Hall's estimates are very close to the modern values, but at San Blas in Mexico his error was almost 60 milligal. This site, which was probably chosen in the interests of peace and quiet, was in a monastery built well to the east of the harbour on a steep sided rock about 37 metres above sea level, and height and terrain could account for about ten milligal of the error. The differences at all Sabine's stations in the far north remain high, and some are actually increased by

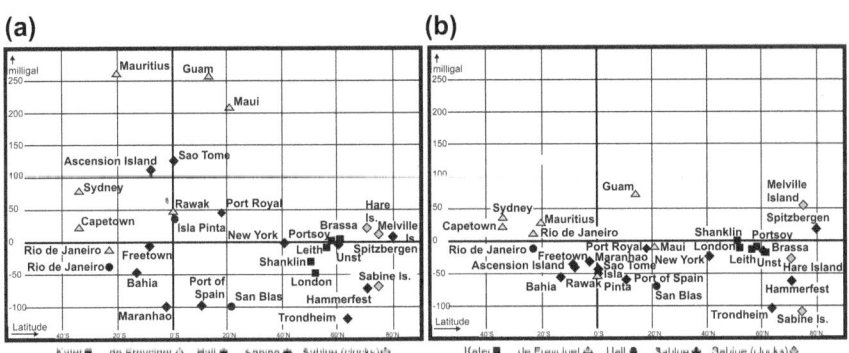

Fig. 14.19 Relative pendulum results, 1818–1823 **a** Comparison with GRS80; **b** Differences from modern estimates of 'g'

the corrections, but even when these measurements are included the overall standard deviation is less than 50 milligal. Considering the conditions under which many of the observations were made, this is impressive.

Coda 7—The Serendipitous Spring of Lucien LaCoste

The spring system suggested by John Herschel would have suffered from an almost fatal drawback that has plagued designers ever since. The spring has a dual function. It has to support the mass (so it has to be strong) but it has to respond to very small changes in weight (so it has to be weak). Lucien LaCoste got round this limitation by designing a spring system that provided the same support to the proof mass over a range of different positions. An essential element in his design was the use of 'zero-length' springs that exert forces proportional to their actual lengths, rather than to their extension beyond some 'natural' length. Equally important, and a feature of most 'astatic' systems, was the idea of supporting the mass via a hinged lever arm, balancing moments about a hinge rather than forces at a point. In combination, these two ideas led to a design that dominated gravity work for more than fifty years.

The system shown in Fig. 14.20 is the very basic one, but with one addition. Not all implementations use a second, weak, measuring spring, and LaCoste's own meters did not. The second spring, however, makes the system easier to explain.

In Fig. 14.20 the downward force on the mass is equal to Mg, and its moment about the hinge is therefore equal to $Mg.A.\cos(e)$, where A is the length of the lever arm and e is the angle it makes with the horizontal.

The force exerted by the spring is kx, where x is its length and k is the spring constant, and its moment about the hinge is equal to kxp, where p is the length of the perpendicular from the hinge to the spring. But:

$$p = h.\sin(\alpha) \quad \text{from the basic properties of a right - angled triangle}$$

and

$$\sin(\alpha)/a = \sin(90 - e)/x \quad \text{(from the sine rule) and} \quad \sin(90 - e) = \cos(e)$$

$$\sin(\alpha) = a.\cos(e)/x$$

So the moment kxp is: $kx.h.\sin(\alpha) = kx.h.a.\cos(e)/x = kh.a.\cos(e)$

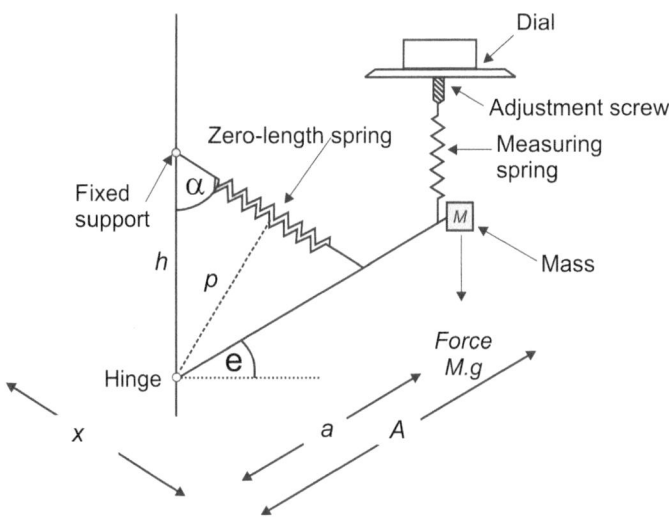

Fig. 14.20 An astatic meter with a zero-length spring and a supplementary measuring spring

The condition for balance is that the two moments are equal, i.e. that:

$$Mg.A.\cos(e) = kha.\cos(e) \quad \text{or} \quad g = kha/MA$$

All the quantities on the right-hand side of the final equation are under the control of the manufacturer, and no angles are involved, so that it is possible to build a system in which, for any chosen value, g_0, of 'g', the moment of the mass M is exactly balanced by the moment of the zero length spring. If this balance is achieved, it is achieved for all angles that the lever arm can make with the horizontal.

If the actual value of 'g' is equal to $g_0 + \delta g$, then the weight to be supported is $M(g_0 + \delta g)$. The zero length spring supports $M.g_0$ of this, leaving the measuring spring to support the additional force $M.\delta g$. Because this is all it has to do, it can be quite weak and therefore sensitive to small changes in δg. It can be raised or lowered by rotating the screw to bring the lever-arm back to the horizontal position, and the rotation needed to do this is a direct measure of δg.

References

Bouguer P (1749) La Figure de la Terre. Jombert, Paris

Balmino G, Vales N, Bonvalot S, Briais A (2011) Spherical harmonic modelling to ultra-high degree of Bouguer and isostatic anomalies. J Geodesy. https://doi.org/10.1007/s00190-011-0533-4

de Freycinet L (1826) Voyage autour du monde par les corvettes l'Uranie et laPhysicienne: 5. Observations du pendule. Pillet Aîné, Paris

Drake S (1990) Galileo: pioneer scientist. University of Toronto Press, Toronto

Galilei G (1632) Dialogo intorno ai due massimi sistemi del mondo: Tolemaio e Copernicano. In: Alberi E (ed) (1842) Le opere de Galileo Galilei, Tomo I. Società Editrice Fiorentina, Firenze

Galilei G (1638) Discorsi e dimostrazioni matematiche intorno a due nuove scienze. Elzevir, Leiden. English edition: Galileo G (1914) Dialogues concerning two new sciences (trans: Crew H, de Salvio A). Macmillan, London

Graney CM (2012) Anatomy of a fall: Giovanni Battista Riccioli and the story of g. Phys Today 65:36–40

Huygens C (1673) Horologium Oscillatorium. Muguet, Paris

Kater H (1818) An account of experiments for determining the length of the pendulum vibrating seconds in the latitude of London. Philos Trans R Soc 108:33–102

Mahoney MS (2000) Huygens and the pendulum: from device to mathematical relation. In: Breger H, Grosholz E (eds) The Growth of Mathematical Knowledge. Kluwer, Amsterdam

Schliesser E, Smith GE (1966) Huygens's 1688 report to the directors of the Dutch East India Company on the measurement of longitude at sea and its implications for the non-uniformity of gravity. De Zeventiende Eeuw 12:198–214

Vollgraf JA (ed) (1929) Oeuvres complètes de Christiaan Huygens. Tome XVI: Mécanique jusqu'à 1666. Martinus Nijhoff, Den Haag

Index

© Springer International Publishing AG, part of Springer Nature 2018
J. Milsom, *The Hunt for Earth Gravity*,
https://doi.org/10.1007/978-3-319-74959-4